AF352877

Advancing Knowledge on
Cyanobacterial Blooms in Freshwaters

Advancing Knowledge on Cyanobacterial Blooms in Freshwaters

Editors

Elisabeth (Savi) Vardaka
Konstantinos Ar. Kormas

MDPI • Basel • Beijing • Wuhan • Barcelona • Belgrade • Manchester • Tokyo • Cluj • Tianjin

Editors
Elisabeth (Savi) Vardaka
International Hellenic University (IHU)
Greece

Konstantinos Ar. Kormas
University of Thessaly
Greece

Editorial Office
MDPI
St. Alban-Anlage 66
4052 Basel, Switzerland

This is a reprint of articles from the Special Issue published online in the open access journal *Water* (ISSN 2073-4441) (available at: https://www.mdpi.com/journal/water/special_issues/cyanobacterial_blooms).

For citation purposes, cite each article independently as indicated on the article page online and as indicated below:

LastName, A.A.; LastName, B.B.; LastName, C.C. Article Title. *Journal Name* **Year**, *Article Number*, Page Range.

ISBN 978-3-03943-505-0 (Hbk)
ISBN 978-3-03943-506-7 (PDF)

Cover image courtesy of Matina Katsiapi.

Contents

About the Editors

Elisabeth (Savi) Vardaka is an Associate Professor at the Department of Nutritional Sciences and Dietetics of the International Hellenic University. She received her bachelor's in biology and earned her Ph.D. in aquatic microbiology from the Department of Biology of the Aristotle University of Thessaloniki, Greece. Her research interests are focused on microbial biodiversity (morphological, ecophysiological and molecular approach), ecological assessment of water quality based on phytoplankton, toxic cyanobacteria and/or cyanotoxins, and their effects on public health. Her current research interests include human microbiota and interactive effects on nutrition and health. She has published numerous scientific articles in academic peer-reviewed journals and conferences on these topics. Her published works have received more than 1100 citations, and her h-index is 17.

Konstantinos Ar. Kormas' research interests are focused on the diversity and ecological role of prokaryotes and unicellular eukaryotes in the aquatic environment, including the interactions of symbiotic microorganisms with aquatic animals. His work includes field and laboratory work in marine and freshwater systems with a specialty in molecular approaches, including genomics and metagenomics. He is the author/co-author of 103 peer-reviewed papers, 9 book chapters and 1 book in Greek ("Ecology of aquatic microorganisms"). He is currently involved in two H2020 and four national research projects. His published works have received more than 2100 citations, and his h-index is 27. Link to full CV: https://sites.google.com/site/kkormas/.

Editorial

Advancing Knowledge on Cyanobacterial Blooms in Freshwaters

Elisabeth Vardaka [1],* and Konstantinos Ar. Kormas [2]

[1] Department of Nutritional Sciences and Dietetics, International Hellenic University (IHU), Alexander Campus, 574 00 Thessaloniki, Greece

[2] Department of Ichthyology & Aquatic Environment, School of Agricultural Sciences, University of Thessaly, 384 46 Volos, Greece; kkormas@uth.gr

*   Correspondence: evardaka@ihu.gr; Tel.: +30-231-001-3417

Received: 23 July 2020; Accepted: 14 September 2020; Published: 16 September 2020

Abstract: Cyanobacterial blooms have become a frequent phenomenon in freshwaters worldwide; they are a widely known indicator of eutrophication and water quality deterioration. Information and knowledge contributing towards the evaluation of the ecological status of freshwaters, particularly since many are used for recreation, drinking water, and aquaculture, is valuable. This Special Issue, entitled "Advancing Knowledge on Cyanobacterial Blooms in Freshwaters", includes 11 research papers that will focus on the use of complementary approaches, from the most recently developed molecular-based methods to more classical approaches and experimental and mathematical modelling regarding the factors (abiotic and/or biotic) that control the diversity of not only the key bloom-forming cyanobacterial species, but also their interactions with other biota, either in freshwater systems or their adjacent habitats, and their role in preventing and/or promoting cyanobacterial growth and toxin production.

Keywords: cyanobacteria; microcystin; water quality; human and animal health; nutrients; climate change; eutrophication; model

---

1. Introduction

Cyanobacterial blooms constitute a water quality problem that has been widely acknowledged to cause negative effects on the use, safety, and sustainability of drinking and recreational waters supplies and fisheries. Water eutrophication combined with relative high water temperature promote cyanobacterial growth and water bloom formation. The cyanobacterial bloom season may be of extended or even continuous duration throughout the year, especially in freshwaters experiencing external and internal conditions, such as high temperatures and irradiance, thermal stratification, long water replacement times, high phosphorus and nitrogen concentrations, and low zooplankton grazing [1] or low viral attack [2]. Evidence for this can be obtained from lakes in the Mediterranean region [3,4]. In a human and climatically changing world, cyanobacterial blooms seem to have increased in frequency and longevity globally, and are likely to expand and thrive in water resources [1,5,6]. Moreover, the successful dispersal and colonization of new aquatic habitats or newly formed systems by airborne bloom-forming cyanobacteria, such as the cosmopolitan potentially toxic *Microcystis*, seem to facilitate their global expansion and are of immediate ecological, economic, and health-related interest [7,8]. Of most concern are the cyanotoxins along with the mechanisms that induce their release and fate in the aquatic environment. These secondary metabolites pose a potential hazard for human health and agricultural and aquaculture products directed for animal and human consumption, therefore the strict and reliable control of cyanotoxins is crucial for assessing risk [6]. In this direction, a deeper understanding of the mechanisms that determine cyanobacterial bloom structure and toxin production become a target for managing practices [5,6].

This Special Issue, entitled "Advancing Knowledge on Cyanobacterial Blooms in Freshwaters", includes 11 research papers [9–19] and aims to bring together the recent research of multi- and interdisciplinary approaches from the field to the laboratory and back again, driven by working hypotheses based on any aspect from ecological theory to applied research on mitigating cyanobacterial blooms.

2. Contributions

Despite the fact that the interest in cyanobacterial blooms is focused on lakes and reservoirs, where the cyanobacterial bloom presence is most intense, the role of Cyanobacteria in other natural or engineered ecosystemsis also attracting scientific interest. One such case is presented by Li et al. [13], who provided a study where potentially toxic cyanobacteria dominate in aquaculture ponds and, thus, might affect the overall aquaculture production.

The issue of deciphering the role of cyanobacterial blooms at the ecosystem level requires investigations both at the organism level (autoecology) and the community level (synecology), which can be approached by field and/or experimental works. Zhang et al. [10] provided an improved modeling approach on the role of turbulence mixing on the growth of *Microcystis aeruginosa* as a means of affecting differential nutrient transfer between this cyanobacterium and *Scenedesmus quadricauda*, frequently occurring chlorophyte in eutrophic freshwaters. Another autoecological approach is given by Muhetaer et al. [17]. In this study, the authors conducted experiments in order to evaluate the comparative effects of different light intensities on the growth and stress responses of two bloom-forming cyanobacterial species, *Pseudanabaena galeata* (strain NIES 512) and *Microcystis aeruginosa* (strain NIES 111) for a few days.

Apart from the role of Cyanobacteria in previously understudied ecosystems, new ecophysiological roles are gaining momentum in cyanobacteria research, such as the role of cyanobacteria in methane cycling in the oxic waters of lakes [20]. In the current Special Issue, Khatun et al. [16] found that *Synechococcus* was the most likely methane producer in nine Japanese lakes.

The human, animal, and plant health risks associated with low ecological lake water quality and its subsequent toxic cyanobacterial blooms are showcased by Tokodi et al. [15] in a Serbian lake with 50 years of persistent blooms of toxic cyanobacteria and considerable concentrations of toxins. This renders imperative not only the need for mitigating the existing problems but also the importance of righteous predictive approaches for future cyanobacterial blooms. Towards this direction, an integrated approach for the management of cyanobacterial blooms targeting the impacted ecological functions is presented by Hall et al. [11]. In the fight against toxic cyanobacterial blooms, Herrera et al. [9] proved experimentally the oppressive effect of eight phenyl-acyl compounds in the growth of *Microcystis aeruginosa*, and also the non-toxic effect of specific caffeic acid concentrations in non-target zooplankton organisms, expanding, thus, the potential use of novel substances in cyanobacterial biomass reduction.

The importance of hydrogeomorphological features in shaping cyanobacterial communities as subcommunities of the phytoplankton community was showcased in two papers of the current Special Issue. Yao et al. [19] investigated a 2-year succession of cyanobacteria and eukaryotic phytoplankton functional groups in a reservoir, while Katsiapi et al. [14] used a system of two recently interconnected ancient lakes to show that this recent mode of water translocation between the two lakes increased the occurrence of potentially toxic and bloom-forming cyanobacteria. Another study with a dataset covering a whole decade between 2004 and 2014 reported that cyanobacteria constituted up to 60% of the primary producer biovolume, with increased dominance after 2010 [18]. The importance of phosphorus management was showcased by Liu et al. [12], not only for cyanobacteria control but also for the "satellite" bacterial communities of a eutrophic lake, which, in total, might affect the biogeochemical functioning of the whole lake.

Cyanobacteria are among these microorganisms that pose major health risks to humanity, but also, at least for some of them, they seem to be favored by our current climatic instability [21]. Predicting the onset and development of cyanobacterial and algal blooms remains a challenging task [1]. Due to the many aspects of the complex ecophysiology of Cyanobacteria which remain either understudied or even unknown and due to the climatic variability which renders the accurate prediction of the environmental settings favorable to Cyanobacteria rather vague, a truly interdisciplinary and not multidisciplinary scientific effort is imperative. This might require us to reset our current working hypotheses or even generate new ones [22]. Future research should include more frequent samplings with standardized protocols [23], following the best established approaches from field to laboratory analysis to data processing and analysis [24], and not only those dictated by managerial directives [25]. Finally, the interconnectivity between experimental and field experiments which combine classic approaches, such as microscopy analysis, with more modern methodologies such as -omics for acquiring more conclusive insights on the biology, ecology, and mitigation of Cyanobacteria and their freshwater blooms seems to be the most inclusive approach for the time being.

Author Contributions: Writing—original draft preparation, E.V. and K.A.K.; writing—review and editing, E.V. and K.A.K. All authors have read and agreed to the published version of the manuscript.

Funding: This research received no external funding.

Acknowledgments: We would like to thank both the research community for contributing a wide range of valuable papers and the MDPI publisher for their support in this Special Issue.

Conflicts of Interest: The authors declare no conflict of interest.

References

1. Paerl, H.W. Mitigating harmful Cyanobacterial blooms in a human- and climatically-impacted world. *Life* **2014**, *4*, 988–1012. [CrossRef] [PubMed]

2. van Hannen, E.J.; Zwart, G.; van Agterveld, M.P.; Gons, H.J.; Ebert, J.; Laanbroek, H.J. Changes in Bacterial and Eukaryotic Community Structure after Mass Lysis of Filamentous Cyanobacteria Associated with Viruses. *Appl. Environ. Microbiol.* **1999**, *65*, 795–801. [CrossRef] [PubMed]

3. Vardaka, E.; Moustaka–Gouni, M.; Cook, C.M.; Lanaras, T. Cyanobacterial blooms and water quality in Greek waterbodies. *J. Appl. Phycol.* **2005**, *17*, 291–401. [CrossRef]

4. Moustaka-Gouni, M.; Vardaka, E.; Michaloudi, E.; Kormas, A.K.; Tryfon, E.; Mihalatou, H.; Gkelis, S.; Lanaras, T. Plankton food web structure in a eutrophic polymictic lake with a history of toxic cyanobacterial blooms. *Limnol. Oceanogr.* **2006**, *51*, 715–727. [CrossRef]

5. Paerl, H.W. Controlling harmful cyanobacterial blooms in a climatically more extreme world: Management options and research needs. *J. Plankton Res.* **2017**. [CrossRef]

6. Huisman, J.; Codd, G.A.; Paerl, H.W.; Ibelings, B.W.; Verspagen, J.M.H.; Visser, P.M. Cyanobacterial blooms. *Nat. Rev. Microbiol.* **2018**, *16*, 471–483. [CrossRef]

7. Genitsaris, S.; Kormas, K.A.; Moustaka-Gouni, M. Airborne algae and cyanobacteria: Occurrence and related health effects. *Front. Biosci.* **2011**, *3*, 772–787.

8. Curren, E.; Leong, S.C.Y. Natural and anthropogenic dispersal of cyanobacteria: A review. *Hydrobiologia* **2020**. [CrossRef]

9. Herrera, N.; Florez, M.T.; Velasquez, J.P.; Echeverri, F. Effect of Phenyl-Acyl Compounds on the Growth, Morphology, and Toxin Production of *Microcystis aeruginosa* Kützing. *Water* **2019**, *11*, 236. [CrossRef]

10. Zhang, H.; Huang, F.; Li, F.; Gu, Z.; Chen, R.; Zhang, Y. An Improved Logistic Model Illustrating *Microcystis aeruginosa* Growth under Different Turbulent Mixing Conditions. *Water* **2019**, *11*, 669. [CrossRef]

11. Hall, E.S.; Hall, R.K.; Aron, J.L.; Swanson, S.; Philbin, M.J.; Schafer, R.J.; Jones-Lepp, T.; Heggem, D.T.; Lin, J.; Wilson, E.; et al. An Ecological Function Approach to Managing Harmful Cyanobacteria in Three Oregon Lakes: Beyond Water Quality Advisories and Total Maximum Daily Loads (TMDLs). *Water* **2019**, *11*, 1125. [CrossRef] [PubMed]

12. Liu, Y.; Qu, X.; Elser, J.J.; Peng, W.; Zhang, M.; Ren, Z.; Zhang, H.; Zhang, Y.; Yang, H. Impact of Nutrient and Stoichiometry Gradients on Microbial Assemblages in Erhai Lake and Its Input Streams. *Water* **2019**, *11*, 1711. [CrossRef]

13. Li, H.; Chen, H.; Gu, X.; Mao, Z.; Zeng, Q.; Ding, H. Dynamics of Cyanobacteria and Related Environmental Drivers in Freshwater Bodies Affected by Mitten Crab Culturing: A Study of Lake Guchenghu, China. *Water* **2019**, *11*, 2468. [CrossRef]

14. Katsiapi, M.; Genitsaris, S.; Stefanidou, N.; Tsavdaridou, A.; Giannopoulou, I.; Stamou, G.; Michaloudi, E.; Mazaris, A.D.; Moustaka-Gouni, M. Ecological Connectivity in Two Ancient Lakes: Impact Upon Planktonic Cyanobacteria and Water Quality. *Water* **2020**, *12*, 18. [CrossRef]

15. Tokodi, N.; DrobacBacković, D.; Lujić, J.; Šćekić, I.; Simić, S.; Đorđević, N.; Dulić, T.; Miljanović, B.; Kitanović, N.; Marinović, Z.; et al. Protected Freshwater Ecosystem with Incessant Cyanobacterial Blooming Awaiting a Resolution. *Water* **2020**, *12*, 129. [CrossRef]

16. Khatun, S.; Iwata, T.; Kojima, H.; Ikarashi, Y.; Yamanami, K.; Imazawa, D.; Kenta, T.; Shinohara, R.; Saito, H. Linking Stoichiometric Organic Carbon–Nitrogen Relationships to planktonic Cyanobacteria and Subsurface Methane Maximum in Deep Freshwater Lakes. *Water* **2020**, *12*, 402. [CrossRef]

17. Muhetaer, G.; Asaeda, T.; Jayasanka, S.M.D.H.; Baniya, M.B.; Abeynayaka, H.D.L.; Rashid, M.H.; Yan, H. Effects of Light Intensity and Exposure Period on the Growth and Stress Responses of Two Cyanobacteria Species: *Pseudanabaenagaleata* and *Microcystis aeruginosa*. *Water* **2020**, *12*, 407. [CrossRef]

18. Chirico, N.; António, D.C.; Pozzoli, L.; Marinov, D.; Malagó, A.; Sanseverino, I.; Beghi, A.; Genoni, P.; Dobricic, S.; Lettieri, T. Cyanobacterial Blooms in Lake Varese: Analysis and Characterization over Ten Years of Observations. *Water* **2020**, *12*, 675. [CrossRef]

19. Yao, L.; Zhao, X.; Zhou, G.-J.; Liang, R.; Gou, T.; Xia, B.; Li, S.; Liu, C. Seasonal Succession of Phytoplankton Functional Groups and Driving Factors of Cyanobacterial Blooms in a Subtropical Reservoir in South China. *Water* **2020**, *12*, 1167. [CrossRef]

20. Bižić, M.; Klintzsch, T.; Ionescu, D.; Hindiyeh, M.Y.; Günthel, M.; Muro-Pastor, A.M.; Eckert, W.; Urich, T.; Keppler, F.; Grossart, H.-P. Aquatic and terrestrial cyanobacteria produce methane. *Sci. Adv.* **2020**, *6*, eaax5343. [CrossRef]

21. Cavicchioli, R.; Ripple, W.J.; Timmis, K.N.; Azam, F.; Bakken, L.R.; Baylis, M.; Behrenfeld, M.J.; Boetius, A.; Boyd, P.W.; Classen, A.T.; et al. Scientists' warning to humanity: Microorganisms and climate change. *Nat. Rev. Microbiol.* **2019**, *17*, 569–586. [CrossRef] [PubMed]

22. Widder, S.; Allen, R.J.; Pfeiffer, T.; Curtis, T.P.; Wiuf, C.; Sloan, W.T.; Cordero, O.X.; Brown, S.P.; Momeni, B.; Shou, W.; et al. Challenges in microbial ecology: Building predictive understanding of community function and dynamics. *ISME J.* **2016**, *10*, 2557–2568. [CrossRef]

23. Kurmayer, R.; Christiansen, G.; Kormas, K.; Vyverman, W.; Verleyen, E.; Ramos, V.; Vasconcelos, V.; Salmaso, N. Sampling and metadata. In *Molecular Tools for the Detection and Quantification of Toxigenic Cyanobacteria*; Kurmayer, R., Sivonen, K., Wilmotte, A., Salmaso, N., Eds.; John Wiley & Sons: Hoboken, NJ, USA, 2017; pp. 19–42.

24. Moustaka-Gouni, M.; Sommer, U.; Katsiapi, M.; Vardaka, E. Monitoring of cyanobacteria for water quality: Doing the necessary right or wrong? *Mar. Freshw. Res.* **2020**, *71*, 717–724. [CrossRef]

25. Moustaka-Gouni, M.; Sommer, U.; Economou-Amilli, A.; Arhonditsis, G.B.; Katsiapi, M.; Papastergiadou, E.; Kormas, K.A.; Vardaka, E.; Karayanni, H.; Papadimitriou, T. Implementation of the Water Framework Directive: Lessons Learned and Future Perspectives for an Ecologically Meaningful Classification Based on Phytoplankton of the Status of Greek Lakes, Mediterranean Region. *Environ. Manag.* **2019**, *64*, 675–688. [CrossRef] [PubMed]

Article

Dynamics of Cyanobacteria and Related Environmental Drivers in Freshwater Bodies Affected by Mitten Crab Culturing: A Study of Lake Guchenghu, China

Hongmin Li [1,2], Huihui Chen [1], Xiaohong Gu [1,*], Zhigang Mao [1], Qingfei Zeng [1] and Huiping Ding [1,2]

[1]   State Key Laboratory of Lake Science and Environment, Nanjing Institute of Geography and Limnology, Chinese Academy of Sciences, Nanjing 210008, China; hongmin1059@163.com (H.L.); hhchen@niglas.ac.cn (H.C.); zgmao@niglas.ac.cn (Z.M.); qfzeng@niglas.ac.cn (Q.Z.); cynthiadhp@126.com (H.D.)

[2]   University of Chinese Academy of Sciences, Beijing 100049, China

*   Correspondence: xhgu@niglas.ac.cn; Tel.: +86-25-86882006

Received: 11 October 2019; Accepted: 22 November 2019; Published: 23 November 2019

Abstract: Mitten crab aquaculture is prevalent in China, however, knowledge about the threat of cyanobacteria in mitten crab aquaculture-impacted water bodies is limited. Here, seasonal variations of cyanobacteria and their relationships with environmental factors were investigated for Lake Guchenghu area. Results suggested the changes of cyanobacteria community in crab ponds distinguished from the adjacent lake. In the lake, cyanobacterial biomass (3.86 mg/L, 34.6% of the total phytoplankton) was the highest in autumn with the dominance of *Oscillatoria*, *Aphanocapsa* and *Pesudanabaena*. By contrast, in crab ponds, cyanobacteria (46.80 mg/L, 97.2% of the total phytoplankton biomass) were the most abundant in summer when *Pesudanabaena* and *Raphidiopsis* were the dominant species. Of particular note was that obviously higher abundance of filamentous and potentially harmful species (e.g., *Raphidiopsis raciborskii* and *Dolichospermum circinale*) were observed in ponds compared to the lake. Specifically, water depth (WD), permanganate index (COD$_{Mn}$), total phosphorus (TP), N:P ratio, and NO$_2^-$-N were the key environmental variables affected cyanobacteria composition. For crab ponds, N:P ratio, water temperature (WT) and TP were the potential environmental drivers of cyanobacteria development. This study highlighted the fact that mitten crab culture had non-negligible influences on the cyanobacteria community and additional attention should be paid to the cyanobacteria dynamics in mitten crab culture-impacted water bodies, especially for those potentially harmful species.

Keywords: mitten crab culture; cyanobacteria community; seasonal variation; environmental factors; potentially harmful species; Lake Guchenghu

---

1. Introduction

Cyanobacteria are ubiquitous in diverse aquatic environments throughout the world, including lakes, reservoirs, ponds, etc. With accelerating industrial and agricultural development, the rising nutrient pollution intensified the proliferation of cyanobacteria in freshwater bodies worldwide [1–3]. The impact of aquaculture on water quality and the expansion of harmful cyanobacteria is a major environmental issue. Given the increase in intensive aquaculture, many environmental challenges emerged, among which environmental degradation associated with cyanobacterial blooms appeared to be the most alarming [4]. The noxious cyanobacterial blooms would cause various problems associated with unpleasant tastes, odors and toxic metabolite production which could threaten aquatic products and human health [2,4–7].

Over the past few years, mitten crab culture has become one of the most economically important freshwater aquacultures in China, with current yield of this species increasing dramatically (from 8.4×10^3 tons in 1991 to 8.2×10^5 tons in 2015) [8]. Whilst previous studies suggested that crab farming was less detrimental to water environments than fish or shrimp farming, the concentrations of total nitrogen (TN) and total phosphorus (TP) in crab culturing ponds were higher than those in neighboring lake areas [9]. In addition, most crab ponds are shallow, small, and stagnant, which are likely to deteriorate the water environment. Thus, the influence of crab culture on the cyanobacteria dynamics should not be neglected because crab culture, to some extent, changes the physical and chemical properties of water bodies, and biotic community structure [10]. Although many studies investigated cyanobacteria dynamics in fish or shrimp ponds, little is known about the ecology of cyanobacteria in crab ponds. Furthermore, crab culture is also regarded as one of the significant sources of surface water pollution because most of the effluents from crab ponds are eventually discharged into their surrounding lakes [9–11]. Hence, cyanobacteria, as a key group in response to nutrient pollution processes, should be paid attention in crab aquaculture-impacted water bodies.

To mitigate the risk of cyanobacterial blooms, knowledge about the environmental drivers of cyanobacteria dynamics is essential for establishing appropriate management strategies. Until recently, known driving factors on cyanobacteria variations include physical and chemical parameters like nutrient conditions, water temperature, pH, dissolved oxygen, water depth and so on [12–14]. Most investigators have advocated for the consideration of nitrogen and phosphorus as the major causes of eutrophication and cyanobacterial blooms [1,15–17]. Nonetheless, the general consensus is that cyanobacterial blooms are complex events, typically caused by multiple factors occurring simultaneously instead of a single environmental factor [18–21]. Due to the complexity of growth regulators, investigation concerning the effects of environmental factors on the abundance of cyanobacteria requires further efforts.

Lake Guchenghu, a significant water body in the middle-lower area of the Yangtze River in China, is well-known for Chinese mitten crab culture and also provides drinking water for nearby residents. Mitten crab ponds are located in the reclamation area around the lake; with the production of mitten crab increasing rapidly since the 1990s. It has become one of the commercially successful industries, with the largest aquaculture area and the highest output in China [10]. However, environmental pollution problems in Lake Guchenghu have become serious over time [11]. In recent years, water quality of Lake Guchenghu has declined, and it exhibited symptoms of eutrophication where crab culture is one of the key pollution sources [11,22]. An earlier research reported that cyanobacteria dominated in this lake in April and July, accounting for 56%–80% of the total phytoplankton [22]. The dominance of cyanobacteria might be further enhanced if the water quality deteriorates in the future. Notwithstanding the Lake Guchenghu and adjacent mitten crab ponds are located in the same watershed and the ponds withdraw water from the lake, the abundance and diversity of cyanobacteria in crab ponds might be different from those in lake. Nevertheless, there is still a lack of fundamental research about the characteristics of cyanobacteria composition and seasonal variation, and its driving factors in this crab aquaculture-impacted subtropical lake area.

Understanding the dynamics of cyanobacteria and how they relate to the environmental factors in mitten crab culture-impacted water bodies are crucial for unraveling the response of cyanobacteria to mitten crab culture activity, and for helping future management to select precise strategies to ensure aquatic product safety and human health. To this end, the objectives of the current study are to: (1) Determine the composition and abundance of cyanobacteria and related environmental variables in Lake Guchenghu and surrounding mitten crab ponds in different seasons; (2) reveal the influence of mitten crab culture on seasonal variations of the cyanobacteria community; (3) clarify the major environmental factors that affect the development of cyanobacteria; and (4) assess the threat of cyanobacteria in mitten crab culture-impacted water bodies.

2. Materials and Methods

2.1. Study Area

The Lake Guchenghu (Latitude: 31°14′–31°18′ N, Longitude: 118°53′–118°57′ E, Area: 3300 hm^2; Figure 1) located in Nanjing City (Jiangsu Province, China) is a crucial water source for the commercial aquaculture of mitten crab, and provides drinking water to part of the local residents. The lake is in the subtropical monsoonal climate zone and the annual average precipitation is 1105.1 mm, of which the summer rainfall is 428.6 mm (i.e., about 40% of the annual precipitation). Mitten crab ponds are located in the reclamation area around the Lake Guchenghu, pumping water from the lake in spring and discharging wastewater back into the lake in autumn. Due to the reclamation, the lake area is only 39 km^2, while the storage volume is 65 million m^3.

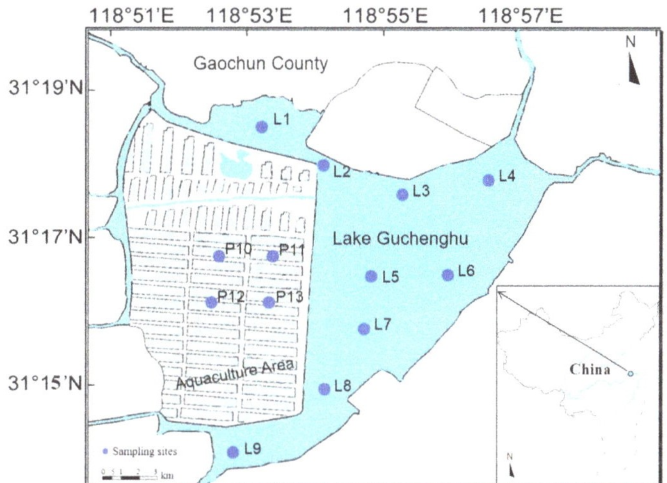

Figure 1. Location of sampling points in Lake Guchenghu and adjacent mitten crab ponds.

2.2. Sample Collection

Water samples were collected from nine sites (L1–L9) in Lake Guchenghu and four sites (P10-P13) in mitten crab ponds (Figure 1). These water samples were gathered in May (spring), August (summer) and November (autumn), 2017 and February (winter), 2018. At each site, about 2 L of water sample was collected from the surface layer (0–0.5 m) with a 2.5 L plexiglass water sampler. The phytoplankton sample (1 L) was extracted from the surface water sample, stored in clean 1 L plastic containers and 15 mL of Lugols iodone solution was instantly added. All samples were stored in a portable refrigerator (about 4 °C) and then immediately transported to the laboratory. To analyze the composition of phytoplankton, it was concentrated to 30 mL to obtain a quantitative sample of phytoplankton after standing for 48 h.

2.3. Analysis Methods

The basic physicochemical parameters, i.e., water temperature (WT), pH, turbidity and dissolved oxygen (DO), were measured using a YSI multiparameter water-quality monitor (HORIBA U-5000, HORIBA, Ltd., Kyoto, Japan). Water depth (WD) was determined using a precision bathometer (SpeedTech SM-5, SpeedTech, Ltd., USA). The standard analysis methods of total nitrogen (TN), total phosphorus (TP), ammonia nitrogen (NH$_4^+$-N), nitrate nitrogen (NO$_3^-$-N), nitrite nitrogen (NO$_2^-$-N), permanganate index (COD$_{Mn}$) and suspended solids (SS) concentrations adopted in current study are described in Jin and Tu (1990). Phytoplankton samples were identified and counted using a microscope (Eclipse Ni-U, Nikon Co., Tokyo, Japan) 10 × 40 magnification with a 0.1 mL counting chamber following earlier studies [23,24]. Each sample was counted three times. The cell density was recorded as cells/L. The measure of algal abundance was based on cell counts. The mean cell volumes of the algae were calculated based on appropriate geometric configurations [25] and converted into biomass with the algal cell density (1 mm^3 ≈ 1 mg fresh weight).

2.4. Statistical Analysis

One-way analysis of variance (ANOVA) was performed to analyze the differences in environmental factors between lakes and ponds in different months using SPSS 16.0 software (SPSS Inc., Chicago, IL, USA). Based on the cyanobacteria relative abundance dataset, the taxa richness, dominance, evenness and Shannon, Simpson and Margalef indices were determined using PAST 3.0 (Øyvind Hammer, Natural History Museum, University of Oslo [ohammer(at)nhm.uio.no], http://folk.uio.no/ohammer/past/). The Kruskal–Wallis non-parametric test was conducted to assess discrepancies in environmental variables among the thirteen sampling sites. To understand the temporal variation of cyanobacteria community structure in the lake and ponds, non-metric multidimensional scaling (NMDS) was performed using PAST 3.0 according to Gower dissimilarity. Analysis of similarities (ANOSIM) was conducted by PRIMER 5.0 software (PRIMER-E, Ltd., Plymouth, United Kingdom) in order to evaluate whether there is a significant difference in cyanobacteria community structure among different months and sampling sites. We adopted redundancy analysis (RDA) to identify the physicochemical parameters affecting the dynamics of cyanobacteria. Before the forward-selection process, all cyanobacteria and water quality data were transformed into $\log_{10}(x + 1)$ format (except pH). All the statistical analyses (RDA) were conducted using CANOCO 4.5. According to the value of the first gradient length of standard deviations, canonical correspondence analysis (CCA) analyses were selected. The RDA results were visualized in the form of ordination diagrams. Species scores are shown as triangles. Each water quality variable is represented by an arrow, which determines an axis. The projection of a taxon on this axis indicates the level of the variable where the taxon is most abundant. Variables with lines that are close to each other and oriented in the same direction are highly positively correlated, while those oriented in opposite directions are highly negatively correlated. Two lines at a 90° angle demonstrate that the corresponding variables are uncorrelated. Pearson correlations between cyanobacteria abundance and environmental factors were performed using SPSS 18.0 software.

3. Results

3.1. Environmental Parameters

The temporal variations of physicochemical parameters in Lake Guchenghu and the crab ponds are presented in Table 1. During the study period, the pH values in Lake and ponds were all above 7.0, ranging from 7.06 to 9.15. WT varied widely throughout the year with ranges of 3.10–29.47 °C and 3.12–31.01 °C in the lake and ponds, respectively. The lowest WT was recorded in February, while the highest WT was observed in August. In May and August, the WT of ponds the was significantly higher than that of the lake ($P < 0.05$). DO varied considerably over time in the ponds, with an average of 11.9 mg/L (lake) and 14.12 mg/L (ponds) with no obvious temporal pattern. The WD of the lake fluctuated (2.82–4.19 m) across different seasons, while the ponds exhibited small changes in WD all year round. The turbidity of the lake peaked in August, while the higher turbidity in the ponds occurred in November and February.

Regarding the nutrient concentrations, NO_2^--N and NO_3^--N in the lake were higher than those in adjacent ponds for most of the months. By contrast, the concentrations of NH_4^+-N, TN and TP in the ponds were evidently higher compared with those in the lake; their concentrations in November and February were higher than in May and August. This could be because the harvest season for crab is in autumn with the discharge of farmed tail water increasing, resulting in higher nitrogen and phosphorus concentrations. The N:P ratio in the ponds (from 4.39 to 23.02) was lower than in the lake (from 5.75 to 62.54) because TP concentration in the pond was significantly higher than that of the lake throughout the four seasons ($P < 0.05$). Moreover, the COD_{Mn} levels in the ponds were higher than in the lake during studied period, with ranges of concentrations 4.64–9.19 mg/L (ponds) and 2.86–4.37 mg/L (lake), respectively. These results indicate that most of the environmental factors of the ponds are significantly different from the lake in May and August.

Table 1. Mean ± standard deviation (SD) value of environmental parameters in thirteen sampling stations of Lake Guchenghu and adjacent mitten crab ponds measured from May 2017 to February 2018.

Variable	May, 2017		August, 2017		November, 2017		February, 2018	
	Lake	Ponds	Lake	Ponds	Lake	Ponds	Lake	Ponds
pH	8.23 ± 0.19	8.76 ± 0.42 *	9.15 ± 0.17	8.24 ± 0.41 *	7.61 ± 0.23	7.06 ± 0.25 *	7.43 ± 0.13	7.45 ± 0.06
WT/(°C)	21.15 ± 0.75	24.93 ± 1.81 *	29.47 ± 0.31	31.01 ± 1.54 *	18.37 ± 0.25	18.40 ± 0.18	3.10 ± 0.14	3.12 ± 1.53
DO/(mg/L)	12.43 ± 2.04	10.23 ± 0.75 *	11.19 ± 0.39	16.49 ± 4.82 *	10.37 ± 0.76	12.89 ± 4.75	13.68 ± 7.33	16.90 ± 2.67 *
WD/(m)	2.82 ± 0.89	1.11 ± 0.02 *	4.12 ± 0.55	1.13 ± 0.01 *	4.19 ± 0.33	0.98 ± 0.02 *	4.05 ± 0.31	0.74 ± 0.27 *
Turb/(NTU)	7.02 ± 4.84	19.5 ± 7.81 *	26.3 ± 12.08	27.9 ± 6.93	18.9 ± 1.62	45.2 ± 24.8 *	4.46 ± 1.10	45.9 ± 19.71 *
NO_2^--N/(mg/L)	0.03 ± 0.00	0.02 ± 0.01 *	0.04 ± 0.01	0.025 ± 0.01	0.02 ± 0.01	0.01 ± 0.01	0.07 ± 0.05	0.03 ± 0.03
NO_3^--N/(mg/L)	0.56 ± 0.07	0.32 ± 0.28 *	0.45 ± 0.14	0.23 ± 0.18 *	0.13 ± 0.04	0.16 ± 0.13	1.87 ± 1.11	0.67 ± 1.01
NH_4^+-N/(mg/L)	0.05 ± 0.02	0.34 ± 0.30 *	0.05 ± 0.02	0.17 ± 0.14 *	0.11 ± 0.04	0.17 ± 0.14	0.15 ± 0.03	2.51 ± 3.81
TN/(mg/L)	1.02 ± 0.11	2.00 ± 0.54 *	1.18 ± 0.17	1.57 ± 0.48 *	1.25 ± 0.07	1.66 ± 0.75	2.47 ± 1.49	3.04 ± 2.62
TP/(mg/L)	0.02 ± 0.01	0.09 ± 0.02 *	0.03 ± 0.01	0.18 ± 0.10 *	0.22 ± 0.02	0.36 ± 0.11 *	0.04 ± 0.00	0.14 ± 0.08 *
N:P	59.6 ± 17.6	22.5 ± 9.62 *	35.5 ± 4.74	11.4 ± 5.75 *	5.75 ± 0.34	4.39 ± 0.87 *	62.5 ± 41.8	23.0 ± 16.8
COD_{Mn}/(mg/L)	4.14 ± 0.30	9.19 ± 1.31 *	2.86 ± 0.19	6.22 ± 1.28 *	4.37 ± 0.18	4.64 ± 2.07	4.05 ± 0.61	7.62 ± 3.03

WT—water temperature; DO—dissolved oxygen; WD—water depth; TN—total nitrogen; Turb—turbidity; TP—total phosphorous; N:P—ratio of total nitrogen to total phosphorus; COD_{Mn}—permanganate index (chemical oxygen demand). Asterisks beside the data of ponds indicate significant differences with respect to the value of lake area (* $P < 0.05$).

3.2. Seasonal Variations of Cyanobacteria

In the study period, a total of 52 cyanobacteria species belonging to 21 genera were detected in Lake Guchenghu; while 16 genera and 32 cyanobacteria species were identified in ponds (Table S1). The genera *Aphanothece*, *Planktothrix*, *Romeria*, *Phormidium* and *Spirulina* were only found in the lake, whereas the genus *Anabaenopsis* were only recorded in ponds. Cyanobacteria richness and species diversity in August and November were greater than those in May and February in both lake and ponds (Table S2).

The relative abundance of cyanobacteria exhibited distinct seasonal changes. Cyanobacteria abundance demonstrated wide variations throughout the year with ranges from 1.12×10^5 cells/L to 2.62×10^7 cells/L in the lake, and the biomass of cyanobacteria varied from 0.05 mg/L to 3.86 mg/L (Figure 2(A1,A2)). In May, cyanobacteria abundance reached the maximum value of 2.62×10^7 cells/L (accounting for more than 94.5% of the total phytoplankton), where *Merismopedia* was the dominant cyanobacteria taxa with 97.7% of relative abundance. Nonetheless, the maximum biomass of cyanobacteria appeared in November (3.86 mg/L, 34.6% of the total phytoplankton) instead of May (Figure 2(A2)). The difference may be due to the fact that the size of the cells of different species can vary widely. *Merismopedia* spp. prevalent in May are tiny sized species which despite being present in very large cell numbers actually contributed little to the total cyanobacterial biomass present in the lake at that time. From Figure 3A, *Oscillatoria* increased and became the dominant genus in August (19.1%) and November (23.8%). In addition, *Dolichospermum* (15.9%) and *Pesudanabaena* (14.0%) were abundant in August; whereas *Aphanocapsa* (13.8%) and *Pesudanabaena* (13.0%) were flourishing in November (Figure 3A).

For the ponds, cyanobacteria abundance and biomass both peaked in August (Figure 2(B1,B2)), with 6.51×10^7 cells/L and 46.80 mg/L (i.e., 92.8% of the total phytoplankton abundance and 97.2% of the total phytoplankton biomass), where *Pesudanabaena* (48.4%) and *Raphidiopsis* (12.4%) were the dominant species (Figure 3B). The maximum value of cyanobacteria biomass in the ponds was much greater than that of the lake. In November, the abundance of cyanobacteria decreased rapidly with less than 2×10^6 cells/L (about 39.1% of the total phytoplankton). The dominant cyanobacteria were *Planktolyngbya*, *Aphanizomenon* and *Oscillatoria*, accounting for 33.9%, 24.4% and 15.1% of the total phytoplankton, respectively (Figure 3B). Notably, cyanobacteria taxa occurred in examined water bodies with the high abundance were filamentous forms (except in May for lake) (e.g., *Oscillatoria, Pesudanabaena, Raphidiopsis, Planktolyngbya,* and *Aphanizomenon*). The proportion of filamentous cyanobacteria in the ponds was higher than that in the lake over the study period, in May and November in particular ($P < 0.01$) (Figure 4).

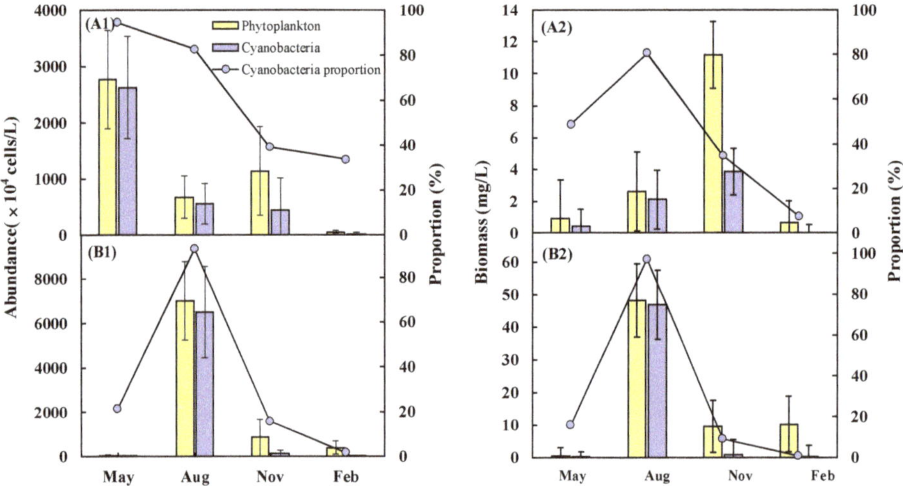

Figure 2. Changes of total phytoplankton abundance and biomass, cyanobacteria abundance and biomass, and cyanobacteria proportion in Lake Guchenghu and adjacent mitten crab ponds from May 2017 to February 2018 ((**A1,A2**): Lake; (**B1,B2**): Ponds).

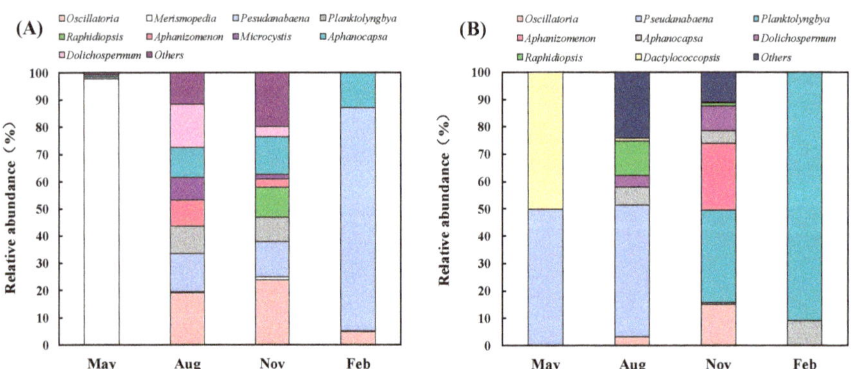

Figure 3. Seasonal variation of cyanobacteria community in Lake Guchenghu and adjacent mitten crab ponds from May 2017 to February 2018 (**A**): Lake; (**B**): Ponds.

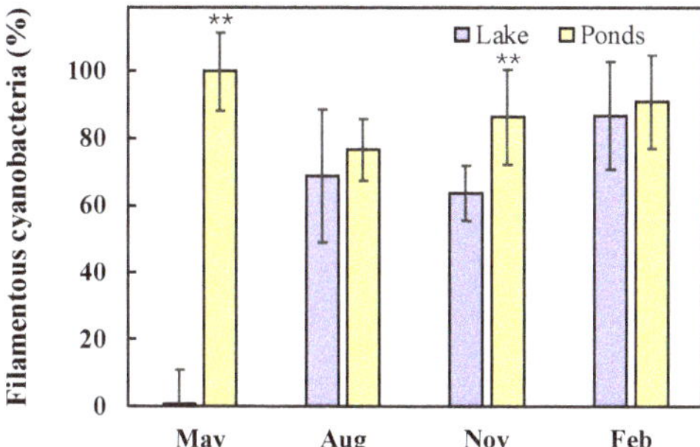

Figure 4. The proportion of filamentous cyanobacteria to total cyanobacteria in Lake Guchenghu and adjacent mitten crab ponds from May 2017 to February 2018. Asterisks above the bars indicate significant differences with respect to the lake area (* $P < 0.05$, ** $P < 0.01$).

During the study periods, five potentially harmful species, all occurred in August and/or November, were identified in the lake and ponds (Table 2). In August, the maximum abundances of the potential toxin producing species in the ponds were much higher than those in lake area. Especially, the abundances of *Dolichospermum circinale* and *Raphidiopsis raciborskii* hit 3.6×10^6 and 8.2×10^6 cells/L in August, respectively. In November, the abundance of potentially harmful cyanobacteria species in the ponds significantly declined.

Table 2. Potential toxin producing species recorded in Lake Guchenghu and adjacent mitten crab ponds.

Species [a]	Family	Maximum Abundance	
		Lake	Ponds
Planktothrix agardhii	Oscillatoriaceae	1.1×10^5 cells/L in November	–
Dolichospermum circinale (synonym: *Anabaena circinale*)	Nostocaceae	1.2×10^5 cells/L in August	3.6×10^6 cells/L in August
Raphidiopsis raciborskii (synonym: *Cylindrospermopsis raciborskii*)	Nostocaceae	–	8.2×10^6 cells/L in August
Cuspidothrix issatschenkoi (synonym: *Aphanizomenon issatschenkoi"*)	Nostocaceae	–	4.5×10^4 cells/L in November
Aphanizomenon flos-aquae	Nostocaceae	1.9×10^5 cells/L in August	–

[a] Potential toxin producing species were identified based on [26] –Nonexistence.

3.3. Quantitative Seasonal Changes in Cyanobacteria Communities

Seasonal variations of cyanobacteria communities at the genera level in the lake and ponds were evident (Figure 5). A remarkable temporal shift in cyanobacteria taxa was observed at the sampling sites in both lake and ponds (Figure 5A,B). All stress values were less than or slightly above 0.1, indicating a good fit by using NMDS analysis [27]. The results of ANOSIM further confirmed distinct discrepancies in the cyanobacteria communities in the lake and ponds among different seasons ($P < 0.01$, Table S3). Figure 5C–F shows that the similarity in cyanobacterial communities between lake and ponds was relatively high in November, but the similarities were not obvious in May and February. In August, the characteristic of cyanobacterial community in the ponds was significantly different from that of the lake ($P < 0.01$, Table S3).

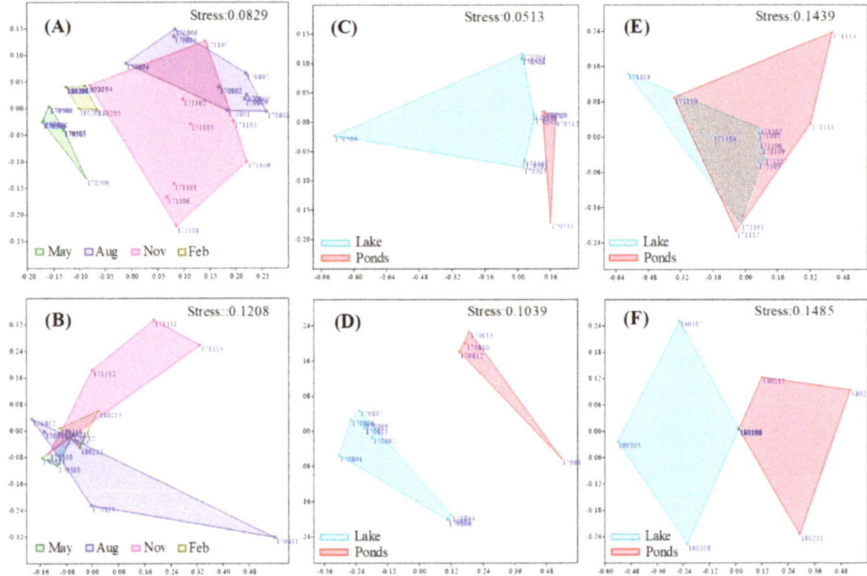

Figure 5. Non-metric multidimensional scaling (NMDS) ordination of cyanobacteria community in Lake Guchenghu and adjacent mitten crab ponds among different months. ((**A**): Lake; (**B**): Ponds; (**C**): May 2017; (**D**): August 2017; (**E**): November 2017; (**F**): February 2018. The first four digits of the number represent the sampling time, and the last two digits represent the sampling point.).

According to cluster analysis (Figure 6), the similarity was relatively high in the same habitat and the same season. In August, the difference in cyanobacterial composition among the sampling sites was significant, especially in the ponds. Consistent with the NMDS analysis (above), Figure 6 reveals that the cyanobacterial composition for the lake and ponds was similar in November, and was significantly discrepant in August.

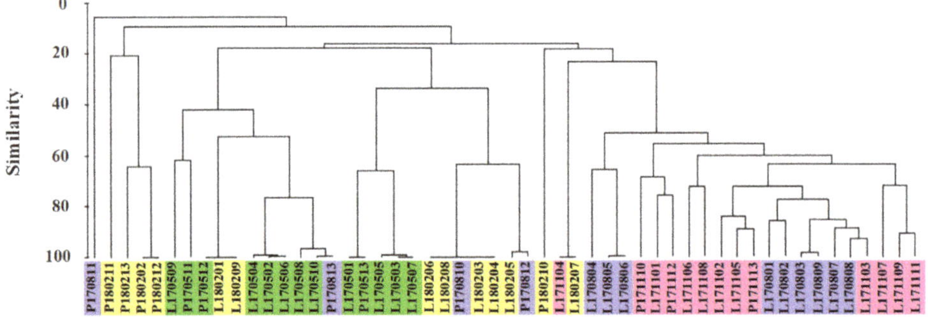

Figure 6. Similarity analysis for cyanobacteria community in Lake Guchenghu and adjacent mitten crab ponds among different months by cluster analysis. (Green for May; Blue for August; Pink for November; and Yellow for February.) ("L" represents Lake; "P" represents Ponds; the first four digits of the number represent the sampling time, and the last two digits represent the sampling point.).

3.4. Relationships between Cyanobacteria Dynamics and Environmental Variables

RDA analysis revealed the relationship between cyanobacteria communities and environmental factors (Figure 7). The environmental variables (details in Table S4) explained 41.8% (54.6%) of the total variance of species distribution in the lake (ponds). In the lake, from 2017 to 2018, the eigenvalues (λ) for RDA axis 1 and axis 2 were 0.356 and 0.136, respectively. The interpretation rate of species-environment relationship for the first two axes was 71.0% (Table S4), which can better explain the relationship between the cyanobacteria community and environmental factors. In crab ponds, the eigenvalues for RDA axis 1 (λ: 0.562) and axis 2 (λ: 0.379) explained a total of 54.6% of the species-environment relation.

The RDA ordination indicated that WD, TN:TP ratio, COD_{Mn}, TP and NO_2^--N are the dominant environmental factors for cyanobacterial composition. For the ponds, N:P ratio, WT, and TP were significant factors of cyanobacteria development from May 2017 to February 2018 (Figure 7; Tables S5 and S6). Our results show that different cyanobacteria taxa contain different environment adaptation abilities. In the lake (Figure 7A), *Oscillatoria* preferred higher TP and NH_4^+-N concentrations; but the opposite was observed in genera like *Chroococcus*, *Pseudanabaena* and *Merismopedia*. WD and turbidity were positively correlated with *Aphanizomenon*, *Dolichospermum*, *Planktolyngbya* and *Oscillatoria*; but were negatively correlated with *Merismopedia*. *Raphidiopsis* abundance was positively correlated with TP and COD_{Mn}, and negatively correlated with N:P ratio. In the ponds (Figure 7B), WT and pH were positively correlated with *Raphidiopsis* and *Pesudanabaena*; but negatively correlated with *Planktolyngbya*, *Aphanizomenon* and *Aphanocapsa*. Besides, many genera clusters (including *Planktolyngbya Aphanizomenon*, *Raphidiopsis* and *Dolichospermum*) were positively correlated with TP; negatively correlated with N:P ratio. On the other hand, the same environmental factors could cause different effects in different habitats. *Planktolyngbya* and *Dolichospermum* were positively correlated with WT and pH in the lake, but negatively associated in the ponds. *Oscillatoria* had a positive correlation with NH_4^+-N in the lake, while the opposite was found in the ponds.

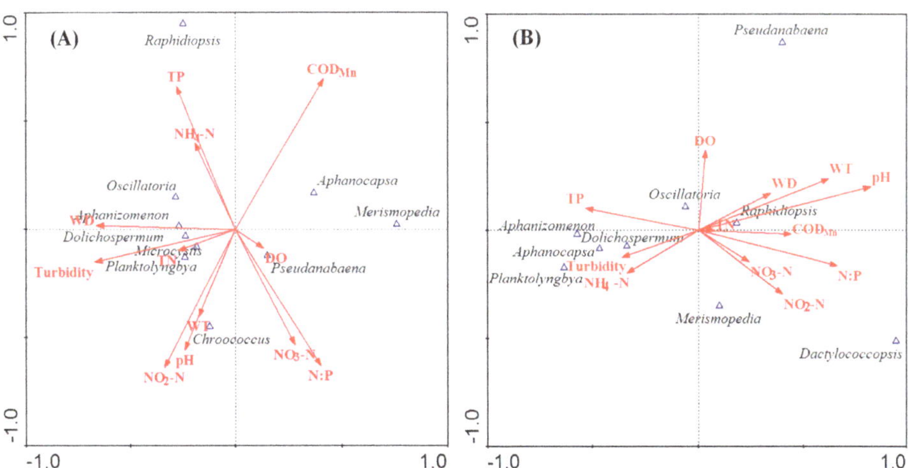

Figure 7. Ordination diagram of redundancy analysis (RDA) for cyanobacteria community associated with environmental factors in Lake Guchenghu and adjacent mitten crab ponds ((**A**): Lake; (**B**): Ponds).

In addition to the RDA results, Pearson correlation coefficients revealed several statistically significant correlations (Figure 8; Table S7). In the lake, cyanobacteria abundance experienced significant positive correlation with TP, but it was negatively correlated with WD, NH_4^+-N, NO_2^--N and TN. Regarding the predominant species, *Merismopedia* abundance correlated positively with the N:P ratio, but negatively with the WD, turbidity, NH_4^+-N, NO_2^--N and TP. *Oscillatoria* abundance correlated positively with pH, WT, turbidity and TP, but negatively with the N:P ratio. *Dolichospermum* abundance exhibited a positive correlation with WD, pH, WT and turbidity, but a negative correlation with COD_{Mn} and N:P. *Raphidiopsis* abundance only demonstrated significant positive correlation with TP. In ponds, whilst cyanobacteria abundance was positively correlated with WT, WD, pH and TP, such correlation was insignificant. *Planktolyngbya* abundance correlated positively with TP, but negatively with pH. *Aphanizomenon* abundance correlated positively with TP, but negatively with COD_{Mn}. The Pearson correlation analysis results (above) are in good agreement with the RDA analysis.

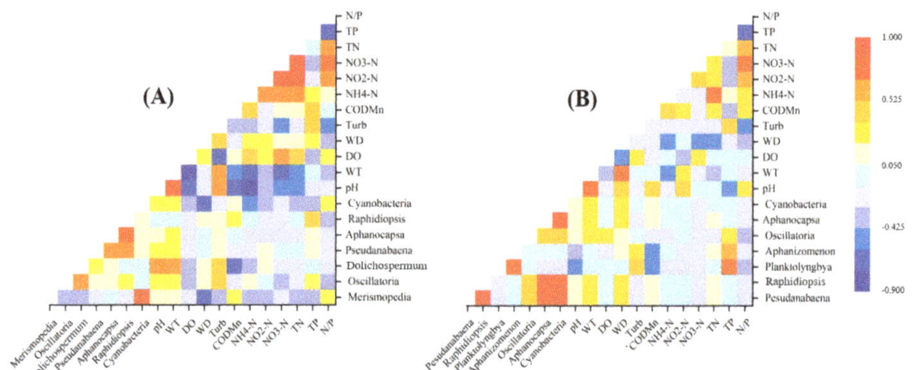

Figure 8. The results of Pearson correlation between predominant cyanobacteria species and environmental parameters in Lake Guchenghu and adjacent mitten crab ponds ((**A**): lake; (**B**): ponds).

4. Discussion

The present study revealed that crab aquaculture has an adverse effect on water quality based on the fact that concentrations of nutrients in crab ponds were evidently higher than those in the adjacent lake area (Table 1). The discharge from the crab ponds (with these nutrients) would increase the risk of pollution in the nearby lake. Regional nutrient enrichment within watersheds might exert a diverse and complex impact on the prevalence and persistence of cyanobacteria over time [13,28]. NMDS analysis indicated that the cyanobacteria community structure had significant seasonal variation in Lake Guchenghu area. Seasonal succession of cyanobacteria has been recorded in many previous studies [29,30]. Physical indicators, limited nutrients, overwintering, and ecological competition all contribute to the temporal variation of the cyanobacteria community.

Our observations revealed that mitten crab culturing had non-negligible impacts on the development of cyanobacteria. The cyanobacterial community in the ponds was different from that of the lake (Figures 5 and 6), especially in August, which was primarily due to the disparities in environmental factors of the two distinctive systems (Table 2). The results of the RDA analysis indicated that WT, WD, turbidity, TP and NH_4^--N played an important role in structuring cyanobacteria assemblages and had a significant influence on cyanobacteria abundance in the lake area. The highest cyanobacteria abundance with the dominance of *Merismopedia* in the lake in spring (May) could have been ascribed to the relative low turbidity and water depth. The low turbidity in May might have contributed to the development of *Merismopedia*, simply because *Merismopedia* was usually the dominant species in clear lakes [31]. Furthermore, Yang et al. (2016) suggested that the water level could alter the patterns of phytoplankton communities and a decline in water level could boost cyanobacteria dominance [32]. The Pearson correlation analysis in the current study also confirmed that a significant negative correlation existed between cyanobacteria abundance and water depth. In fact, WD is never constant in the lake. WD can influence the ecological processes and patterns of waterbodies through water column mixing, light transmission change and nutrient availability [33,34]. However, the species of cyanobacteria are unitary in May, and this change in species composition followed an increase in phosphorus in August and November, which likely promoted the increase in filamentous species such as *Oscillatoria*, *Dolichospermum*, *Pesudanabaena* and *Planktolyngbya* (Figure 3). Most of them, similarly to Nostocales, are nitrogen-fixing species [35]. The dominance of nitrogen-fixing cyanobacteria was mainly due to its significant positive correlation with TP, and negative correlation with TN, NH_4^+-N and NO_2^--N (Figures 7 and 8). With an increase in TP and decrease in N:P ratio from August to November 2017, the abundance of nitrogen-fixing species increased dramatically. Another reason may be that the precipitation increases in summer and autumn in this subtropical monsoonal climate zone, and the water level fluctuates frequently, resulting in many benthic species like *Oscillatoria* being mobilised and becoming entrained within the water column due to the water in the lake being shallow.

By contrast, WT, TP, pH, and N:P ratio were important environmental variables that influenced cyanobacteria development in the ponds, which were different compared to the lake. This confirmed that similar set of environmental factors might not always drive the cyanobacteria development in different types of habitats (lake and ponds). Unlike the lake, the greatest cyanobacteria abundance was observed in ponds where *Pesudanabaena* and *Raphidiopsis* dominated in August (Figure 3), and the maximum abundance was 2.5 times the lake's greatest abundance over the studied period. The WT and TP concentrations in the ponds were evidently higher compared to those in the lake. Especially, in August, the TP concentration in the ponds was six times higher than that in the lake because nutrient control in ponds was more difficult given the need to regularly feed farmed animal and the excretion of waste. Cyanobacteria are known to flourish with relatively high WT and nutrient levels, conditions prevailing in crab culture ponds. Furthermore, crab ponds with distinctly lower values of TN:TP ratios were characterized by a higher proportion of filamentous cyanobacteria compared to the lake, which may be attributed to the fact that most of filamentous cyanobacteria have the capacity of nitrogen fixation [35].

Whilst some studies suggested that cyanobacteria can dominate in both low and high phosphorus conditions [36], our study found that the total phosphorus was the major positive driving factor in governing the cyanobacteria abundance, and nitrogen was not a limiting factor for cyanobacteria proliferation. Numerous studies revealed that increased phosphorus concentrations could lead to high cyanobacteria abundance [37,38]. For example, Andersson et al. (2015) reported the increase of Nostocales was caused by the rise of total phosphorus and decrease of dissolved inorganic nitrogen in the coastal Bothnian Sea, which could support our results. Furthermore, the abundance of cyanobacteria was also related to the alkaline conditions. For instance, the dominant species *Oscillatoria* and *Dolichospermum* were positively correlated with pH in the lake (Figures 7 and 8). However, some of the cyanobacteria taxa (e.g., *Planktolyngbya*, *Aphanizomenon*, *Raphidiopsis*, etc.) in the aquaculture ponds were negatively correlated with pH, where the water was alkaline on average. On one hand, high pH is the result of photosynthesis of cyanobacteria, where the consumption of carbon dioxide would lead to a rise in the pH of the water [39]. On the other hand, such alkaline conditions foreordained cyanobacteria being dominant owing to the reduced efficiency of other phytoplankton species to utilize carbon at high pH [40].

Despite the growing interest in investigating the dynamics of cyanobacteria in freshwater bodies, little effort has been made to account for the composition of cyanobacteria and its threat in crab aquaculture-impacted water bodies, especially crab-culturing ponds. From the results of the present study, the known potential cyanotoxin producers, species like *R. raciborskii* (synonym: *Cylindrospermopsis raciborskii*) and *D. circinale* (synonym: *Anabaena circinale*) [26,41], were abundant taxa found in the crab ponds and the peak abundances of these species occurred in summer. In the ponds, *R. raciborskii* and *D. circinale* were positively correlated with WT, WD, COD_{Mn} and TP; negatively correlated with NO_3^--N and N:P ratio (Figures 7 and 8). Crab ponds with the higher values of WT, WD, TP and COD_{Mn} concentration and the lower concentration of NO_3^--N and N:P ratio would benefit these species in summer. Their excessive expansion suggests a high risk of potential cyanotoxins existing in the ponds and associated cyanotoxins to humans through intake of aquatic products, which could cause serious social and economic losses simultaneously [42,43]. With intensifying eutrophication and global warming, *R. raciborskii* as an invasive species began to dominate many lakes and reservoirs in

China, and the known concentration of CYNs could hit 8.25 µg/L in reservoirs [44,45]. Nonetheless, the current study did not evaluate cyanotoxin levels in freshwater and "seafood" products. Hence, future studies should investigate cyanotoxin, taste and odor levels associated with the abundance of harmful cyanobacteria.

The current study reflected the importance of establishing rational management of the crab culturing industry and controlling the excessive expansion of cyanobacteria in crab culture ponds. Moreover, it is necessary to strengthen the treatment of crab culture wastewater before being discharged into the neighboring lakes in order to reduce the pollution in the lake. According to the survey results of pollution sources and pollution load of Lake Guchenghu in 2012, the main pollution sources were aquaculture culture, industrial wastewater, domestic sewage and crop farming. Aquaculture culture has become the primary source of pollutants in the Lake Guchenghu area [11]. Regular monitoring of cyanobacteria and cyanotoxin in these mitten crab culture-impacted water bodies is recommended in order to avoid the large-scale cyanobacteria proliferation, and control their harmful impact on freshwater "seafood" products and human health.

5. Conclusions

The present study demonstrated that the activities of mitten crab culture played a certain role in the change of cyanobacterial community structure. In mitten crab ponds, whilst cyanobacteria could be found throughout the year, the period related to the highest risk was August (summer) because cyanobacteria (including several potentially harmful species) were most abundant during this time of the year with biomass reaching 46.80 mg/L. By contrast, the cyanobacteria biomass (3.86 mg/L) peaked in November in the nearby lake, accounting for a smaller proportion of the total phytoplankton biomass which have relatively low risk. RDA and Pearson correlation analyses indicated that N:P ratio, COD_{Mn}, TP and $NO_2{}^-$-N had significant influence on cyanobacteria abundance in the lake, whereas WT, N:P ratio and TP played an important role in cyanobacteria dynamics in mitten crab culture ponds. This also confirmed that cyanobacteria development was not always affected by the same set of environmental factors in different types of habitats. The results of this study suggested that regular monitoring on cyanobacteria dynamics in aquaculture-impacted water bodies is necessary to accurately evaluate cyanobacteria species' responses to environmental changes, and predict the risk of harmful cyanobacterial blooms.

Supplementary Materials: The following are available online at http://www.mdpi.com/2073-4441/11/12/2468/s1, Table S1: The list of cyanobacteria species in Lake Guchenghu and adjacent mitten crab ponds (May 2017–February 2018); Table S2: Cyanobacteria community metrics (mean ± standard deviation) observed over four study seasons in Lake Guchenghu basin. Values were calculated from all sites for each season; Table S4: Summary of redundancy analysis between environmental factors and cyanobacteria abundance in Lake Guchenghu and adjacent mitten crab ponds; Table S5: The correlation matrix between RDA axes and environmental factors in Lake Guchenghu from May 2017 to February 2018; Table S6: The correlation matrix between RDA axes and environmental factors in mitten crab ponds from May 2017 to February 2018; Table S7: Results of Pearson correlation between dominant cyanobacteria species and environmental parameters in Lake Guchenghu and adjacent mitten crab ponds.

Author Contributions: Conceptualization, X.G. and H.L.; Methodology, H.C. and Z.M.; Formal analysis, H.L., H.C. and H.D.; Investigation, H.C., Z.M. and Q.Z.; Writing-original draft preparation, H.L.; Writing-review and editing, H.L., H.C. and Z.M.

Funding: This work was supported by the National "Twelfth Five-Year" Plan for Science & Technology (No. 2015BAD13B06), National Natural Science Foundation of China (No. 31870449), Water Conservancy Science and Technology Project of Jiangsu Province (No. 2017013), Water environment governance projects of Taihu (TH2018303), the National Key R & D Program of China (No. 2018YFD0900804) and the Major Science and Technology Program for Water Pollution Control and Treatment (2017ZX07204005).

Acknowledgments: We thank Yuting Wang for her guidance in the data analysis phase.

Conflicts of Interest: The authors declare no conflict of interest.

References

1. Cremona, F.; Tuvikene, L.; Haberman, J.; Nõges, P.; Nõges, T. Factors controlling the three-decade long rise in cyanobacteria biomass in a eutrophic shallow lake. *Sci. Total Environ.* **2018**, *621*, 352–359. [CrossRef]
2. Havens, K.E.; Ji, G.; Beaver, J.R.; Fulton, R.S.; Teacher, C.E. Dynamics of cyanobacteria blooms are linked to the hydrology of shallow Florida lakes and provide insight into possible impacts of climate change. *Hydrobiologia* **2019**, *829*, 43–59. [CrossRef]
3. Lenard, T.; Ejankowski, W.; Poniewozik, M. Responses of Phytoplankton Communities in Selected Eutrophic Lakes to Variable Weather Conditions. *Water* **2019**, *11*, 1207. [CrossRef]
4. Hu, X.; Zhang, R.; Ye, J.; Wu, X.; Zhang, Y.; Wu, C. Monitoring and research of microcystins and environmental factors in a typical artificial freshwater aquaculture pond. *Environ. Sci. Pollut. Res.* **2018**, *25*, 5921–5933. [CrossRef] [PubMed]
5. Ndlela, L.L.; Oberholster, P.J.; Van, W.J.H.; Cheng, P. An overview of cyanobacterial bloom occurrences and research in Africa over the last decade. *Harmful Algae* **2016**, *60*, 11–26. [CrossRef]
6. Liao, J.; Zhao, L.; Cao, X.; Sun, J.; Gao, Z.; Wang, J.; Jiang, D.; Fan, H.; Huang, Y. Cyanobacteria in lakes on Yungui Plateau, China are assembled via niche processes driven by water physicochemical property, lake morphology and watershed land-use. *Sci. Rep.* **2016**, *6*, 36357. [CrossRef]
7. Alvarez, S.; Lupi, F.; Solís, D.; Thomas, M. Valuing Provision Scenarios of Coastal Ecosystem Services: The Case of Boat Ramp Closures Due to Harmful Algae Blooms in Florida. *Water* **2019**, *11*, 1250. [CrossRef]
8. Zeng, Q.; Jeppesen, E.; Gu, X.; Mao, Z.; Chen, H. Cannibalism and Habitat Selection of Cultured Chinese Mitten Crab: Effects of Submerged Aquatic Vegetation with Different Nutritional and Refuge Values. *Water* **2018**, *10*, 1542. [CrossRef]
9. Cai, C.; Gu, X.; Huang, H.; Dai, X.; Ye, Y.; Shi, C. Water quality, nutrient budget, and pollutant loads in Chinese mitten crab (Eriocheir sinensis) farms around East Taihu Lake. *Chin. J. Oceanol. Limn.* **2012**, *30*, 29–36. [CrossRef]
10. Zeng, Q.; Gu, X.; Chen, X.; Mao, Z. The impact of Chinese mitten crab culture on water quality, sediment and the pelagic and macrobenthic community in the reclamation area of Guchenghu Lake. *Fish. Sci.* **2013**, *79*, 689–697. [CrossRef]
11. Gu, X.K.; Gu, X.H.; Zeng, Q.F.; Mao, Z.G.; Li, X.G.; Wang, Y.P.; Wang, W.X. Spatial-Temporal Variation and Developing Tendency of Water Quality in Gucheng Lake and Inlets and Outlets of the Lake. *J. Ecol. Rural Environ.* **2016**, *32*, 68–75.
12. Fortin, N.; Aranda-Rodriguez, R.; Jing, H.; Pick, F.; Bird, D.; Greer, C.W. Detection of microcystin-producing cyanobacteria in Missisquoi Bay, Quebec, Canada, using quantitative PCR. *Appl. Environ. Microbiol.* **2010**, *76*, 5105–5112. [CrossRef] [PubMed]
13. Brasil, J.; Attayde, J.L.; Vasconcelos, F.R.; Dantas, D.D.; Huszar, V.L. Drought-induced water-level reduction favors cyanobacteria blooms in tropical shallow lakes. *Hydrobiologia* **2016**, *770*, 145–164. [CrossRef]
14. Yang, C.; Nan, J.; Li, J. Driving factors and dynamics of phytoplankton community and functional groups in an estuary reservoir in the Yangtze River, China. *Water* **2019**, *11*, 1184. [CrossRef]
15. Bullerjahn, G.S.; McKay, R.M.; Davis, T.W.; Baker, D.B.; Boyer, G.L.; Anglada, L.V.; Doucette, G.J.; Ho, J.C.; Irwin, E.G.; Kling, C.L. Global solutions to regional problems: Collecting global expertise to address the problem of harmful cyanobacterial blooms. A Lake Erie case study. *Harmful Algae* **2016**, *54*, 223–238. [CrossRef]
16. Funari, E.; Manganelli, M.; Buratti, F.M.; Testai, E. Cyanobacteria blooms in water: Italian guidelines to assess and manage the risk associated to bathing and recreational activities. *Sci. Total Environ.* **2017**, *598*, 867–880. [CrossRef]
17. Conley, D.J.; Paerl, H.W.; Howarth, R.W.; Boesch, D.F.; Seitzinger, S.P.; Havens, K.E.; Lancelot, C.; Likens, G.E. Controlling eutrophication: Nitrogen and phosphorus. *Science* **2009**, *323*, 1014–1015. [CrossRef]
18. Heisler, J.; Glibert, P.M.; Burkholder, J.M.; Anderson, D.M.; Cochlan, W.; Dennison, W.C.; Dortch, Q.; Gobler, C.J.; Heil, C.A.; Humphries, E. Eutrophication and harmful algal blooms: A scientific consensus. *Harmful Algae* **2008**, *8*, 3–13. [CrossRef]
19. Chan, F.; Pace, M.L.; Howarth, R.W.; Marino, R.M. Bloom formation in heterocystic nitrogen-fixing cyanobacteria: The dependence on colony size and zooplankton grazing. *Limnol. Oceanogr.* **2004**, *49*, 2171–2178. [CrossRef]

20. Lee, T.A.; Rollwagen-Bollens, G.; Bollens, S.M.; Faber-Hammond, J.J. Environmental influence on cyanobacteria abundance and microcystin toxin production in a shallow temperate lake. *Ecotoxicol. Environ. Saf.* **2015**, *114*, 318–325. [CrossRef]

21. Woodhouse, J.N.; Kinsela, A.S.; Collins, R.N.; Bowling, L.C.; Honeyman, G.L.; Holliday, J.K.; Neilan, B.A. Microbial communities reflect temporal changes in cyanobacterial composition in a shallow ephemeral freshwater lake. *ISME J.* **2016**, *10*, 1337–1351. [CrossRef] [PubMed]

22. Zeng, Q.; Gu, X.; Mao, Z.; Sun, M.; Gu, X. Assessment of trophic levels and phytoplankton variation in Guchenghu Lake and canal route. *Chin. Environ. Sci.* **2012**, *32*, 1487–1494.

23. Jin, X.C.; Tu, Q.Y. *The Standard Methods for Observation and Analysis in Lake Eutrophication*; China Environmental Science Press: Beijing, China, 1990; p. 240.

24. Hu, H. *The Freshwater Algae of China: Systematics, Taxonomy and Ecology*; Science Press: Beijing, China, 2006.

25. Hillebrand, H.; Dürselen, C.; Kirschtel, D.; Pollingher, U.; Zohary, T. Biovolume calculation for pelagic and benthic microalgae. *J. Phycol.* **1999**, *35*, 403–424. [CrossRef]

26. Carmichael, W.W. Health Effects of Toxin-Producing Cyanobacteria: "The CyanoHABs". *Hum. Ecol. Risk Assess. Int. J.* **2001**, *7*, 1393–1407. [CrossRef]

27. Clarke, K.R. Non-parametric multivariate analyses of changes in community structure. *Aust. J. Ecol.* **1993**, *18*, 117–143. [CrossRef]

28. Walls, J.T.; Wyatt, K.H.; Doll, J.C.; Rubenstein, E.M.; Rober, A.R. Hot and toxic: Temperature regulates microcystin release from cyanobacteria. *Sci. Total Environ.* **2018**, *610*, 786–795. [CrossRef]

29. Pilkaitytė, R.; Razinkovas, A. Seasonal changes in phytoplankton composition and nutrient limitation in a shallow Baltic lagoon. *Boreal Environ. Res.* **2007**, *12*, 551–559.

30. Dalu, T.; Wasserman, R.J. Cyanobacteria dynamics in a small tropical reservoir: Understanding spatio-temporal variability and influence of environmental variables. *Sci. Total Environ.* **2018**, *643*, 835–841. [CrossRef]

31. Qin, X.B.; Huang, P.Y.; Liu, M.H.; Ma, C.X.; Yu, H.X. Phytoplankton assemblage classification in Anbang river wetland in summer. *South Chin. Fish. Sci.* **2007**, *3*, 1–7.

32. Yang, J.; Lv, H.; Liu, L.; Yu, X.; Chen, H. Decline in water level boosts cyanobacteria dominance in subtropical reservoirs. *Sci. Total Environ.* **2016**, *557*, 445–452. [CrossRef]

33. Liu, L.; Liu, D.F.; Johnson, D.M.; Yi, Z.Q.; Huang, Y.L. Effects of vertical mixing on phytoplankton blooms in Xiangxi Bay of Three Gorges Reservoir: Implications for management. *Water Res.* **2012**, *46*, 2121–2130. [CrossRef] [PubMed]

34. Wantzen, K.M.; Rothhaupt, K.O.; Mörtl, M.; Cantonati, M.; László, G.; Fischer, P. Ecological effects of water-level fluctuations in lakes: An urgent issue. *Hydrobiologia* **2008**, *613*, 1–4. [CrossRef]

35. Bothe, H.; Schmitz, O.; Yates, M.G. Newton WE Nitrogen fixation and hydrogen metabolism in cyanobacteria. *Microbiol. Mol. Biol. Rev.* **2010**, *74*, 529–551. [CrossRef]

36. Carey, C.C.; Ibelings, B.W.; Hoffmann, E.P.; Hamilton, D.P.; Brookes, J.D. Eco-physiological adaptations that favour freshwater cyanobacteria in a changing climate. *Water Res.* **2012**, *46*, 1394–1407. [CrossRef] [PubMed]

37. An, W.C.; Li, X.M. Phosphate adsorption characteristics at the sediment—Water interface and phosphorus fractions in Nansi Lake, China, and its main inflow rivers. *Environ. Monit. Assess.* **2009**, *148*, 173–184. [CrossRef]

38. Andersson, A.; Höglander, H.; Karlsson, C.; Huseby, S. Key role of phosphorus and nitrogen in regulating cyanobacterial community composition in the northern Baltic Sea. *Estuar. Coast. Shelf Sci.* **2015**, *164*, 161–171. [CrossRef]

39. Paerl, H.W.; Paul, V.J. Climate change: Links to global expansion of harmful cyanobacteria. *Water Res.* **2012**, *46*, 1349–1363. [CrossRef]

40. Dokulil, M.T.; Teubner, K. Cyanobacterial dominance in lakes. *Hydrobiologia* **2000**, *438*, 1–12. [CrossRef]

41. Antunes, J.T.; Leão, P.N.; Vasconcelos, V.M. *Cylindrospermopsis raciborskii*: Review of the distribution, phylogeography, and ecophysiology of a global invasive species. *Front. Microbiol.* **2015**, *6*, 473. [CrossRef]

42. Kozak, A.; Celewicz-Gołdyn, S.; Kuczyńska-Kippen, N. Cyanobacteria in small water bodies: The effect of habitat and catchment area conditions. *Sci. Total Environ.* **2019**, *646*, 1578–1587. [CrossRef]

43. Drobac, D.; Tokodi, N.; Lujić, J.; Marinović, Z.; Subakov-Simić, G.; Dulić, T.; Važić, T.; Nybom, S.; Meriluoto, J.; Codd, G.A. Cyanobacteria and cyanotoxins in fishponds and their effects on fish tissue. *Harmful Algae* **2016**, *55*, 66–76. [CrossRef] [PubMed]

44. Lei, L.; Peng, L.; Huang, X.; Han, B. Occurrence and dominance of *Cylindrospermopsis raciborskii* and dissolved cylindrospermopsin in urban reservoirs used for drinking water supply. *South China Environ. Monit. Assess.* **2014**, *186*, 3079–3090. [CrossRef] [PubMed]
45. Liu, L.; Chen, H.; Liu, M.; Yang, J.R.; Xiao, P.; Wilkinson, D.M.; Yang, J. Response of the eukaryotic plankton community to the cyanobacterial biomass cycle over 6 years in two subtropical reservoirs. *ISME J.* **2019**, *13*, 2196–2208. [CrossRef] [PubMed]

Article

An Improved Logistic Model Illustrating *Microcystis aeruginosa* Growth Under Different Turbulent Mixing Conditions

Haiping Zhang [1], Fan Huang [1], Feipeng Li [2,*], Zhujun Gu [2], Ruihong Chen [1] and Yuehong Zhang [1]

[1] College of Environmental Science and Engineering, Tongji University, Shanghai 200092, China; hpzhang@tongji.edu.cn (H.Z.); huangfan__fan@163.com (F.H.); chen_ruihong@foxmail.com (R.C.); 1710798@tongji.edu.cn (Y.Z.)
[2] School of Environment and Architecture, University of Shanghai for Science and Technology, Shanghai 200093, China; chenghao_1009@163.com
* Correspondence: leefp@126.com; Tel.: +86-021-65980757

Received: 24 January 2019; Accepted: 22 March 2019; Published: 31 March 2019

Abstract: To overcome the limitations of the normal logistic equation, we aimed to improve the logistic model under hydrodynamic conditions for the examination of the responses of cyanobacterium, coupled turbulence mixing, and growth of cyanobacterium in population dynamics models. Selecting *Microcystis aeruginosa* and experimenting with the ideal conditions in a laboratory beaker, the chlorophyll-a concentration reached the corresponding maximum under each turbulent condition compared with the control. According to the experiment results, the theory of mass transfer, turbulence mixing, and the logistic equation are organically combined. The improved logistic growth model of *Microcystis aeruginosa* and competition growth model in the symbiont *Scenedesmus quadricauda* under turbulent conditions were established. Using the MATLAB multi-parameter surface fitting device, both models produced good fitting effects, with $R > 0.95$, proving that the results fit the models, and demonstrating the relationship of the unity of nutrient transfer and algae growth affected by turbulence mixing. With continuous increases in turbulent mixing, the fitted curve became smoother and steadier. Algae stimulated by turbulence accelerate reproduction and fission to achieve population dominance. The improved logistic model quantitatively explains the *Microcystis aeruginosa* response to turbulence and provides a basis to represent ecological and biogeochemical processes in enclosed eutrophic water bodies.

Keywords: *Microcystis aeruginosa*; logistic equation; max algal population; hydrodynamic; mass transfer

1. Introduction

In hypertrophic freshwater lakes, turbulent mixing is increasingly thought to be a key factor in the growth, blooms, and composition of the cyanobacterium *Microcystis aeruginosa* [1–4]. Turbulent mixing preventing nuisance blooms of cyanobacterium has been reported at the population and ecosystem levels in a few cases [5–7]. Bronnenmeier and Märkl [8] found that higher turbulence can result in algal cell wall breakage, and Regel et al. [9] found that turbulence has little effect on the algae under low turbulence intensity, whereas high shear turbulence can cause cell deformation or breakage, suggesting an inhibitory effect of higher turbulence intensities on algae growth. However, due to the complex interactions between algae physiology and behavior, nutrient dynamics, and the light environment, many studies failed to accurately characterize these inhibitory effects in nature [10,11]. To qualitatively describe algae dynamics and correlation with turbulent mixing, researchers have conducted series of laboratory tests or enclosed experiments to produce more controlled environments [12–14]. The velocity factor has been proposed as the factor for determining the algal growth rate [15]. However, the growth responses of cyanobacterium compared to the flow velocity and corresponding nutrient uptake obtained from laboratory experiments are often contradictory [11]. These conflicting results are not only due to many studies using qualitative analyses, but also due to the adopted flow velocity, which can only represent the average algal environment rather than the environment surrounding the cells.

Several studies have reported the algae physiological responses to small-scale fluid motion and corresponding nutrients uptake at the cellular level. Grobbelaar [16] found that increased turbulence, which increased exchange rates of nutrients and metabolites between algae cells and their growth medium, together with increased light/dark frequencies, increased productivity and photosynthetic efficiency. Arin et al. [17] suggested that small-scale turbulence may affect the community response differently when nutrient conditions vary. Increased turbulence is relatively important for phytoplankton when nutrients are plentiful, whereas in nutrient-poor conditions, the effect of turbulence is negligible. A possible mechanism could be that small-scale turbulence accelerates the flowing of ions toward the surface of the algae cells; hence, the nutrients can be absorbed by algae under eutrophic conditions. However, Iversen et al. [18] found that small-scale turbulence appeared to affect diatoms and primary production rates regardless of nutrient condition. In terms of nutrient uptake of algae cells, turbulence may have a hysteresis effect on algae cell growth [19]. A reduction in turbulence affects the thickness of the cell laminar boundary layer, which increases the ion flux to the cell with increasing turbulence intensity. Hondzo and Wuest [20] proposed a conceptual model describing how to integrate fluid motion in Monod-type kinetics, accounting for the effect of turbulence on microorganism growth and nutrients uptake. However, the responses of algae to turbulence flow are species-specific. These conclusions from red tide algae, green algae, and diatoms may not be useful in explaining the effect of turbulence on cyanobacterium, which have specific physical characteristics. Xiao et al. [21] found that both *Microcystis flos-aquae* and *Anabaena flos-aquae* adjust their growth rates in response to turbulence levels via the asynchronous cellular stoichiometry of C, N, and P, and especially the phosphorus strategy, to improve the nutrient use efficiency. However, the current understanding of the direct effects of small-scale turbulence on *M. aeruginosa* in eutrophic water bodies remains limited. Missaghi et al. [22] quantified the influence of fluid motion on the growth rate and vertical cell distribution of *M. aeruginosa*.

Shallow lakes are strongly influenced by wind-driven wave turbulence [23]. Microbial communities are strongly influenced by wind waves, often resulting in the formation and persistence of cyanobacterial blooms in the surface waters [24,25]. Nutrient supply is also regarded as a main factor influencing algal mass propagation during bloom periods [26]. Although a number of studies have investigated the impacts of turbulence and mixing on the growth of several cyanobacteria, to date, the direct physical effect on the production of microcystins is still unclear [27]. In a six-day mesocosm experiment in Lake Taihu that is situated in the Changjiang (Yangtze) delta, Zhou et al. [28] demonstrated that turbulence promoted the growth of Microcystis spp., especially toxic Microcystis spp., and increased the proportion of toxic Microcystis spp. In the short term (approx. three days). Zhang [29] conducted a series of laboratory experiments in which nutrients were assumed to be plentiful and light available was identical, and found that different turbulence intensities (7.40×10^{-3} m^{2}s^{-3} < energy dissipation rate ε < 1.50 m^{2}s^{-3}) promoted the growth of *Microcystis aeruginosa* compared with the stationary condition, despite some differences in the degree of promotion. Turbulent mixing increases diffusion rate on cell surface and enhances uptake of nutrients [20,28]. The corresponding largest algae biomass was also found in each experiments, and the algae growth rate increased with decreased biological saturation as a result of the relationship between nutrition and space for population growth [29]. Although all the trends of the growth of *Microcystis aeruginosa* were in agreement with the logistic growth, existing logistic models cannot characterize the effects of the different types of bloom outbreaks under different turbulence conditions.

The aim of present study was to introduce turbulence mixing into the examination of the responses of *Microcystis aeruginosa* to coupled turbulence flow and the growth of *Microcystis aeruginosa* in the models of population dynamics. In this study, we found: (1) the hybrid hydraulics can increase the space capacity of algae growth, and (2) the increased rate of algal population is directly proportional to the space amount of population. On this basis, a beaker experiment with controlled variables was conducted to study the influence of hydrodynamics on algae growth. Light, temperature, and nutrient conditions were fixed and the average water mixing rate of the beaker hydrodynamics was used as the variable in the experiments. An improved logistic model of *Microcystis aeruginosa* and a competition growth model with the symbiont *Scenedesmus quadricauda* were established based on the experimental results, among which the data for the competition model were obtained from our research team member Zhang's masters thesis [29].

2. Materials and Methods

2.1. Experimental Procedure

Microcystis aeruginosa (FACHB-905) and *Scenedesmus quadricauda* (FACHB-507) were purchased from the Institute of Hydrobiology, Chinese Academy of Sciences (Wuhan, China). The algae were cultivated in BG11 culture medium for 1 week until the logarithmic growth phase at 4000 lux light intensity and a constant temperature of 26 °C. Experimental medium and utensils were all sterilized at 121 °C for 30 min and then cooled to room temperature before use. Glassware used in the experiments was prepared by soaking in 0.1 M hydrochloric acid overnight and rinsed. The experiments were conducted in an artificial climate chamber at constant temperature and light conditions. The specific volume of the medium was added to a 2 L beaker and then the *M. aeruginosa* in the logarithmic growth phase were added to 1800 mL of supplemented media. The date of inoculation was recorded, and algae density and chlorophyll-a (Chl-a) concentrations were measured every day. Each experiment was repeated 3 times. The experiment period was 20 days.

2.2. Sample Analysis and Date Processing

Algae growth was quantified using the number of algae cells and Chl-a concentration. Algae cells were counted using a hem cytometer under an optical microscope (Olympus CX21, Olympus Corporation, Tokyo, Japan). The Chl-a concentration was determined by phytoplankton classification using PHYTO-PAM fluorescence analyzer (Walz, Effeltrich, Germany). When the Chl-a concentration in the sample was higher than 300 μg/L, fluorescence resorption from the chlorophyll molecule could interfere with the results. To avoid this effect, when the measured Chl-a content was higher than 300 μg/L, the sample was diluted and re-measured.

A magnetic stirrer created circulation flow in the water to form a closed turbulent environment. Turbulent intensity was controlled by adjusting the speed of the magnetic stirrer. The experiments included one control and three turbulent conditions with speeds of 100 rpm, 150 rpm, and 200 rpm. The relationship between water turbulent mixing capabilities and flow velocity were tested throughout the experiments. A three-dimensional model was established using Pro Engineer wildfire 5.0 (Parametric Technology Corporation, Boston, MA, USA). The Mesh under Ansys14.0 software package (ANSYS, America) was used to calculate the unstructured mesh. Numerical analysis was fitted using the cure fitting tool in MATLAB R2013a (MathWorks, Natick, MA, USA). Image data were processed using Origin Pro 8.0 (OriginLab, Northampton, MA, USA).

2.3. Test for Hydraulic Parameter Description

A high definition camera (Sony NEX-5N, Sony Corporation, Tokyo, Japan) with an EF S18–55 mm lens that is not used for full frame was used to observe the water mixing characteristics in the beaker. The parameter was set to 25 frames/s, shutter 1/40 s, and aperture F8. The camera focal plane was parallel to the container axis. Algae were added to 2 L of water in the test vessel and then the speed of the magnetic stirrer was adjusted. The camera was turned on when the rotational speed was stable, and 40 μL dye was added to the water on the surface in the center. The RGB color (Red-green-blue) gradient was used to record the mixing of the dye. As shown in Figure 1, the dye was completely mixed into the water. The central vortex flow path controlled the water mixture and resulted in an exchange between the upper and lower water. For example, the RGB color gradient over time in the beaker under 100 rpm stirring is shown in Figure 2.

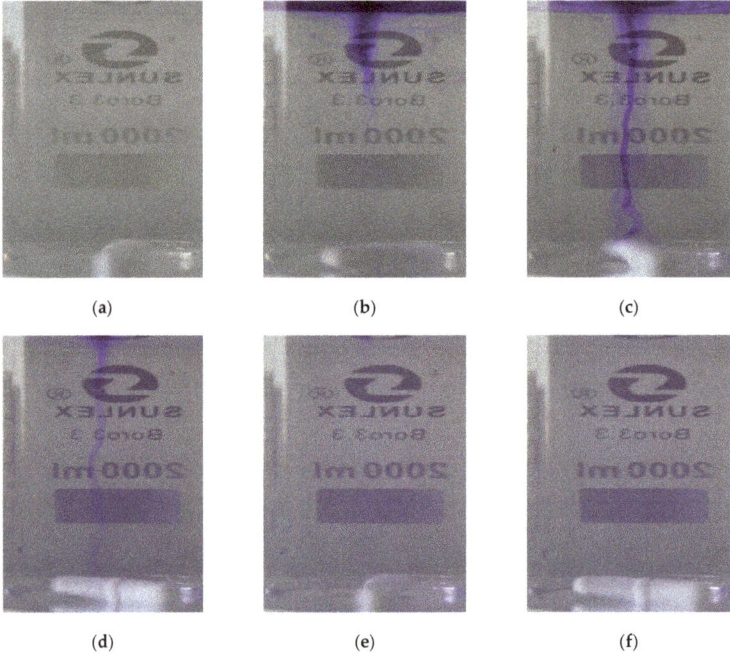

Figure 1. The dye around the beaker achieved complete mixing due to the stirring paddle, among which Fig a-f are RGB color gradient at the time of 0 s (**a**), 1 mim (**b**), 2 min (**c**), 4 mim (**d**), 6 min (**e**) and complete mixing (**f**), respectively.

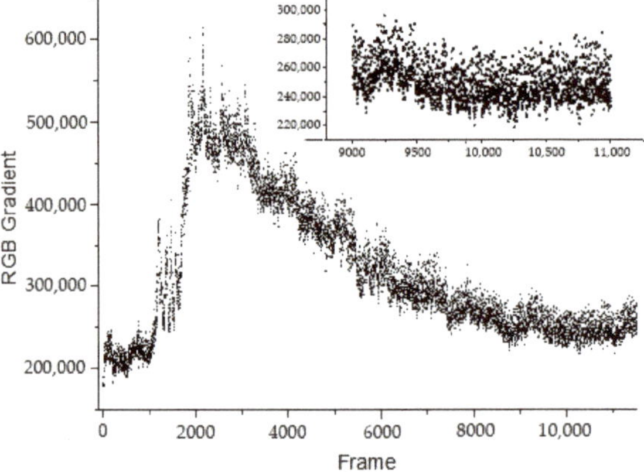

Figure 2. Variation of RGB color gradient with time in a beaker under 100 rpm stirring, among which the inset image is an enlargement of local data.

Polynomial was used for fitting discrete data to eliminate noise. The maximum value was taken as the demarcation point, and the variation of the RGB color gradient over time was divided into 2 parts. The lowest point at the front of the RGB gradient was set as a starting point and the lowest point at the tail of the RGB gradient value was set as the end point. The interval between the 2 points was the mixed frame. The mixing time was determined by frame rate. For example, for the beaker under 100 rpm stirring, the end data fitting is shown in Figure 3. According to the above numerical method, the mixing time under different stirrer speeds was calculated and the reciprocal mixing time was taken as the water mixing rate. Figure 4 shows the significant correlation (R^2 = 0.9940) between the mixing time and stirrer speed. So, the mixing disturbance capacity could be described by the reciprocal mixing time in these beakers.

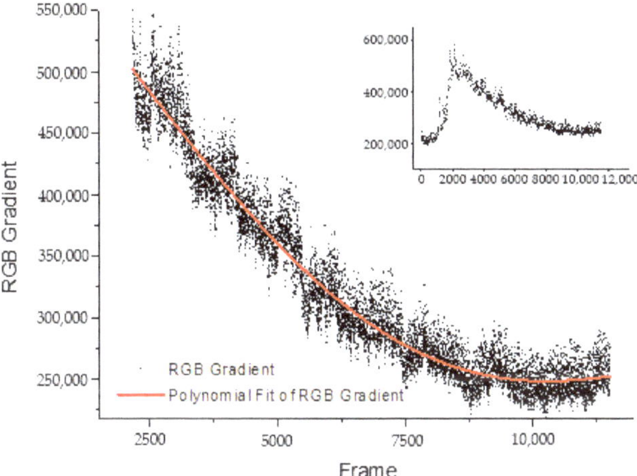

Figure 3. The end data fitting under 100 rpm stirring, among which the inset image is the variation of RGB color gradient with time in a beaker under 100 rpm stirring.

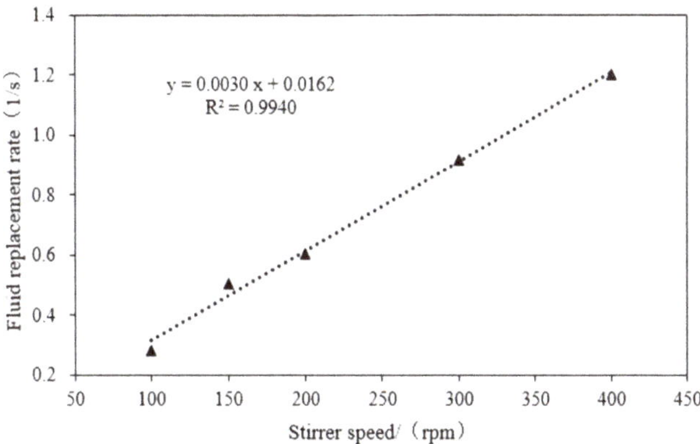

Figure 4. Correlation between the mixing time and stirrer speed.

3. Experimental Results and Model Constructing

3.1. Effect of Turbulence Mixing on Growth of Microcystis aeruginosa

Figure 5a,b show the growth of Cyanobacteria (*Microcystis aeruginosa*) under different turbulence intensities. Compared to the control condition, turbulence promoted the growth of

Microcystis aeruginosa. The Chl-a concentration change in *Microcystis aeruginosa* and the change in the number of algae cells under different turbulent mixing conditions were significantly related ($r > 0.96$). As the curves show, the growth trend between different turbulence conditions were approximately same: the lag phase occurs, then the logarithmic phase is entered, and finally stability is achieved, in agreement with the logistic growth model. Under the same light intensity and nutrient level conditions, this kind of algae is assumed to have the same r value (maximum growth rate). At the beginning of cultivation, *M. aeruginosa* was in the adaptive phase, in which the synthesis rate and concentration of Chl-a were lower than during the other phases. After six days, the growth in *Microcystis aeruginosa* entered into the logarithmic phase, in which the concentration of Chl-a increased logarithmically and differences between groups appeared. Based on the microscopic examination of the blood plate count to determine the cell damage rate, there was no algal cell damage. Under low hydraulic shear strengths, the hydraulic shear effect does not destroy the algae cell.

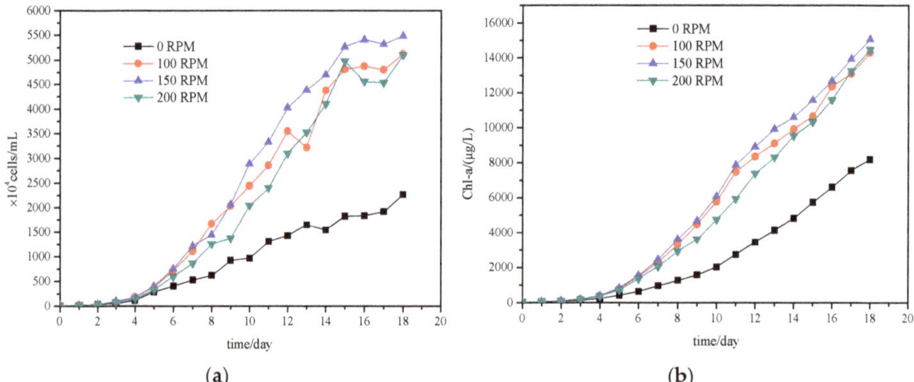

Figure 5. (**a**) Cell number density and(**b**) Chl-a concentration changes in *Microcystis aeruginosa* under different turbulence conditions.

3.2. Construction of Logistic Model under Water Disturbance

3.2.1. Hypothesis

The maximum biomass or maximum abundance (M) is related to the combined homogenization effects between the nutrients and harmful substances in the water. The homogenization effect of the concentration field is based on the turbulent mixing. Based on the linear relationship between the maximum concentrations and turbulence mixing, and combining the maximum space capacity of algae growth under turbulent conditions, the logistic model can be optimized in terms of hydrodynamic force.

3.2.2. Modified Equation Deduction Based on Active M

Due to the continuous nutrient consumption and the waste discharge of the cell, the size of the biological growth space in the surroundings was affected by the mass transfer efficiency between the tiny water body (the laminar boundary layer) around the cell and the large water body (diffusion rate of nutrient and self-produced waste of the cell). This mass transfer efficiency was directly proportional to the concentration gradient between tiny water body (laminar boundary layer) and the large water body around the cell. Under eutrophication conditions, the beaker can be regarded as a large water body. Turbulent water mixing can prevent hazardous substances around the cell from accumulating and replenish nutrients around the tiny water body (Figure 6).

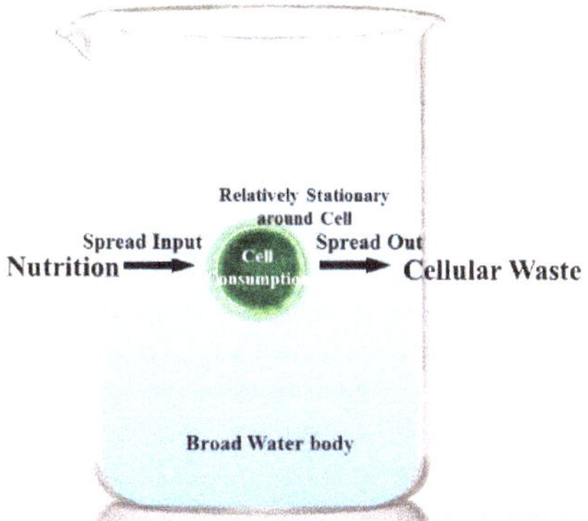

Figure 6. Water mixing can prevent hazardous substances around cell from accumulating and replenish nutrients around the tiny water body.

As described above, the material transport capacity can be characterized by the fluid replacement rate to which nutrient transport by microfluidics is related. Assuming that infinity concentration is *c*, complete mixture is represented as the speed at which infinite distance fluid replaces the fluid block of nutrient depletion and water stacking in the laminar boundary layer.

$$v_{replace} = \frac{V}{T} \tag{1}$$

where V is the volume of reactor, T is the complete mixture time, and $v_{replace}$ is the mixing rate.

The biochemical reaction equation is assumed to be $A + B \rightarrow C + D$. Then the growth limitation has the following relationship:

$$K(ABCD) \approx \frac{c(A)c(B)}{c(C)c(D)} \tag{2}$$

where K($ABCD$) is reaction rate, A and B are the nutrients, and C and D are the metabolites.

This relationship reflects the effect of the concentration of substances surrounding the cell on the growth of the cell in the period of balance or dynamic balance. Mass transfer is the key to a well-distributed concentration field and the limitation on increasing algae density. Under ideal growth conditions, the concentration of extracellular substances has a significant influence on the growth of algae. In the impounded body, concentration gradients can reach an equilibrium state within infinite time. However, cell biological processes are dynamic equilibrium processes, and waste accumulation and nutrient decrease lead to cell death. So, delayed extracellular diffusion of substances A, B, C, and D influence the maximum biomass, whereas the hydraulic condition decreases C and D and increases A and B by homogeneous mixing, which pushes the reaction toward the right, that is, promotes the growth of algae.

The fluid replacement ratio affects the concentration of the laminar boundary layer material, which is directly proportional to the nutrient concentration with which the cell is in contact. The fluid replacement ratio is directly proportional to the concentration of the substances in the area surrounding of the cell; therefore, the fluid replacement ratio is directly proportional to the mass of cell growth (maximum biomass or abundance).

The differential form of the logistic function is described as the following: growth speed k is directly proportional to the residual biomass $(M - P)$:

$$\frac{dP}{dt} = kP = r(M - P)P \tag{3}$$

where P is biomass or abundance, t is time, k is growth rate of the algae, and M is maximum biomass or maximum abundance.

The influence of mass transfer efficiency on maximum biomass (abundance) results in M being optimized into a linear relationship as follows:

$$M = M_0 + K_M v_{replace} \tag{4}$$

where M_0 is the maximum biomass (abundance) in static moment and K_M is constant.

Then, the integral form of the logistic function is:

$$P(t) = \frac{M}{1 + e^{-rM(t-t^*)}} \tag{5}$$

where r is proportional constant of growth rate and space, and t^* is the time to reach half of maximum biomass (abundance).

The relationship between mass transfer efficiency and M is substituted into an integral form of the logistic function:

$$P\left(t, U\right) = M_0 + KMvreplace1 + e - (M_0 + KMvreplace)(t - t^*) \tag{6}$$

As such, two-dimensional fitting target formula was achieved. The parameter optimized hydrodynamic force logistic function can be achieved using the *N-t-P* relational formula for parameter fitting.

3.2.3. Model Verification

Multi-parameter estimation adopts least squares estimation based on the regional optimization idea. In the initial value area, the best function matching of data is found by minimizing the sum of squared errors. Parameter c is selected to minimize the weighted sum of squares of residuals (or deviations) between fitting data and actual observed values at each point. The curve is the fitting curve of data under the meaning of weighted least square. There are many methods that have been successfully used to solve fitting curves. For linear models, parameters can generally be calculated by establishing and solving equations and then a fitted curve can be obtained.

The largest iteration number was 400. The calculated cutoff point is the change of residual sum of squares between fitting data and measured data on the former and latter iteration, which is less than 10^{-6}. The calculated cutoff point is the change of the sum of squared residuals of parameters on the former and latter iteration was less than 10^{-6}.

3.2.4. Optimized Parameter Estimation of M

The Curve Fitting Tool in MATLAB was used for fitting data. Using a nonlinear least squares method, Robust is Bisqure and the Levenberg-Marquardt was used. Since the larger sensitivity of surface fitting to initial value of parameter, the numerical values are estimated.

For the maximum Chl-a concentration, M_0 was estimated at 6000 µg/mL and k_M was estimated at 8000. The specific growth rate constant was estimated to be 1.0×10^{-5} at $\frac{1}{2} M_{max}$ of 10 days. The parameter estimation of the abundance of the maximum algae cells number and the calculated fitting results are shown in the online Supplementary Materials.

3.2.5. Calculation Fitting Results

For the maximum Chl-a concentration, Table 1 shows the parameter estimation based on a 95% confidence interval and Table 2 shows the fitting evaluation.

Table 1. The parameter estimation based on 95% confidence intervals (CIs) for the maximum chlorophyll-a (Chl-a) concentration.

Parameter	Calculated Value	Estimated Lower Limit	Estimated Higher Limit
M_0 (µg/mL)	9471	8192	1.075×10^4
k_m	9779	7527	1.203×10^4
r	2.899×10^{-5}	2.059×10^{-5}	3.738×10^{-5}
t^* (d)	11.53	10.72	12.33

Table 2. The fitting evaluation parameters for the maximum Chl-a concentration.

Parameter	Value
SSE (sum of squares due to error)	9.682×10^7
R^2	0.9411
Adjusted R^2	0.9386
RMSE (Root mean square error)	1160

Considering the universality of the function, the $V_{replace}$-*t*-*P* relation was employed for drawing to better show the relationships among turbulent mixing, time, and Chl-a concentration calculated by the fitting function (Figure 7).

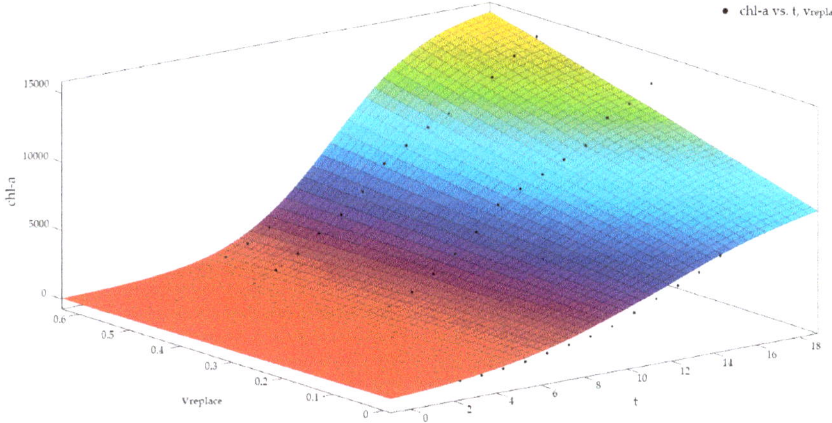

Figure 7. Relationship among hydrodynamic impact factor, time, and Chl-a concentration.

3.3. Deduction of Improved Logistic Function Facing Population Competition

Assuming that there are *n* kinds of hydrobiontic algae in water, due to the fixed biological capacity of water body, the corresponding cell differential equation is:

$$\frac{dP_i}{dt} = k_i P_i = r_i (M - P_1 - P_2 \cdots - P_i \cdots - P_n) P_i \tag{7}$$

The equation set of logistic function is:

$$
\begin{cases}
P_1(t) = \dfrac{M-P_2-P_3\cdots-P_n}{1+e^{-r_1\left(M-P_2-P_3\cdots-P_n\right)\left(t-t_1^*\right)}} \\[2ex]
P_2(t) = \dfrac{M-P_1-P_3\cdots-P_n}{1+e^{-r_2\left(M-P_1-P_3\cdots-P_n\right)\left(t-t_2^*\right)}} \\[1ex]
\quad\vdots \\[1ex]
P_i(t) = \dfrac{M-P_1-P_2\cdots-P_{i-1}-P_{i+1}\cdots-P_n}{1+e^{-r_i\left(M-P_1-P_2\cdots-P_{i-1}-P_{i+1}\cdots-P_n\right)\left(t-t_i^*\right)}} \\[1ex]
\quad\vdots \\[1ex]
P_n(t) = \dfrac{M-P_1-P_2\cdots-P_{n-1}}{1+e^{-r_n\left(M-P_1-P_2\cdots-P_{n-1}\right)\left(t-t_n^*\right)}}
\end{cases}
\tag{8}
$$

where M is the maximum total biomass or maximum total abundance, r_i is the specific growth rate of i algae, and t_i^* is the time that i algae requires to reach half of maximum biomass (abundance).

Combined with N-t-P relationship, the known biomass of *Microcystis aeruginosa* and the competition growth formula based on the logistic function in the symbiosis with *Microcystis aeruginosa* and *Scenedesmus quadricauda* is:

$$
P_1(t) = \frac{M_0 + k_M V_{replace} - kk_1 P_2}{1 + e^{-r_1\left(M_0 + k_M V_{replace} - kk_1 P_2\right)\left(t-t_1^*\right)}}
\tag{9}
$$

where P_1 reflects *Scenedesmus quadricauda*, in µg/mL, P_2 reflects *Microcystis aeruginosa* in µg/mL, and kk_1 is the biological plot ratio.

According to the experimental results, the average Chl-a concentration of *Scenedesmus quadricauda* was 33.91 µg/10^4 ind (individual) and the average Chl-a concentration of *Microcystis aeruginosa* was 2.40 µg/10^4 ind. Under a flow rate of 100 rpm, the maximum biomass and specific growth rate are listed in Table 3. The adjusted R^2 was greater than 0.6.

Table 3. The maximum biomass and specific growth rate under 100 rpm stirring.

Rotation Rate (rpm)	Maximum Biomass (µg/mL)	Hybrid Effect Coefficient	Biological Plot Ratio (*Microcystis aeruginosa* Is Unit 1)	Specific Growth Rate	Half-Saturation Time (d)
100	2620	−2586	0.9205	1.025×10^{-4}	21.21

The analysis of the fitted parameters suggested that the competition of different species causes a decrease in maximum biomass. The transportation and transfer due to water disturbance promotes growth space competition among cells. For instance, increasing diffusion of *Microcystis aeruginosa* toxins further reduces the growth space. Algae stimulated by external factors accelerate reproduction and fission to achieve population dominance. However, the results showed that algae in the same ecological niche (*Microcystis aeruginosa* and *Scenedesmus quadricauda* have similar biological plot ratios: *Microcystis aeruginosa*:*Scenedesmus quadricauda* = 1:0.9205) compete with each other and slow down the maximum growth rate. The results obtained from the model agree with observed results.

4. Discussion

Without considering the velocity of mechanical damage, small-scale turbulence strongly influences the growth of *Microcystis aeruginosa*. Many studies found that turbulence through water mixing affects the physico-chemical properties of the water column, increases the diffusion rate on the cell surface, and enhances the uptake of nutrients [20,30–32], which then promotes the growth of Microcystis spp., especially the toxic Microcystis spp. In controlled laboratory experiments, where nutrients are plentiful and available light is identical, different turbulent conditions have promoted the growth of *Microcystis aeruginosa*. Compared with the control condition, the Chl-a concentration reached the corresponding maximum under each turbulence condition. The growth rate of *Microcystis aeruginosa* decreased with increasing biological saturation in the limited beaker space, reflecting the restriction of nutrition and space on algal population growth. The population dynamics was consistent with the logistic function under each turbulence condition. The improved logistic growth model of *M aeruginosa* and the competition growth model with the symbiont S. *quadricauda* under turbulent conditions were established based these experimental results. We introduced turbulence for the examination of physiological responses of *Microcystis aeruginosa*, coupled turbulence, and growth of *Microcystis aeruginosa* in the models of population dynamics. The *R* value of the fitting curve was greater than 0.9, confirming the results fit the model and the unity of the nutrient and algae growth relationship affected by hydrodynamic force. With the increase in turbulence intensity, turbulence has a limited influence on algal growth space.

Phytoplankton population dynamics are controlled by the relative rather than absolute timescales of mixing [33]. Zhou et al. [28] found the short-term (approx. three days') turbulence favored the growth of toxic Microcystis species. In an ideal environment, they assumed that same algae species have similar *r* values. The initial algae biomass and the *r* value dominate the beginning of the growth curve. The well-fitted algae curves and similar growth rates support the former hypothesis. At the end of the growth period, the algal population limit influences the size of the algal population. Algal population has a maximum biomass that can change with the turbulence conditions, which proves the feasibility of our conjecture. In the improved logistic growth model of *M. aeruginosa*, compared with the static conditions, water turbulence can promote the growth of *M. aeruginosa* and increase the concentration of chl-a under certain turbulence intensity. This conclusion fits the experiment data that showed that algae growth curves are similar to each other under the conditions of high rotational speed. Water turbulence may provide an advantage for *M. aeruginosa* in allowing access to more nutrients and preventing the accumulation of hazardous material around cell by changing the thickness of the extracellular laminar boundary layer. The same phenomenon has also been observed in other studies [34,35].

Under competition conditions, the fitted results confirm the maximum biomass conjecture that similar maximum biomasses occur under different hydraulic conditions and the influence of the flow rate on maximum algae biomass is similar to a linear relationship. The growth rates of algae under competition compared from single algae is larger, which conforms to the explanation that *Microcystis aeruginosa* toxin has a carcinogenic mechanism. Cancer due to biotoxicity leads to a decrease in total biomass but improved multiplication capacity and increased death rates. Moreover, high intensity turbulence mixing plays an important role in *Scenedesmus quadricauda* becoming a dominant species, in agreement with other observed data that showed that potentially toxic cyanobacteria predominate under low turbulence, whereas diatoms and green algae dominate growth under high turbulence [10]. Compared with other phytoplankton, many previous studies have suggested that Cyanobacteria can adjust their position in the water by changing their buoyancy to obtain more nutrients, causing Cyanobacteria blooms. When the turbulence intensity increases, the absorption and photosynthetic efficiency of *Scenedesmus quadricauda* increases in the growth space, and the competitive advantage of *Microcystis aeruginosa* decreases. Cyanobacteria are more sensitive to small-scale turbulence than green algae, resulting in inhibition of growth under high intensity mixing [36].

5. Conclusions

(1) According to the experimental results, the timing of the lag and logarithmic growth phases were similar between turbulence treatments. The process starts from the lag phase, then enters a logarithmic phase, and finally reaches a stable phase, which fits the logistic growth curve. However, the abundance differed between treatments, showing that increases in turbulence can increase growth rate.

(2) Based on the logistic basic function, a planktonic algae model under hydrodynamic conditions was established. When $R > 0.95$, well-fitted results were obtained, in which the algae growth better matched an s-shaped curved surface growth.

(3) Under the same conditions of adequate nutrients and light, the same type of planktonic algae have a logistic growth rate ratio constant r. Water changes the extracellular environment of planktonic algae through the mass transfer effect and then changes the growth limitation of algae, thus affecting the algal group size. However, water flow enhances the algae biomass. With a continuous increase in flow speed, the fitted curved surface becomes smoother and steadier.

Supplementary Materials: The following are available online at http://www.mdpi.com/2073-4441/11/4/669/s1. Table 1. The parameters estimation based on 95% confidence interval for maximum cell number of abundance. Table 2. The fitting evaluation parameters for maximum cell number of abundance.

Author Contributions: Conceptualization, H.Z.; Formal Analysis, H.Z. and F.L.; Methodology, R.C. and Y.Z.; Software, R.C. and Z.G.; Data Curation, Y.Z.; Writing-Original Draft Preparation, F.L., Z.G., and F.H.; Writing-Review & Editing, F.L. and F.H.

Funding: This research has been supported by National Natural Science Foundation of China (51409190, 51379146).

Acknowledgments: Comments and suggestions from anonymous reviewers, the Associate Editor, and the Editor are greatly appreciated.

Conflicts of Interest: The authors declare no conflict of interest.

References

1. Mischke, U. Cyanobacteria associations in shallow polytrophic lakes: Influence of environmental factors. *Acta Oecol.* **2003**, *24*, S11–S23. [CrossRef]
2. Peeters, F.; Straile, D.; Lorke, A.; Ollinger, D. Turbulent mixing and phytoplankton spring bloom development in a deep lake. *Limnol. Oceanogr.* **2007**, *52*, 286–298. [CrossRef]
3. Paerl, H.W.; Huisman, J. Climate-Blooms like it hot. *Science* **2008**, *320*, 57–58. [CrossRef]
4. Visser, P.M.; Ibelings, B.W.; Bormans, M.; Huisman, J. Artificial mixing to control cyanobacterial blooms: A review. *Aquat. Ecol.* **2016**, *50*, 423–441. [CrossRef]
5. Berman, T.; Shteinman, B. Phytoplankton development and turbulent mixing in Lake Kinneret, 1992–1996. *J. Plankton Res.* **1998**, *20*, 709–726. [CrossRef]
6. Imteaz, M.A.; Asaeda, T. Artificial mixing of lake water by bubble plume and effects of bubbling operations on algal bloom. *Water Res.* **2000**, *34*, 1919–1929. [CrossRef]
7. Liu, L.; Liu, D.F.; Johnson, D.M.; Yi, Z.Q.; Huang, Y.L. Effects of vertical mixing on phytoplankton blooms in Xiangxi Bay of Three Gorges Reservoir: Implications for management. *Water Res.* **2012**, *46*, 2121–2130. [CrossRef]
8. Bronnenmeier, R.; Märkl, H. Hydrodynamic stress capacity of microorganisms. *Biotechnol. Bioeng.* **1982**, *24*, 553–578. [CrossRef] [PubMed]
9. Regel, R.H.; Brookes, J.D.; Ganf, G.G.; Griffiths, R. The influence of experimentally generated turbulence on the Mash01 unicellular Microcystis aeruginosa strain. *Hydrobiologia* **2004**, *517*, 107–120. [CrossRef]
10. Huisman, J.; Sharples, J.; Stroom, J.M.; Visser, P.M.; Kardinaal, W.E.A.; Verspagen, J.M.H.; Sommeijer, B. Changes in turbulent mixing shift competition for light between phytoplankton species. *Ecology* **2004**, *85*, 2960–2970. [CrossRef]
11. Li, F.P.; Zhang, H.P.; Zhu, Y.P.; Xiao, Y.H.; Chen, L. Effect of flow velocity on phytoplankton biomass and composition in a freshwater lake. *Sci. Total Environ.* **2013**, *447*, 64–71. [CrossRef]
12. O'Brien, K.R.; Meyer, D.L.; Waite, A.M.; Ivey, G.N.; Hamilton, D.P. Disaggregation of Microcystis aeruginosa colonies under turbulent mixing: Laboratory experiments in a grid-stirred tank. *Hydrobiologia* **2004**, *519*, 143–152. [CrossRef]
13. Zhang, H.P.; Chen, R.H.; Li, F.P.; Chen, L. Effect of flow rate on environmental variables and phytoplankton dynamics: Results from field enclosures. *Chin. J. Oceanol. Limn.* **2015**, *33*, 430–438. [CrossRef]
14. San, L.; Long, T.Y.; Liu, C.C.K. Algal bioproductivity in turbulent water: An experimental study. *Water* **2017**, *9*, 304. [CrossRef]

15. Long, T.Y.; Wu, L.; Meng, G.H.; Guo, W.H. Numerical simulation for impacts of hydrodynamic conditions on algae growth in Chongqing Section of Jialing River, China. *Ecol. Model.* **2011**, *222*, 112–119. [CrossRef]
16. Grobbelaar, J.U. Turbulence in Mass Algal Cultures and the Role of Light/Dark Fluctuations. *J. Appl. Phycol.* **1994**, *6*, 331–335. [CrossRef]
17. Arin, L.; Marrase, C.; Maar, M.; Peters, F.; Sala, M.M.; Alcaraz, M. Combined effects of nutrients and small-scale turbulence in a microcosm experiment. I. Dynamics and size distribution of osmotrophic plankton. *Aquat. Microb. Ecol.* **2002**, *29*, 51–61. [CrossRef]
18. Iversen, K.R.; Primicerio, R.; Larsen, A.; Egge, J.K.; Peters, F.; Guadayol, O.; Jacobsen, A.; Havskum, H.; Marrase, C. Effects of small-scale turbulence on lower trophic levels under different nutrient conditions. *J. Plankton Res.* **2010**, *32*, 197–208. [CrossRef]
19. Hondzo, M.; Warnaars, T.A. Coupled effects of small-scale turbulence and phytoplankton biomass in a small stratified lake. *J. Environ. Eng-Asc.* **2008**, *134*, 954–960. [CrossRef]
20. Hondzo, M.; Wuest, A. Do Microscopic Organisms Feel Turbulent Flows? *Environ. Sci. Technol.* **2009**, *43*, 764–768. [CrossRef]
21. Xiao, Y.; Li, Z.; Li, C.; Zhang, Z.; Huo, J.S. Effect of small-scale turbulence on the physiology and morphology of two bloom-forming cyanobacteria. *PLoS ONE* **2016**, *11*, e0168925. [CrossRef] [PubMed]
22. Missaghi, S.; Hondzo, M.; Sun, C.; Guala, M. Influence of fluid motion on growth and vertical distribution of cyanobacterium *Microcystis aeruginosa*. *Aquat. Ecol.* **2016**, *50*, 639–652. [CrossRef]
23. G.-Tóth, L.; Parpala, L.; Balogh, C.; Tátrai, I.; Baranyai, E. Zooplankton community response to enhanced turbulence generated by water-level decrease in Lake Balaton, the largest shallow lake in Central Europe. *Limnol. Oceanogr.* **2011**, *56*, 2211–2222.
24. Moreno-Ostos, E.; Cruz-Pizarro, L.; Basanta, A.; George, D. The influence of wind-induced mixing on the vertical distribution of buoyant and sinking phytoplankton species. *Aquat. Ecol.* **2009**, *43*, 271–284. [CrossRef]
25. Wu, T.; Qin, B.; Zhu, G.; Luo, L.; Ding, Y.; Bian, G. Dynamics of cyanobacterial bloom formation during short-term hydrodynamic fluctuation in a large shallow, eutrophic, and wind-exposed Lake Taihu, China. *Environ. Sci. Pollut. Res.* **2013**, *20*, 8546–8556. [CrossRef]
26. Huang, C.C.; Li, Y.M.; Yang, H.; Sun, D.Y.; Yu, Z.Y.; Zhang, Z.; Chen, X.; Xu, L.J. Detection of algal bloom and factors influencing its formation in Taihu Lake from 2000 to 2011 by MODIS. *Environ. Earth Sci.* **2014**, *71*, 3705–3714. [CrossRef]
27. Meissner, S.; Fastner, J.; Dittmann, E. Microcystin production revisited: Conjugate formation makes a major contribution. *Environ. Microbiol.* **2013**, *15*, 1810–1820. [CrossRef] [PubMed]
28. Zhou, J.; Qin, B.Q.; Han, X.X.; Zhu, L. Turbulence increases the risk of microcystin exposure in a eutrophic lake (Lake Taihu) during cyanobacterial bloom periods. *Harmful Algae* **2016**, *55*, 213–220. [CrossRef] [PubMed]
29. Zhang, Y.H. The Effect of Turbulence on *Microcystis Aeruginosa* and *Scenedesmus Quadricauda*. Master's Thesis, Tongji University, Shanghai, China, 2014. (In Chinese with English abstract)
30. Bergstedt, M.S.; Hondzo, M.M.; Cotner, J.B. Effects of small scale fluid motion on bacterial growth and respiration. *Freshw. Biol.* **2004**, *49*, 28–40. [CrossRef]
31. Guasto, J.S.; Rusconi, R.; Stocker, R. Fluid mechanics of planktonic microorganisms. *Annu. Rev. Fluid Mech.* **2012**, *44*, 373–400. [CrossRef]
32. Prairie, J.C.; Sutherland, K.R.; Nickols, K.J.; Kaltenberg, A.M. Biophysical interactions in the plankton: A cross-scale review. *Limnol. Oceanogr Fluids Environ.* **2012**, *2*, 121–145. [CrossRef]
33. O'Brien, K.R.; Ivey, G.N.; Hamilton, D.P.; Waite, A.M. Simple mixing criteria for the growth of negatively buoyant phytoplankton. *Limnol Oceanogr.* **2003**, *48*, 1326–1337. [CrossRef]
34. KarpBoss, L.; Boss, E.; Jumars, P. Nutrient fluxes to planktonic osmotrophs in the presence of fluid motion. *Oceanogr. Mar. Biol.* **1996**, *34*, 71–107.
35. Warnaars, T.A.; Hondzo, M. Small-scale fluid motion mediates growth and nutrient uptake of Selenastrum capricornutum. *Freshw. Biol.* **2006**, *51*, 999–1015. [CrossRef]
36. Thomas, W.H.; Gibson, C.H. Effects of small-scale turbulence on microalgae. *J. Appl. Phycol.* **1990**, *2*, 71–77. [CrossRef]

Article

Effects of Light Intensity and Exposure Period on the Growth and Stress Responses of Two Cyanobacteria Species: *Pseudanabaena galeata* and *Microcystis aeruginosa*

Guligena Muhetaer [1,*], Takashi Asaeda [2,3,4], Senavirathna M. D. H. Jayasanka [1], Mahendra B. Baniya [1,5], Helayaye D. L. Abeynayaka [1], M. Harun Rashid [6] and HongYu Yan [1]

[1]   Graduate School of Science and Engineering, Saitama University, 255 Shimo-okubo, Sakura-ku, Saitama 338-8570, Japan; jayasanka@mail.saitama-u.ac.jp (S.M.D.H.J.); baniyam57@gmail.com (M.B.B.); hdlakmali@yahoo.com (H.D.L.A.); en.k.943@ms.saitama-u.ac.jp (H.Y.)

[2]   Institute for studies of the Global Environment, 7-1 Sophia University, Kioicho, Chiyoda, Tokyo 102-0094, Japan; asaeda@mail.saitama-u.ac.jp

[3]   Institute Hydro Technology Institute, 4-3-1 Shiroyama Trust Tower, Tranomon, Minato, Tokyo 105-0001, Japan

[4]   Research and Development Center, Nippon Koei, 2304 Inarihara, Tsukuba, Ibaraki 300-1259, Japan

[5]   Provincial Government, Ministry of Physical Infrastructure Development, Gandaki Province, Pokhara 33700, Nepal

[6]   Department of Agronomy, Bangladesh Agricultural University, Mymensingh 2202, Bangladesh; mhrashid@bau.edu.bdmailto

*   Correspondence: gulgina1112@gmail.com; Tel.: +81-09064936640

Received: 29 November 2019; Accepted: 31 January 2020; Published: 3 February 2020

Abstract: Light is an important factor that affects cyanobacterial growth and changes in light can influence their growth and physiology. However, an information gap exists regarding light-induced oxidative stress and the species-specific behavior of cyanobacteria under various light levels. This study was conducted to evaluate the comparative effects of different light intensities on the growth and stress responses of two cyanobacteria species, *Pseudanabaena galeata* (strain NIES 512) and *Microcystis aeruginosa* (strain NIES 111), after periods of two and eight days. The cyanobacterial cultures were grown under the following different light intensities: 0, 10, 30, 50, 100, 300, and 600 μmol m^{-2} s^{-1}. The optical density (OD$_{730}$), chlorophyll a (Chl-a) content, protein content, H$_2$O$_2$ content, and the antioxidative enzyme activities of catalase (CAT) and peroxidase (POD) were measured separately in each cyanobacteria species. *P. galeata* was negatively affected by light intensities lower than 30 μmol m^{-2} s^{-1} and higher than 50 μmol m^{-2} s^{-1}. A range of 30 to 50 μmol m^{-2} s^{-1} light was favorable for the growth of *P. galeata*, whereas *M. aeruginosa* had a higher tolerance for extreme light conditions. The favorable range for *M. aeruginosa* was 10 to 100 μmol m^{-2} s^{-1}.

Keywords: cyanobacterial growth; stress responses; *Pseudanabaena galeata*; *Microcystis aeruginosa*; oxidative stress; antioxidative enzymes

---

1. Introduction

Eutrophication and global warming have promoted the growth of cyanobacteria in freshwater systems worldwide, and this trend is expected to increase in the future [1]. Cyanobacterial blooms are a serious issue in fresh, brackish, and marine water, as they decrease light penetration through the water and deplete dissolved oxygen, causing mortality for aquatic life [2]. Cyanobacteria are oxygenic autotrophs, constituting the largest and most diverse community of photosynthetic prokaryotes [3].

They play imperative roles in carbon and nutrient cycling in aquatic systems, as well as in degrading water quality and safety, causing numerous problems with water sources and ecosystem management [4]. From an environmental perspective, cyanobacterial blooms and their effects have been reported in the scientific literature for more than a century, and their probability and severity have both escalated over time [5]. The mass development of cyanobacteria has resulted in various negative consequences including eutrophication, ecosystem imbalances, and scenic impairments [6]. Their ability to produce toxic secondary metabolites is becoming an increasingly important environmental issue, which is threatening human, animal, and plant health. Cyanobacterial blooms can, thereby, lead to ecosystem destruction [7,8]. Given these problems, gaps in our knowledge of cyanobacteria must be filled.

It is important to understand the environmental factors associated with cyanobacterial growth and metabolism to address real world problems. This subject has been extensively mentioned in connection to climate change [9,10], and anthropogenic activities associated with land use changes are also being experienced worldwide and have been widely discussed in the literature [11,12]. Cyanobacteria have several environmental drivers, such as water temperature, water column irradiance, and stratification of the water column coupled with long residence time, the availability of nitrogen (N) and phosphorus (P), and the turbidity and salinity of water. Of these drivers, light intensity is a particularly important determinant of the growth of cyanobacterial blooms [13]. The growth patterns of cyanobacteria respond to different light intensities via morphological and physiological changes [14], and detailed investigation of these changes is vital for the formulation of both proactive and reactive planning for freshwater resources management.

Pseudanabaena galeata and *Microcystis aeruginosa* are two cyanobacteria species that are known to degrade water quality and safety in many parts of the world [15]. *P. galeata* is a non-bloom-forming species, however, it forms odorous 2-methylisoborneol (2-MIB) which causes operational issues with water supplies [16]. *M. aeruginosa* is a bloom forming species that synthesizes secondary toxic metabolites, namely microcystins, which affect water safety, and therefore human health, thereby posing serious social and ecological hazards [17,18]. In addition to these concerns that arise with both species, their worldwide occurrences have increased as eutrophication and global warming which together have provided suitable conditions for vigorous growth [19,20].

As the frequency of cyanobacteria blooms continues to increase, various control measures are being applied to control their growth [21]. Methods such as the mixing of water in lakes or reservoirs mainly focus on exposing cyanobacteria to low levels of light. However, the photosynthetic organisms of cyanobacteria can also be stressed by high-light exposure, and the excess energy produces reactive oxygen species (ROS), which cause severe photodamage to their cellular components [22,23]. The growth rates of *Microcystis* and *Anabaena* species are high under low-light levels (25 μmol m^{-2} s^{-1}), however, these growth rates decrease under high-light levels (200 μmol m^{-2} s^{-1}) [22,24].

Although high-light levels can negatively affect cyanobacteria species, the responses and relationships of their oxidative stress, antioxidant, pigmentation, and protein contents are currently not well understood. Therefore, in this study, we aimed to explore the responses of *P. galeata* and *M. aeruginosa* to low-light and high-light stressors. We measured hydrogen peroxide (H_2O_2), antioxidant enzymes (catalase (CAT) and peroxidase (POD)), chlorophyll a (Chl-a), protein contents, and optical densities (OD_{730}) under different light conditions. The growth performances of these two species were calculated and validated using cyanobacteria growth models proposed by Steele [25], Platt and Jassby [26], and Peeters and Eilers [27], to determine the applicability of our findings to cyanobacteria control.

2. Materials and Methods

2.1. Cyanobacterial Cultures and Incubation

Strains of the cyanobacterial species *P. galeata* and *M. aeruginosa* were obtained from the National Institute for Environmental Studies (NIES) at Ibaraki, Japan. The samples were cultured in BG-11 medium [28] and acclimatized for 14 days inside an incubator (MIR-254, Sanyo, Tokyo, Japan) at 20 °C; they underwent manual shaking three to five times per day. The samples were cultured under controlled light conditions, with photon flux levels of 20 to 30 μmol m^{-2} s^{-1} emitted from cool white fluorescent lamps (5000 K color temperature). The cycle of light conditions to dark conditions was maintained at 12 h light and 12 h dark [29], using an automatic setup time device (REVEX PT7, Saitama, Japan).

2.2. Growth Experimental Setup

All of the experiments were conducted inside an incubator (MIR-254, Sanyo, Tokyo, Japan) at a constant temperature of 20 °C throughout the experimental period. To characterize the growth response to different light conditions, incubated *P. galeata* (NIES 512) and *M. aeruginosa* (NIES 111) cells were separately subjected to seven different photon flux levels (0, 10, 30, 50, 100, 300, and 600 μmol m^{-2} s^{-1}); a constant temperature of 20 °C was used to calculate the saturation light intensities. Light was supplied from cold white fluorescent lamp sources. The cultures were maintained in an illumination cycle of 12 h light and 12 h dark (12L:12D). Zero μmol m^{-2} s^{-1} light intensity means that there was no light source; these cultures were maintained in 24 h of darkness. The light intensities were measured using a quantum sensor (ml-020P, EKO Instruments Co., Ltd., Tokyo, Japan), and were read as voltage output (mV) by a voltage logger (LR5041, HIOKI, Nagano, Japan). The cells that were cultured under different light levels were sampled for Chl-a and enzyme analysis in two-day intervals. To ensure the homogenous exposure of cells to the light environment, the cyanobacteria cultured flasks were shaken gently five times a day. Each treatment was performed in triplicate.

2.3. Measuring Protein

The concentration of protein was measured using the Bradford method [30]. A crude protein extract was discolored with Coomassie (G-250). After incubation at room temperature (25 ± 2 °C) for 10 min, the absorbance was measured at 595 nm using an ultraviolet visible (UV-Vis) spectrometer (UVmini-1240, Shimadzu, Kyoto, Japan). Protein was diluted with the same buffer and was stained with Coomassie (G-250) dye and used to prepare the standard curve; deionized water was used as the blank.

2.4. Measuring OD$_{730}$

To estimate the growth of cyanobacteria, the OD$_{730}$ was measured by taking 1 mL of sample from each flask. The OD$_{730}$ was measured with a UV-Vis spectrophotometer (UVmini-1240, Shimadzu, Kyoto, Japan) at an optical absorption wavelength of 730 nm, using a previously proposed methodology [31,32].

2.5. Measuring Chlorophyll a Content

The concentration of Chl-a in the cyanobacteria samples was measured according to the method described by Holm and Romo [33,34]. The 1 mL cell suspensions of the two species were centrifuged separately at 10,000× *g* for 10 min at 4 °C, and the supernatant was removed. Each cell pellet was washed once with Milli-Q water and then extracted in 1 mL of 80% acetone. The mixture was shaken vigorously and maintained in darkness overnight at room temperature (25 ± 2 °C). Then, each sample was again centrifuged at 10,000× *g*, and the supernatant was measured with a UV-Vis spectrophotometer (UVmini-1240, Shimadzu, Kyoto, Japan) at absorption wavelengths of 660 and 645 nm. To correct for the absorbance for pheophytin a, the samples were acidified with 0.1 N HCL, and their absorbance was measured again. The chlorophyll a content was calculated by using Equation (1) [33]:

$$Chl - a = (9.76 \times A_{660}) - (0.99 \times A_{645}) \tag{1}$$

where $Chl - a$ is the content of chlorophyll a (expressed in μg per mL), and A_{660} and A_{645} are the absorbance values 660 nm and 645 nm, respectively.

2.6. Measuring H_2O_2 Content

The method specified by Jana [35] was employed to measure the H_2O_2 concentration in the cultured cyanobacteria samples. The *P. galeata* (NIES 512) and *M. aeruginosa* (NIES 111) cell pellets were obtained by centrifuging at $10,000\times g$ for 10 min and removing the supernatant. The cell pellets were washed once with Milli-Q water and homogenized in 1 mL of 0.1 M pH 6.5 phosphate buffer to extract the internal H_2O_2. The homogenate was then centrifuged at $10,000\times g$ for 10 min at 4 °C and the extract was used for H_2O_2 estimation. A reaction mixture of 0.1% titanium chloride in 20% H_2SO_4 (v/v) was added to the supernatant, and after a 1 min incubation period, the mixture was centrifuged at room temperature (25 ± 2 °C) and the absorbance was measured at 410 nm with a UV-Vis spectrophotometer (UVmini-1240, Shimadzu, Kyoto, Japan). The H_2O_2 concentration was determined using the pre-prepared standard curve for known concentration series. An extinction coefficient of 0.28 mmol^{-1} cm^{-1} was used to calculate the concentration of H_2O_2 in µmol mL^{-1}.

2.7. Measuring CAT Activity

The CAT activity was measured using the method proposed by Aebi [36]. The cyanobacteria cells were homogenized in a phosphate buffer (pH 7.0), supernatant liquid was, then, taken as an enzyme extract, after being centrifuged at $12,000\times g$ at 4 °C for 10 min. The decrease in absorbance at 240 nm was recorded for 3 min. The CAT activity was calculated using an extinction coefficient of 39.4 mM^{-1} cm^{-1}.

2.8. Measuring POD Activity

The POD activity was measured based on guaiacol oxidation, as proposed by MacAdam [37]. The reaction mixture contained 920 µL of 100 mM potassium phosphate buffer (pH 6.8), 15 µL of 0.6% H_2O_2, and 65 µL of enzyme extract. The increase in absorbance was measured at 420 nm every 10 s for 3 min.

2.9. Cell Growth Measurement and Different Model Fitting

The cell growth of cyanobacteria was measured using OD_{730} measurements. OD_{730} was measured using a UV-Vis spectrophotometer (UVmini-1240, Shimadzu, Kyoto, Japan).

The cell growth measurements were compared with three cell growth models that have been proposed for cyanobacteria growth [25–27]. The cell growth of cyanobacteria was measured using OD_{730} measurements, which were applied to the growth rate in Equation (2) [38]. Then, the calculated growth rates were compared with the three model outputs obtained for the different light intensities of the present experiment.

$$\mu = \frac{(logOD_t - logOD_0)}{t} \times 3.32 \tag{2}$$

where μ is the cell growth rate, t is the time in days, OD_t is the optical density after t days, and OD_t is the optical density at the beginning of the experiment zero time. Different models proposed by Steele [25], Platt and Jassby [26], and Peeters and Eilers [27] were fitted to our experimental observations [39].

The model proposed by Steele [25], named Model I, is written as:

$$\mu_{T,I} = \mu_{maxT} \times I/I_{optT} \times \exp\left(I - I/I_{optT}\right) \tag{3}$$

where $\mu_{T,I}$ is the growth rate at light intensity I, and μ_{maxT} and I_{optT} are the estimated maximal growth rates and the optimal light intensity at temperature T, respectively.

The model proposed by Platt and Jassby [26], named Model II, is written as:

$$\mu_{T,I} = \mu_{maxT} \times \tan h[\alpha \times (I - I_c)/\mu_{maxT}] \tag{4}$$

where α is the growth efficacy, I_c is the estimated light intensity without growth ($I_c \geq 0$), and tanh is the hyperbolic tangential function.

The model proposed by Peeters and Eilers [27], named Model III, is written as:

$$\mu_{T,I} = 2 \times \mu_{maxT} \times (1 + \beta) \times \left(I/I_{optT}\right) / \left[\left(I/I_{optT}\right)^2 + 2 \times \left(I/I_{optT}\right) \times \beta + 1\right] \tag{5}$$

where β is the attenuation coefficient, which allows for consideration of the photoinhibition phenomenon.

2.10. Statistical Analysis

All the presented results are expressed as the mean ± SD (n = 3). A two-way analysis of variance (ANOVA) followed by Tukey's post-hoc tests were performed to examine the statistical significance of variations among the means between the light exposure period and light intensity combinations used for *P. galeata* and *M. aeruginosa*. Statistical analyses were performed by using IBM SPSS Statistics for Windows, Version 25.0. (Armonk, NY, USA: IBM Corp).

3. Results

The protein concentrations of *P. galeata* and *M. aeruginosa* species after two- and eight-day periods of exposure to light of different intensities are presented in Figure 1a,b. After the two-day period of exposure, the protein concentrations of each exposure condition were nearly the same as the initial values, 61.6 ± 3.0 µg mL^{-1} for *P. galeata* ($F_{6,14} = 1.250$ and $p = 0.34$) and 54.5 ± 0.70 µg mL^{-1} for *M. aeruginosa* ($F_{6,14} = 3.960$ and $p = 0.016$). However, for *P. galeata* ($F_{6,14} = 32.795$ and $p < 0.0001$) the concentration substantially increased after eight days. It increased rapidly up to 100 µmol m^{-2} s^{-1}, and then continued to increase at a slower rate up to 600 µmol m^{-2} s^{-1}. For *M. aeruginosa* ($F_{6,14} = 34.824$ and $p < 0.0001$), the protein concentration exhibited a decreasing trend with respect to increasing light intensity, which is an almost opposite trend to that of *P. galeata*. The protein concentration of *M. aeruginosa* decreased rapidly up to a light intensity of 100 µmol m^{-2} s^{-1} and, then, continued to decrease at a slower rate up to 600 µmol m^{-2} s^{-1}.

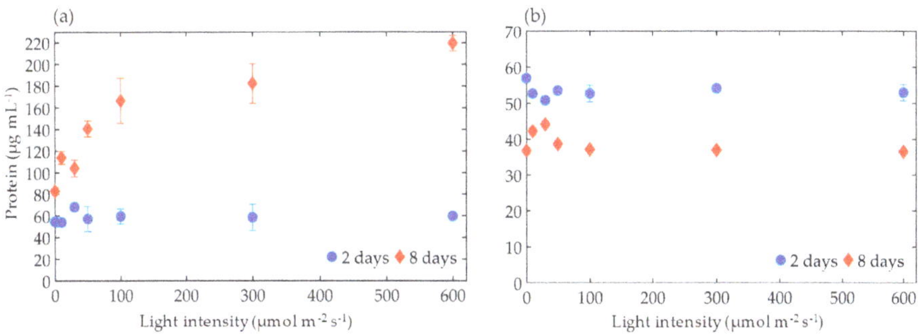

Figure 1. Changes in protein concentration with respect to light intensity after two and eight days for two species. (**a**) *Pseudanabaena galeata*; (**b**) *Microcystis aeruginosa*.

The variations in OD$_{730}$ with respect to light intensity are presented in Figure 2a,b. *P. galeata* and *M. aeruginosa* both showed a statistically significant difference between the two- and eight-day periods of exposure (*P. galeata* $F_{6,28} = 188.811$, $p < 0.0001$ and *M. aeruginosa* $F_{6,28} = 145.041$, $p < 0.0001$). After two days, relatively constant OD$_{730}$ values were observed under all of the light intensities, for both *P. galeata* and *M. aeruginosa*. The eight-day OD$_{730}$ value of *P. galeata* increased under 30 and 50 µmol m^{-2} s^{-1} light intensities and, then, decreased under higher light intensities over the two-day period. However, the OD_{730} values of *M. aeruginosa* after eight days were approximately double the OD_{730} values of *P. galeata*.

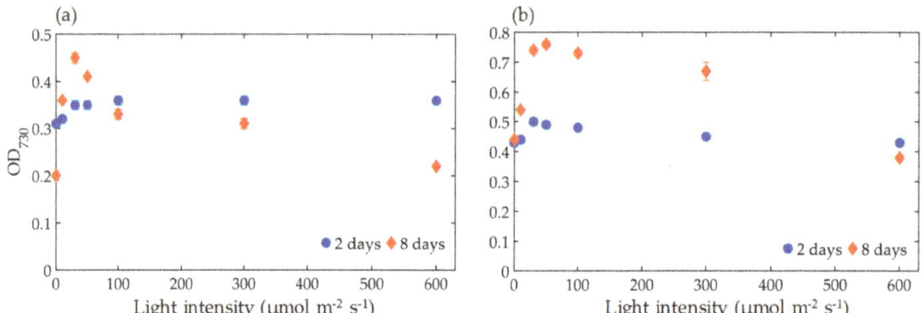

Figure 2. Changes in OD_{730} with respect to light intensity after two and eight days. (**a**) *P. galeata*; (**b**) *M. aeruginosa*.

The Chl-a concentrations of *P. galeata* and *M. aeruginosa* under different light intensities are presented in Figure 3a,b. The Chl-a concentrations of *P. galeata* and *M. aeruginosa* showed a statistically higher significant difference after eight than after two days (*P. galeata* $F_{6,28} = 21.876$, $p < 0.0001$ and *M. aeruginosa* $F_{6,28} = 9.036$, $p = 0.0001$). After the two-day exposure, the Chl-a concentrations of both species showed a slightly decreasing trend after 50 μmol m^{-2} s^{-1}. After eight days, the Chl-a concentration increased under 30 and 50 μmol m^{-2} s^{-1} for *P. galeata*, however, it decreased more under the high-light intensity than that after the two-day exposure. After eight days, the Chl-a content of *M. aeruginosa* also followed the same trend as *P. galeata*, but with different values. Chl-a gradually increased up to 50 μmol m^{-2} s^{-1} light intensity, and then started to degrade under higher light intensities.

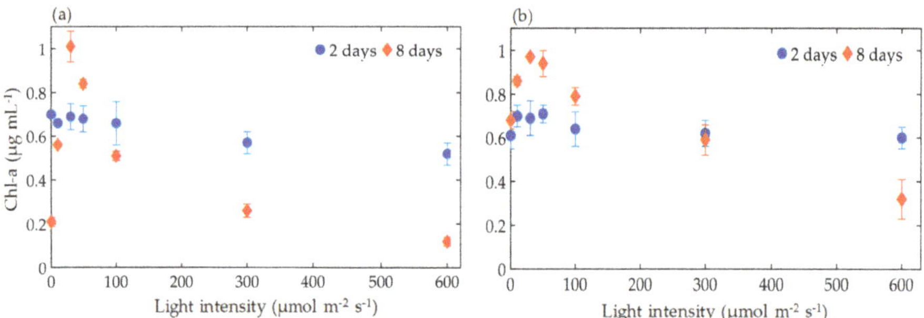

Figure 3. Changes in chlorophyll a (Chl-a) content with respect to culture light intensity. (**a**) *P. galeata*; (**b**) *M. aeruginosa* under different light intensities (0 to 600 μmol photons m^{-2} s^{-1}) after two and eight days.

The ratio of H_2O_2 per protein (H_2O_2/protein) over different light intensities is shown in Figure 4a,b. After two days, *P. galeata's* H_2O_2/protein decreased under 30 and 50 μmol m^{-2} s^{-1}, increased gradually up to 300 μmol m^{-2} s^{-1}, and then remained level under 600 μmol m^{-2} s^{-1}. After eight days, the H_2O_2/protein was steady up to 100 μmol m^{-2} s^{-1} and, then, followed a decreasing trend under 300 and 600 μmol m^{-2} s^{-1}. The H_2O_2/protein of *M. aeruginosa* showed an increasing trend up to 100 μmol m^{-2} s^{-1} under both two and eight days of exposure. Then, under 300 and 600 μmol m^{-2} s^{-1}, we observed a slightly increasing trend for the two-day exposure and a decreasing trend for the eight-day exposure. Both *P. galeata* and *M. aeruginosa* showed a statistically significant difference in H_2O_2/protein after eight days as compared with after two days (*P. galeata* $F_{6,28} = 9.036$, $p < 0.0001$ and *M. aeruginosa*: $F_{6,28} = 10.864$, $p < 0.0001$).

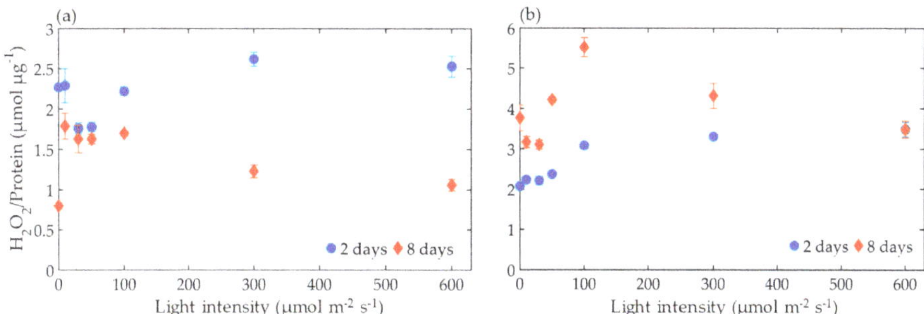

Figure 4. Changes in H_2O_2/protein with respect to culture light intensity. (**a**) *P. galeata*; (**b**) *M. aeruginosa*, after two and eight days.

The relationship between Chl-a content and H_2O_2 is shown in Figure 5a,b. We found a clear negative correlation between Chl-a content and H_2O_2 for *P. galeata* (correlation coefficient, r = 0.94) and *M. aeruginosa* (r = 0.71) after the two-day period only. The Chl-a content and H_2O_2 results were scattered beyond the two-day period. Both species *P. galeata* ($F_{6,28}$ = 40.569 and p < 0.0001) and *M. aeruginosa* ($F_{6,28}$ = 16.026 and p < 0.0001) showed statistically significant differences.

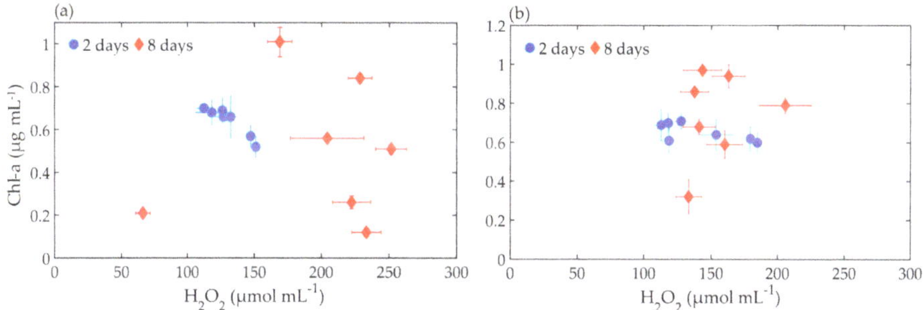

Figure 5. Chl-a content and H_2O_2. (**a**) *P. galeata*; (**b**) *M. aeruginosa*, after two and eight days.

Figure 6a,b depicts the H_2O_2 per OD_{730} (H_2O_2/OD_{730}) variation with respect to light intensity for both species after two and eight days. After two days, P. galeata H_2O_2/OD_{730} remained the same across the light intensities. However, for *M. aeruginosa*, H_2O_2/OD_{730} decreased for 10 and 30 µmol m^{-2} s^{-1}, and then increased with further increasing light intensities. After eight days, P. galeata H_2O_2/OD_{730} increased more significantly than after two days of exposure, except for under 30 µmol m^{-2} s^{-1} ($F_{6,28}$ = 69.894, p < 0.0001). For *M. aeruginosa*, H_2O_2/OD_{730} decreased significantly from a light intensity of 30 µmol m^{-2} s^{-1} ($F_{6,28}$ = 13.964, p < 0.0001). We identified a clearer positive correlation between H_2O_2/OD_{730} and light intensity for *P. galeata* after two days of exposure (r = 0.74) and after eight-day exposure (r = 0.88), as well as for M. aeruginosa after two (r = 0.91) and eight (r = 0.54) days of exposure.

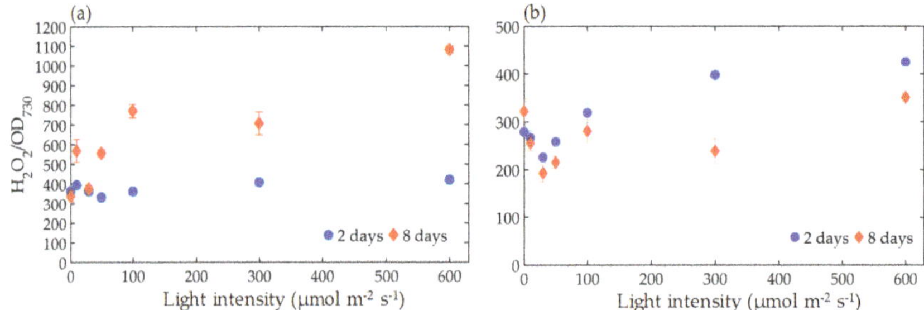

Figure 6. H_2O_2/OD_{730} with respect to culture light intensity. (**a**) *P. galeata*; (**b**) *M. aeruginosa*, under different light intensities (0 to 600 µmol m^{-2} s^{-1}) after two and eight days.

The CAT activity of two cyanobacteria species were found to be a function of H_2O_2/protein, as shown in Figure 7a,b. *P. galeata* showed a statistically significant difference ($F_{6,28}$ = 2.870, p = 0.026), as did *M. aeruginosa* ($F_{6,28}$ = 2.881, p = 0.026). The CAT activity was highest for the two-day period for the highest H_2O_2/protein values; after that, its value decreased with decreasing H_2O_2/protein for *P. galeata*. In contrast, H_2O_2/protein increased with time for *M. aeruginosa*, therefore, a higher CAT activity was generated.

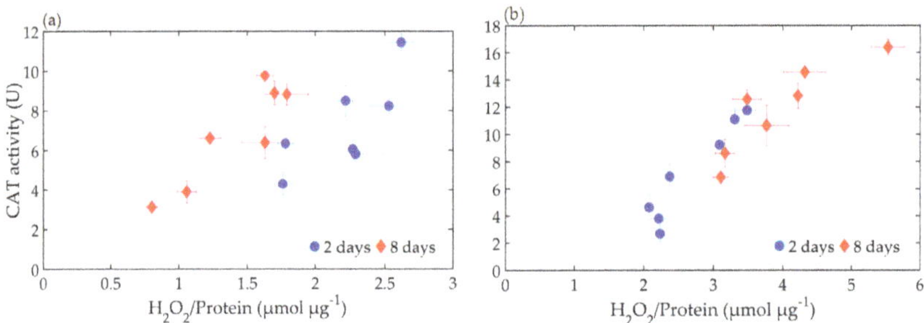

Figure 7. Catalase (CAT) activity with respect to H_2O_2/protein. (**a**) *P. galeata*; (**b**) *M. aeruginosa*, under different light intensities (0 to 600 µmol photons m^{-2} s^{-1}) after two and eight days.

The POD activity of the two cyanobacteria species was also found to be a function of H_2O_2/protein, as shown in Figure 8a,b. *P. galeata* showed a statistically significant difference ($F_{6,28}$ = 16.452, p < 0.0001), as did *M. aeruginosa* ($F_{6,28}$ = 3.640, p = 0.009). In *P. galeata*, POD activity after two days was higher than that in the eight-day experiment. In contrast, in *M. aeruginosa*, the POD activity increased with increasing H_2O_2/protein during both experiments.

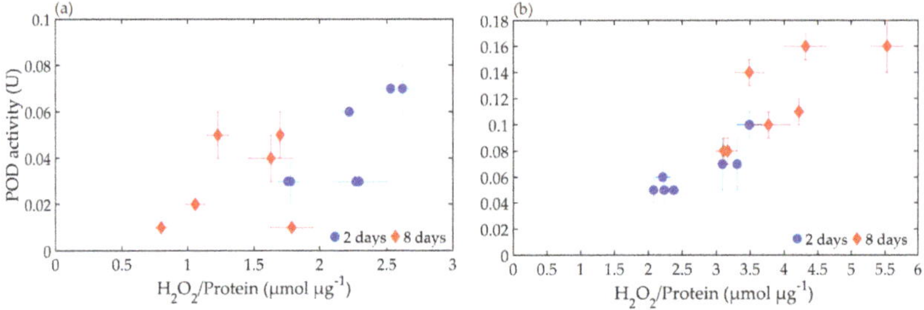

Figure 8. Peroxidase (POD) activity in relation to H_2O_2/protein. (**a**) *P. Galeata*; (**b**) *M. aeruginosa*, under different light intensities (0 to 600 µmol m^{-2} s^{-1}) after two and eight days.

The trend was also observed in the color of the sample, which changed after both periods under different light intensities, as shown in Figure 9.

Figure 9. Photographs showing growth during the two- and eight-day periods. (**a**) *P. galeata*; (**b**) *M. aeruginosa*, under different light intensities (0 to 600 µmol m^{-2} s^{-1}).

The observed growth rates were fitted to the models proposed by Steele (Model I) [25], Platt and Jassby (Model II) [26], and Peeters and Eilers (Model III) [27]. The observed two-day growth rate of *P. galeata* fitted Model II, and the eight-day growth rate fitted Model I (Figure 10). The observed two- and eight-day data of *M. aeruginosa* fitted Model III, and the eight-day data fitted model III. *P. galeata* attained a maximum growth rate of 0.13 day^{-1} within the first two days under a light intensity of 100 µmol m^{-2} s^{-1}, whereas *M. aeruginosa* attained a maximum growth rate of 0.15 day^{-1} within the first two days under a light intensity of 30 µmol m^{-2} s^{-1} (Figure 11). The mean growth rates of *P. galeata* and *M. aeruginosa* were 0.08 day^{-1} and 0.12 day^{-1}, respectively, achieved under a light intensity of 30 µmol m^{-2} s^{-1}.

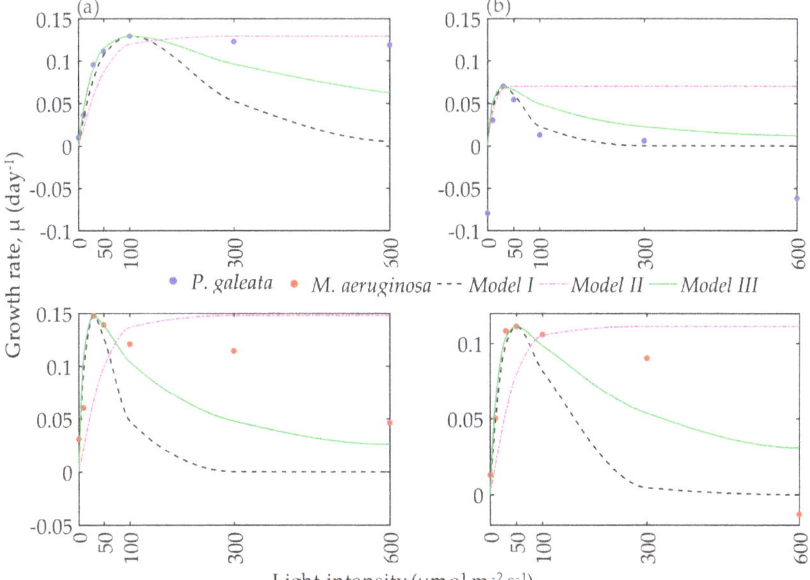

Figure 10. Growth rates of *P. galeata* and *M. aeruginosa* as a function of light intensity at different time intervals. (**a**) Two days; (**b**) eight days. The observed data were fitted with the models proposed by Steele (Model I) [25], Platt and Jassby (Model II) [26], and Peeters and Eilers (Model III) [27].

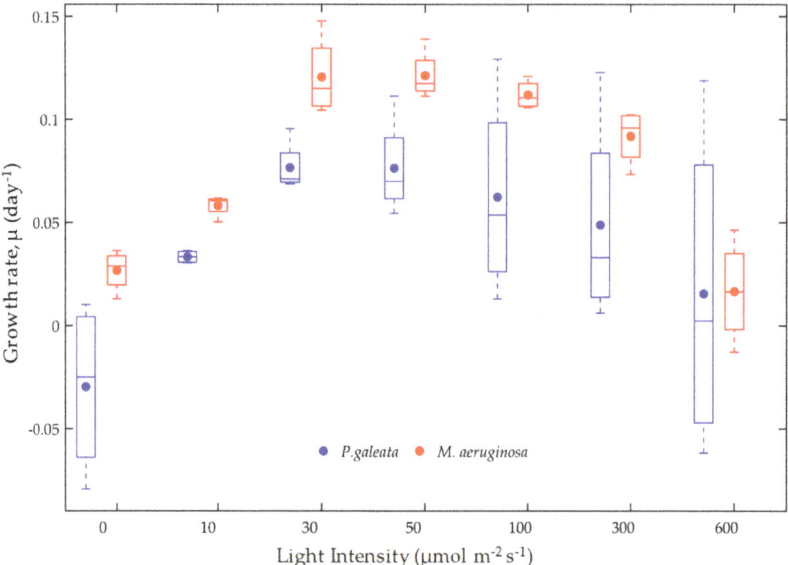

Figure 11. Comparison of growth rates of *P. galeata* and *M. aeruginosa* under different light intensities. The central lines indicate the median, and the bottom and top edges of each box indicate the 25th and 75th percentiles, respectively. The whiskers extend to the most extreme data points that were not considered outliers, and the circles indicate the mean value.

4. Discussion

Two cyanobacteria species, *P. galeata* and *M. aeruginosa*, exhibited different responses under different light intensities, and with respect to increasing light conditions, particularly when exceeding 50 μmol m^{-2} s^{-1}. *P. galeata* was negatively affected when the light intensity exceeded 100 μmol m^{-2} s^{-1}, and *M. aeruginosa* was also negatively affected. However, under two days of exposure, neither species was influenced, even by the extreme light conditions (300 and 600 μmol m^{-2} s^{-1}), suggesting that under a short exposure to high-light intensity, both species would survive. However, the extended exposure (eight days) increased the stress on both species. Therefore, both species have the capacity to tolerate light stress for shorter durations but lose this tolerance after extended exposure periods. Light intensities between 30 to 50 μmol m^{-2} s^{-1} can be considered preferable light conditions for *P. galeata* growth; with a corresponding range of 10 to 100 μmol m^{-2} s^{-1} for *M. aeruginosa* (Figure 11). These findings show similar trends to previously published results, as the growth of *Microcystis* and *Anabaena* species increased under low light (25 μmol m^{-2} s^{-1}) but decreased under high light (200 μmol m^{-2} s^{-1}) [22,24,34]. However, we confirmed that light intensities exceeding 200 μmol m^{-2} s^{-1} further intensified stress on both cyanobacteria species. As the only parameter that was varied in the present experiment was light, this suggests that cells become stressed mainly due to photosystem-produced H_2O_2, even at light intensities under 600 μmol m^{-2} s^{-1}, as in most cyanobacteria species, photoinhibition occurred when the light intensity exceeded 1000 μmol m^{-2} s^{-1} [40–42].

The reduced protein content, which reflects increased stress or vice versa for *M. aeruginosa* and most cyanobacteria species [43], decreased more after the eight-day period than the two-day period, independent from the light intensity. This reveals that *M. aeruginosa* has a defense mechanism to prevent cell damage from light (Figure 1a,b). The observed decreases in protein content with increasing light intensity in *M. aeruginosa* are a result of decreased phycobiliprotein synthesis, which protects against the absorption of excess light energy and the increased degradation of protein by proteases [44–46]. The increased OD$_{730}$ of *M. aeruginosa*, except at 0 and 600 μmol m^{-2} s^{-1} (two- and eight-day differences in *M. aeruginosa* OD_{730} under 600 µmol m^{-2} s^{-1} minimum as compared with *P. galeata*), evidences the survival and continuous cell proliferation of *M. aeruginosa*. *P. galeata*, which appeared to be relatively weak, strictly preferred 30 and 50 µmol m^{-2} s^{-1}, and experienced high stress under lower or higher light intensities, considering the Chl-a content and OD_{730}. However, the observed increases in protein content with increasing light stress suggest that there was a deviated stress response for *P. galeata*. The increased protein content can be associated with the upregulation of stress-related protein, as the oxidative stress was enforced due to the elevated H_2O_2 content with increasing light intensity [47]. However, further research focused on the upregulation of stress proteins is necessary to confirm this phenomenon.

The oxidative stress response mechanism, in cyanobacteria, helps to protect from extreme environmental conditions and can trigger antioxidant defense system responses [48]. The balance between the oxidative stress and antioxidative enzymes is disturbed by abiotic stress factors, and cells are subjected to oxidative stress in these conditions [49,50]. In this study, the H_2O_2 concentrations of both species were enhanced under low light (0 and 10 µmol m^{-2} s^{-1}) and high light (100 to 600 µmol m^{-2} s^{-1}) intensities. After two days, the CAT and POD activities of *P. galeata* showed scattered trends with respect to oxidative stress (Figure 7), whereas for *M. aeruginosa*, they showed increasing trends (Figure 8). However, when testing the regression relationships between H_2O_2/protein and antioxidant enzymes, CAT activity was found to have a strong relationship as compared with the POD of both species (CAT: *M. aeruginosa* r = 0.89 and *P. galeata* r = 0.90; POD: *M. aeruginosa* r = 0.75 and *P. galeata* r = 0.45). This confirms that the CAT activity of both cyanobacteria species played a more prominent antioxidant activity role than POD. Under higher stress, the antioxidant balance was also lost.

These findings show that high levels of light exposure can be adopted as a non-chemical method to control *P. galeata* and *M. aeruginosa*. This approach differs from methods such as the artificial mixing of water in lakes and reservoirs, which are based on the hypothesis that low-light exposure suppresses cyanobacteria growth [21]. However, we suggest that, in the case of controlling of *P. galeata* and *M. aeruginosa*, both low- and high-light exposure can be used effectively. Practical methods should be further studied to determine the field implications of high-light exposure for controlling these species. Particular focus should be given to methods that would keep the water column illuminated, thereby exceeding the light levels tolerable for *P. galeata* and *M. aeruginosa*. The growth performance of these two cyanobacteria species was fitted with one or more mathematical models tested (Figure 10), which confirmed the fit of the present data for the evaluation of growth responses of *P. galeata* and *M. aeruginosa* under low- and high-light conditions.

5. Conclusions

The availability of 30 to 50 µmol m^{-2} s^{-1} light was found to be a favorable illumination range for *P. galeata*, with the corresponding range for *M. aeruginosa* being 10 to 100 µmol m^{-2} s^{-1}. Beyond the optimal light intensities, the growth of the two cyanobacteria species was reduced. *M. aeruginosa* demonstrated higher tolerance to higher light intensities than *P. galeata*. High-light intensities, at which growth was lowest, could be used to develop control mechanisms, or to improve the present methods based on low-light exposure. This could help to effectively control cyanobacteria in water bodies. The fitting of the present results with the cyanobacteria growth models confirms that the growth responses of *P. galeata* and *M. aeruginosa* to different light conditions can be modeled to predict and control their occurrence.

Author Contributions: For conceptualization and methodology, G.M. and T.A.; software, validation, formal analysis, and writing—original draft preparation, G.M.; writing—review and editing, G.M., T.A., S.M.D.H.J., M.B.B., and M.H.R.; supervision, T.A.; data curation, G.M, M.B., H.D.L.A., and H.Y.; funding Acquisition, T.A. and S.J. All authors have read and agreed to the published version of the manuscript.

Funding: This research was funded by the Japan Society for the Promotion of Science (JSPS), JSPS KAKENHI grant numbers 19H02245 and 18K13833.

Acknowledgments: We would like to give special thanks to Saitama University for providing laboratory and also give thanks to Kyoko Endo for administrative role during purchasing chemicals, materials and tools for this research study.

Conflicts of Interest: The authors declare no conflict of interest.

References

1. Trolle, D.; Nielsen, A.; Rolighed, J.; Thodsen, H.; Andersen, H.; Karlsson, I.; Refsgaard, J.C.; Olesen, J.; Bolding, K.; Kronvang, B.; et al. Projecting the future ecological state of lakes in Denmark in a 6 degree warming scenario. *Clim. Res.* **2015**, *64*, 55–72. [CrossRef]

2. Wang, B.; Wang, X.; Hu, Y.; Chang, M.; Bi, Y.; Hu, Z. The combined effects of UV-C radiation and H2O2 on Microcystis aeruginosa, a bloom-forming cyanobacterium. *Chemosphere* **2015**, *141*, 34–43. [CrossRef]

3. Demoulin, C.F.; Lara, Y.J.; Cornet, L.; François, C.; Baurain, D.; Wilmotte, A.; Javaux, E.J. Cyanobacteria evolution: Insight from the fossil record. *Free. Radic. Boil. Med.* **2019**, *140*, 206–223. [CrossRef]

4. Yan, D.; Xu, H.; Yang, M.; Lan, J.; Hou, W.; Wang, F.; Zhang, J.; Zhou, K.; An, Z.; Goldsmith, Y. Responses of cyanobacteria to climate and human activities at Lake Chenghai over the past 100 years. *Ecol. Indic.* **2019**, *104*, 755–763. [CrossRef]

5. Śliwińska-Wilczewska, S.; Cieszyńska, A.; Konik, M.; Maculewicz, J.; Latała, A. Environmental drivers of bloom-forming cyanobacteria in the Baltic Sea: Effects of salinity, temperature, and irradiance. *Estuarine, Coast. Shelf Sci.* **2019**, *219*, 139–150. [CrossRef]

6. Mazur-Marzec, H.; Sutryk, K.; Kobos, J.; Hebel, A.; Hohlfeld, N.; Błaszczyk, A.; Toruńska, A.; Kaczkowska, M.J.; Łysiak-Pastuszak, E.; Kraśniewski, W. Occurrence of cyanobacteria and cyanotoxin in the Southern Baltic Proper. Filamentous cyanobacteria versus single-celled picocyanobacteria. *Hydrobiologia* **2013**, *701*, 235–252. [CrossRef]

7. Codd, G.A.; Morrison, L.F.; Metcalf, J.S. Cyanobacterial toxins: Risk management for health protection. *Toxicol. Appl. Pharmacol.* **2005**, *203*, 264–272. [CrossRef]

8. Pearson, L.; Mihali, T.; Moffitt, M.; Kellmann, R.; Neilan, B. On the Chemistry, Toxicology and Genetics of the Cyanobacterial Toxins, Microcystin, Nodularin, Saxitoxin and Cylindrospermopsin. *Mar. Drugs* **2010**, *8*, 1650–1680. [CrossRef]

9. Paerl, H.W.; Huisman, J. Climate change: A catalyst for global expansion of harmful cyanobacterial blooms. *Environ. Microbiol. Rep.* **2009**, *1*, 27–37. [CrossRef]

10. O'Neil, J.; Davis, T.; Burford, M.; Gobler, C. The rise of harmful cyanobacteria blooms: The potential roles of eutrophication and climate change. *Harmful Algae* **2012**, *14*, 313–334. [CrossRef]

11. Huisman, J.; Codd, G.A.; Paerl, H.W.; Ibelings, B.W.; Verspagen, J.M.H.; Visser, P.M. Cyanobacterial blooms. *Nat. Rev. Genet.* **2018**, *16*, 471–483. [CrossRef]

12. Barros, M.U.; Wilson, A.E.; Leitão, J.I.; Pereira, S.P.; Buley, R.P.; Fernandez-Figueroa, E.G.; Capelo-Neto, J. Environmental factors associated with toxic cyanobacterial blooms across 20 drinking water reservoirs in a semi-arid region of Brazil. *Harmful Algae* **2019**, *86*, 128–137. [CrossRef] [PubMed]

13. Berg, M.; Sutula, M. Factors affecting the growth of cyanobacteria with special emphasis on the Sacramento-San Joaquin Delta. South. Calif. Coast. *Water Res. Proj. Tech. Rep.* **2015**, *869*, 7–28.

14. Khatoon, H.; Leong, L.K.; Rahman, N.A.; Mian, S.; Begum, H.; Banerjee, S.; Endut, A. Effects of different light source and media on growth and production of phycobiliprotein from freshwater cyanobacteria. *Bioresour. Technol.* **2018**, *249*, 652–658. [CrossRef] [PubMed]

15. Abeynayaka, H.D.L.; Asaeda, T.; Rashid, M.H. Effects of elevated pressure on Pseudanabaena galeata Böcher in varying light and dark environments. *Environ. Sci. Pollut. Res.* **2018**, *25*, 21224–21232. [CrossRef] [PubMed]

16. Kakimoto, M.; Ishikawa, T.; Miyagi, A.; Saito, K.; Miyazaki, M.; Asaeda, T.; Yamaguchi, M.; Uchimiya, H.; Kawai-Yamada, M. Culture temperature affects gene expression and metabolic pathways in the 2-methylisoborneol-producing cyanobacterium Pseudanabaena galeata. *J. Plant Physiol.* **2014**, *171*, 292–300. [CrossRef] [PubMed]

17. Dang, T.C.; Fujii, M.; Rose, A.L.; Bligh, M.; Waite, T.D. Characteristics of the freshwater cyanobacterium Microcystis aeruginosa grown in iron-limited continuous culture. *Appl. Environ. Microbiol.* **2012**, *78*, 1574–1583. [CrossRef]

18. Princiotta, S.D.; Hendricks, S.P.; White, D.S. Production of Cyanotoxins by Microcystis aeruginosa Mediates Interactions with the Mixotrophic Flagellate Cryptomonas. *Toxins* **2019**, *11*, 223. [CrossRef]

19. Glibert, P.M. Harmful algae at the complex nexus of eutrophication and climate change. *Harmful Algae* **2019**, 101583. [CrossRef]

20. Lürling, M.; Van Oosterhout, F.; Faassen, E. Eutrophication and Warming Boost Cyanobacterial Biomass and Microcystins. *Toxins* **2017**, *9*, 64. [CrossRef]

21. Visser, P.M.; Ibelings, B.W.; Bormans, M.; Huisman, J. Artificial mixing to control cyanobacterial blooms: A review. *Aquat. Ecol.* **2016**, *50*, 423–441. [CrossRef]

22. Islam, M.A.; Beardall, J. Growth and Photosynthetic Characteristics of Toxic and Non-Toxic Strains of the Cyanobacteria Microcystis aeruginosa and Anabaena circinalis in Relation to Light. *Microorganisms* **2017**, *5*, 45. [CrossRef] [PubMed]

23. Muramatsu, M.; Hihara, Y. Acclimation to high-light conditions in cyanobacteria: From gene expression to physiological responses. *J. Plant Res.* **2012**, *125*, 11–39. [CrossRef] [PubMed]

24. Venugopal, V.; Prasanna, R.; Sood, A.; Jaiswal, P.; Kaushik, B.D. Stimulation of pigment accumulation in Anabaena azollae strains: Effect of light intensity and sugars. *Folia Microbiol.* **2006**, *51*, 50–56. [CrossRef] [PubMed]

25. Steele, J.H. Notes on some theoketical pkoblems in pkoduction ecology. Prim. Product. *Aquat. Environ.* **1965**, *18*, 383.

26. Platt, T.; Jassby, A.D. The relationship between photosynthesis and light for natural assemblages of coastal marine phyoplankton. *J. Phycol.* **1976**, *12*, 421–430.

27. Peeters, J.C.H.; Eilers, P. The relationship between light intensity and photosynthesis—A simple mathematical model. *Aquat. Ecol.* **1978**, *12*, 134–136. [CrossRef]

28. Rippka, R.; Stanier, R.Y.; Deruelles, J.; Herdman, M.; Waterbury, J.B. Generic Assignments, Strain Histories and Properties of Pure Cultures of Cyanobacteria. *Microbiol.* **1979**, *111*, 1–61. [CrossRef]

29. Guan, Y.; Deng, M.; Yu, X.; Zhang, W. Two-stage photo-biological production of hydrogen by marine green alga Platymonas subcordiformis. *Biochem. Eng. J.* **2004**, *19*, 69–73. [CrossRef]

30. Bradford, M.M. A rapid and sensitive method for the quantitation of microgram quantities of protein utilizing the principle of protein-dye binding. *Anal. Biochem.* **1976**, *72*, 248–254. [CrossRef]

31. Axler, R.P.; Owen, C.J. Measuring chlorophyll and phaeophytin: Whom should you believe? *Lake Reserv. Manag.* **1994**, *8*, 143–151. [CrossRef]

32. Association, A.P.H.; Association, A.W.W.; Federation, W.P.C.; Federation, W.E. *Standard Methods for the Examination of Water and Wastewater*; American Public Health Association: Washington, DC, USA, 1915; ISBN 8755-3546.

33. Holm, G. Chlorophyll Mutations in Barley. *Acta Agric. Scand.* **1954**, *4*, 457–471. [CrossRef]

34. Romo, S. Growth parameters of Pseudanabaena galeata Böcher in culture under different light and temperature conditions. *Algol. Stud. für Hydrobiol. Suppl.* **1994**, *75*, 239–248.

35. Jana, S.; Choudhuri, M.A. Effects of plant growth regulators on Hill activity of submerged aquatic plants during induced senescence. *Aquat. Bot.* **1984**, *18*, 371–376. [CrossRef]

36. Aebi, H. Catalase in vitro. In *Tumor Immunology and Immunotherapy &ndash*; Cellular Methods Part B; Elsevier: Amsterdam, The Netherlands, 1984; pp. 121–126.

37. Macadam, J.W.; Nelson, C.J.; Sharp, R.E. Peroxidase Activity in the Leaf Elongation Zone of Tall Fescue: I. Spatial Distribution of Ionically Bound Peroxidase Activity in Genotypes Differing in Length of the Elongation Zone. *Plant Physiol.* **1992**, *99*, 872–878. [CrossRef]

38. Foy, R.H.; Gibson, C.E.; Smith, R. V The influence of daylength, light intensity and temperature on the growth rates of planktonic blue-green algae. *Br. Phycol. J.* **1976**, *11*, 151–163. [CrossRef]

39. Sabour, B.; Sbiyyaa, B.; Loudiki, M.; Oudra, B.; Belkoura, M.; Vasconcelos, V. Effect of light and temperature on the population dynamics of two toxic bloom forming Cyanobacteria–Microcystis ichthyoblabe and Anabaena aphanizomenoides. *Chem. Ecol.* **2009**, *25*, 277–284. [CrossRef]

40. Whitelam, G.C.; Cold, G.A. Photoinhibition of photosynthesis in the cyanobacterium Microcystis aeruginosa. *Planta* **1983**, *157*, 561–566. [CrossRef]

41. Machová, K.; Elster, J.; Adamec, L. Xanthophyceaen assemblages during winter-spring flood: Autecology and ecophysiology of Tribonema fonticolum and T. monochloron. *Hydrobiologia* **2008**, *600*, 155–168. [CrossRef]

42. Harel, Y.; Ohad, I.; Kaplan, A. Activation of Photosynthesis and Resistance to Photoinhibition in Cyanobacteria within Biological Desert Crust1[w]. *Plant Physiol.* **2004**, *136*, 3070–3079. [CrossRef]

43. Kaebernick, M.; Neilan, B.A.; Börner, T.; Dittmann, E. Light and the Transcriptional Response of the Microcystin Biosynthesis Gene Cluster. *Appl. Environ. Microbiol.* **2000**, *66*, 3387–3392. [CrossRef] [PubMed]

44. De Oliveira, C.A.; Oliveira, W.C.; Ribeiro, S.M.R.; Stringheta, P.C.; Nascimento, A.G. Effect of light intensity on the production of pigments in Nostoc SPP. *Eur. J. Biol. Med. Sci. Res.* **2014**, *2*, 23–36.

45. Pojidaeva, E.; Zinchenko, V.; Shestakov, S.V.; Sokolenko, A. Involvement of the SppA1 Peptidase in Acclimation to Saturating Light Intensities in Synechocystis sp. Strain PCC 6803. *J. Bacteriol.* **2004**, *186*, 3991–3999. [CrossRef] [PubMed]

46. Rosales-Loaiza, N.; Guevara, M.; Lodeiros, C.; Morales, E. Crecimiento y producción de metabolitos de la cianobacteria marina Synechococcus sp.(Chroococcales) en función de la irradiancia. *Rev. Biol. Trop.* **2008**, *56*, 421–429. [CrossRef]

47. Babele, P.K.; Kumar, J.; Chaturvedi, V. Proteomic De-Regulation in Cyanobacteria in Response to Abiotic Stresses. *Front. Microbiol.* **2019**, *10*, 1315. [CrossRef]

48. Liu, M.; Shi, X.; Chen, C.; Yu, L.; Sun, C. Responses of Microcystis Colonies of Different Sizes to Hydrogen Peroxide Stress. *Toxins* **2017**, *9*, 306. [CrossRef]

49. De Silva, H.C.C.; Asaeda, T. Effects of heat stress on growth, photosynthetic pigments, oxidative damage and competitive capacity of three submerged macrophytes. *J. Plant Interactions* **2017**, *12*, 228–236. [CrossRef]

50. Rastogi, R.P.; Singh, S.P.; Häder, D.-P.; Sinha, R.P. Detection of reactive oxygen species (ROS) by the oxidant-sensing probe 2′,7′-dichlorodihydrofluorescein diacetate in the cyanobacterium Anabaena variabilis PCC 7937. *Biochem. Biophys. Res. Commun.* **2010**, *397*, 603–607. [CrossRef]

Article

Linking Stoichiometric Organic Carbon–Nitrogen Relationships to planktonic Cyanobacteria and Subsurface Methane Maximum in Deep Freshwater Lakes

Santona Khatun [1,*], Tomoya Iwata [2,*], Hisaya Kojima [3], Yoshiki Ikarashi [2], Kana Yamanami [2], Daichi Imazawa [1], Tanaka Kenta [4], Ryuichiro Shinohara [5] and Hiromi Saito [6]

[1] Integrated Graduate School of Medicine, Engineering, and Agricultural Sciences, University of Yamanashi, 4-4-37 Takeda, Kofu 400-8510, Japan; g17lr002@yamanashi.ac.jp

[2] Faculty of Life and Environmental Sciences, University of Yamanashi, 4-4-37 Takeda, Kofu 400-8510, Japan; yoshiki.ikarashi@gmail.com (Y.I.); L14EV033@yamanashi.ac.jp (K.Y.)

[3] Institute of Low Temperature Science, Hokkaido University, Kita-19, Nishi-8, Kita-ku, Sapporo 060-0819, Japan; kojimah@pop.lowtem.hokudai.ac.jp

[4] Sugadaira Research Station, Mountain Science Center, University of Tsukuba, 1278-294 Sugadaira-kogen, Ueda 386-2204, Japan; kenta@sugadaira.tsukuba.ac.jp

[5] National Institute for Environmental Studies, 16-2 Onogawa, Tsukuba 305-8506, Japan; r-shino@nies.go.jp

[6] Department of Marine Biology and Sciences, Tokai University, 1-1 1-Chome 5-Jo Minami-sawa, Minami-ku, Sapporo 005-8601, Japan; saitou@tsc.u-tokai.ac.jp

* Correspondence: santona.khatun@gmail.com (S.K.); tiwata@yamanashi.ac.jp (T.I.)

Received: 29 November 2019; Accepted: 31 January 2020; Published: 2 February 2020

Abstract: Our understanding of the source of methane (CH_4) in freshwater ecosystems is being revised because CH_4 production in oxic water columns, a hitherto inconceivable process of methanogenesis, has been discovered for lake ecosystems. The present study surveyed nine Japanese deep freshwater lakes to show the pattern and mechanisms of such aerobic CH_4 production and subsurface methane maximum (SMM) formation. The field survey observed the development of SMM around the metalimnion in all the study lakes. Generalized linear model (GLM) analyses showed a strong negative nonlinear relationship between dissolved organic carbon (DOC) and dissolved inorganic nitrogen (DIN), as well as a similar curvilinear relationship between DIN and dissolved CH_4, suggesting that the availability of organic carbon controls N accumulation in lake waters thereby influences the CH_4 production process. The microbial community analyses revealed that the distribution of picocyanobacteria (i.e., *Synechococcus*), which produce CH_4 in oxic conditions, was closely related to the vertical distribution of dissolved CH_4 and SMM formation. Moreover, a cross-lake comparison showed that lakes with a more abundant *Synechococcus* population exhibited a greater development of the SMM, suggesting that these microorganisms are the most likely cause of methane production. Thus, we conclude that the stoichiometric balance between DOC and DIN might cause the cascading responses of biogeochemical processes, from N depletion to picocyanobacterial domination, and subsequently influence SMM formation in lake ecosystems.

Keywords: dissolved inorganic nitrogen; dissolved organic carbon; phosphonate; subsurface methane maximum; stoichiometry; *Synechococcus*

---

1. Introduction

In recent decades, considerable efforts have been made to elucidate the source of methane (CH_4) emissions, because the atmospheric concentration of this greenhouse gas has been increasing [1–3].

Among the natural sources of CH_4 emissions, freshwater lakes have recently been recognized as an important contributor, accounting for 6–16% of total natural CH_4 emissions [4,5]. Traditionally, CH_4 production in lakes was believed to occur via anaerobic methanogenesis in oxygen-depleted environments [6,7]. However, recent studies have identified the occurrence of the subsurface methane maximum (SMM) in the oxygenated water columns of many deep freshwater lakes [8–15], as well as in marine ecosystems (known as the methane paradox) [16–18]. As subsurface CH_4 in upper water column layers is closer to the atmosphere than sedimentary CH_4, the formation of the SMM may have a significant impact on CH_4 emissions from freshwater lakes [19].

It has been proposed that SMM might develop as a result of the tributary inflow, and the transport and diffusion of CH_4 from hypolimnetic and littoral sediments [4,9,20–22]. Moreover, anaerobic methanogenesis by archaea present in anoxic microenvironments within algal cell aggregates, zooplankton guts or other particles was also proposed as a possible solution of the methane paradox [10,23]. However, recent studies have added empirical evidence that SMM formation is likely via phosphonate metabolism by planktonic bacteria that carry C-P lyase genes [12,13,15,16,24,25] or via unknown photosynthesis-related processes [26]. For example, planktonic autotrophic and heterotrophic bacteria (e.g., *Trichodesmium*, *Pseudomonas*, *SAR11*, *Synechococcus*) can utilize phosphonate compounds when they are P-starved and produce CH_4 aerobically as a byproduct of phosphonate decomposition. Other methylated compounds, such as dimethylsulfoniopropionate (DMSP), trimethylamine (TMA), and methionine (Met), can also serve as a possible precursor of aerobic CH_4 production by planktonic bacteria and microalgae [17,19,27]. However, most of these studies focused only on specific microbes linked with CH_4 production in oxic water columns. Moreover, few studies have clarified the relationship among biogeochemical processes, microbial communities, and aerobic CH_4 production in lakes.

Our previous studies have shown that SMM starts to develop in early summer in association with the development of stratification, and that it peaks in midsummer around the metalimnion in a deep oligotrophic lake [15]. Then, the SMM disappears in the winter season. It is hypothesized that the seasonal predominance of cyanobacterial populations, such as *Synechococcus*, during the thermal stratification period may be responsible for the development of SMM through organophosphonate metabolism [15]. However, the hypothesis has never been tested in other lake ecosystems. Although a few studies have examined the role of the microbial community in lake SMM formation [12–15], knowledge of the relationship between the vertical profile of dissolved CH_4 concentrations and microbial communities is also lacking. If planktonic cyanobacteria are responsible for SMM formation, then biogeochemical processes other than CH_4 production in metalimnetic waters may also be affected through their metabolic activities, such as primary productivity, nutrient uptake, and the extracellular release of DOC [28,29]. Furthermore, the biogeochemical processes that control SMM development have remained unidentified, even though the degree of the SMM development varies greatly among lakes [15].

The objective of the present study was to clarify the pattern and mechanisms of SMM formation in the aerobic environments of lake ecosystems. We surveyed nine deep freshwater lakes in Japan to determine the pattern of CH_4 supersaturation in the metalimnion and to identify the environmental variables that affect such SMM formation. Moreover, we performed microbial community analyses of bacterioplankton and algae to identify the microbes responsible for aerobic CH_4 production across the lakes. Finally, we analyzed lake physicochemical environments linked with both the microbial community and SMM formation to expand our knowledge of aerobic CH_4 production and emission from oxic layers in deep freshwater lakes.

2. Materials and Methods

2.1. Study Area and Field Survey

The field survey was conducted in nine freshwater lakes of Japan (Figure S1, Tables S1 and S2) during the summer stratification period (from July to September) in 2016–2017. The study lakes are located in temperate and cool-temperate regions (35°11′ N–42°36′ N latitudes) and are mainly classified as caldera lakes, dammed lakes, or tectonic lakes (Table S1). The study lakes are deep stratified lakes, with maximum depths (31–363 m) deeper than the depth of 1% of the irradiance at the surface (i.e., a surrogate for the compensation depth) during the study period. The lake trophy was classified as mesotrophic, oligotrophic, or ultraoligotrophic conditions according to chlorophyll *a* and TP concentrations (Table S2, [30]). Most of the lakes are monomictic or dimictic and rarely freeze over; except Lake Hibara, where the surface water generally freezes during the December–March period. The watersheds of the study lakes are mostly covered with deciduous and coniferous forests (range = 67–99%). Agricultural lands such as rice paddies and farmlands occupy 20% and 11% of the watershed areas of Lake Toya and Lake Nojiri, respectively. Grassland vegetation accounts for 11% of the watershed area of Lake Hibara, while in Lake Ashinoko, residential areas (9%) contribute to the land cover to some extent.

At the deepest point of each study lake, we established a sampling site and measured the vertical profiles of environmental variables such as water temperature (°C), dissolved oxygen (DO) concentration (DO, mgO_2/L) and saturation (%), pH, and photosynthetically active radiation (PAR, $\mu mol\ m^{-2}\ s^{-1}$) by using a multi-parameter sonde (YSI ProDSS, YSI-Nanotech, Tokyo, Japan), a DO meter (HQ40d, Hach, Loveland, CO, USA), a pH meter (Orion3-Star, Thermo Scientific, Chelmsford, MA, USA), and a submersible spherical quantum sensor (LI-193, LI-COR, Inc., Lincoln, NE, USA), respectively. For each sampling site, lake water samples were collected at 10 depths, from the surface to hypolimnetic waters, by using a 6 L Van Dorn sampler. For determination of the dissolved CH_4 concentration, the collected lake water was siphoned into two 30 mL glass vials and sealed with a butyl rubber stopper and an aluminum crimp. Then, 0.2 mL of saturated $HgCl_2$ was added to each vial as a preservative and stored in a dark place at room temperature until analysis. For water quality analyses, the collected lake water was filtered with a 100 μm nylon mesh to remove large particles (e.g., zooplankton), and the filtrates were stored in polypropylene bottles for total nitrogen (TN) and total phosphorus (TP) analyses. The remaining filtrate water was further filtered through pre-combusted GF/F glass fiber filters (Whatman plc., Maidstone, Kent, UK) for dissolved nutrient (i.e., SRP, NH_4, NO_2, and NO_3) and dissolved organic carbon (DOC) analyses; the former samples were stored in polypropylene bottles, while the latter samples were preserved in pre-combusted amber glass bottles. All of the water samples for nutrient and DOC analyses were kept at −20 °C until analyzed. For analyzing dissolved manganese (Mn) concentrations, the 100 μm mesh-filtered lake waters were further filtered with cellulose acetate disposable filters (pore size 0.45 μm), and nitric acid (final concentration = 0.1 M HNO_3) was added to prevent the precipitation and adsorption of metals. Then, the acidified water samples were stored in a refrigerator.

For determination of the vertical profile of phytoplankton biomass and planktonic microbe density, we also collected lake waters from the 10 sampling depths and performed chlorophyll *a* (Chl *a*) measurements and CARD-FISH (catalyzed reporter deposition fluorescence in situ hybridization) analyses. We collected phytoplankton pigment samples by filtering 50–500 mL of lake water with GF/F glass fiber filters, and then stored the samples at −20 °C until analysis. For the CARD-FISH analyses, an aliquot of lake water samples was fixed in a 0.2 μm-filtered paraformaldehyde solution (final concentration = 1%, v/v) and stored at 4 °C for <24 h. To analyze the community structure of planktonic eukaryotic algae and prokaryotic bacteria, we collected additional lake water samples from each epilimnion (0 m depth), metalimnion (7–15 m depth), and hypolimnion (26–60 m depth). The samples for the enumeration of eukaryotic algae were fixed with a Lugol solution (final concentrations, 0.2%–0.4%) and stored in the dark at room temperature, while the samples for bacterial community structure were collected on 0.22 μm filters (Sterivex filter cartridges, Millipore, Billerica, MA, USA) and stored at −20 °C until analysis.

To identify the presence of phosphonic compounds in microbial cells, a two-dimensional NMR (^1H–^{31}P heteronuclear multiple bond correlation, ^1H–^{31}P HMBC) analysis was performed for the suspended particles in lake water samples according to the previous study [31]. A large amount of lake water (68–201 L) was collected from the metalimnion of each lake with the Van Dorn sampler and filtered with a 100 µm nylon mesh to remove large particles. Then, the suspended particulate matter was collected onboard by filtering the collected lake water through a 2 µm cartridge filter (MCP-HX-C10S, Advantec, Inc., Tokyo, Japan) by using a peristaltic pump (Masterflex, L/S VFP002, Cole-Parmer, Vernon Hills, IL, USA). The filter samples were stored at −20 °C until analysis.

We also surveyed all of the major tributary streams identifiable on topographical maps for each study lake (Table S3), except for two inaccessible tributaries (Yanagisawa Stream of Lake Chuzenji and Okotanpe Stream of Lake Shikotsu). Near the downstream end of each tributary inflow, we collected river water samples to analyze the same environmental variables (water temperature, DO, pH, dissolved CH_4, nutrients, DOC, and Mn) as those in the lake survey. In addition, the discharge (m^3/s) of tributary streams was measured by the midsection method using a portable current meter (CR-7WP; Cosumo Riken, Inc., Kashiwara, Japan).

2.2. Chemical Analyses

Dissolved CH_4 concentrations of the collected lake waters were quantified by the headspace equilibration method using a gas chromatograph with an FID detector (GC-FID, GC-8A, Shimadzu Corp., Kyoto, Japan). The headspace phase created by 3 mL of high-purity He gas (>99.9999%) in the vial was equilibrated with the aqueous phase. The pCH_4 in the equilibrated headspace was then analyzed by GC-FID. The dissolved CH_4 concentration (nmol/L) in the water samples were determined from the amounts of gasses in both the equilibrated headspace and the aqueous phase; the latter was calculated by using pCH_4 in the headspace and Henry's law constant.

The DOC concentration of lake water samples was measured using a total organic carbon analyzer (TOC-L CPN, Shimadzu Corp., Kyoto, Japan). The ammonium concentration (NH_4, µM) was quantified spectrophotometrically using the indophenol method [32]. Nitrate (NO_3) and nitrite (NO_2) concentrations were determined through the second-derivative UV spectrophotometric method [33] and Bendschneider and Robinson's method [34], respectively. Dissolved inorganic nitrogen (DIN) concentrations were determined as the sum of NH_4, NO_2, and NO_3. The total nitrogen (TN) concentration of the lake water samples was determined by the alkaline persulfate digestion method, followed by the absorbance measurement (at 220 nm) for the resultant nitrate by UV spectrophotometry. Soluble reactive phosphorus (SRP) concentrations were measured spectrophotometrically using the molybdenum blue method. Likewise, the dissolved total phosphorus (DTP) and total phosphorus (TP) concentrations were quantified using the molybdenum blue method after persulfate digestion for the GF/F-filtered lake waters and unfiltered lake waters, respectively. For the phosphomolybdic acid measurements for SRP, DTP, and TP analyses, we increased the sulfuric acid concentration by a factor of two to minimize the effect of silicate interference [35]. The dissolved organic phosphorus (DOP) concentration was determined as the difference between DTP and SRP concentrations. Dissolved Mn concentrations were analyzed using inductively coupled plasma optical emission spectroscopy (SPS 3520 UV-DD, Hitachi High-Tech Science Corp., Tokyo, Japan).

2.3. Analyses for Phytoplankton Biomass and Community Structure

The Chl *a* sample pigments collected on GF/F glass fiber filters were extracted with N, N-dimethylformamide (6-mL) in the dark for 24 h. The Chl *a* content in lake water (µg-chl *a*/L) was then measured by the Welschmeyer method using a fluorometer (Trilogy, Turner Designs, San Jose, CA, USA). To identify the community structure of eukaryotic algae, we identified phytoplankton species to the lowest taxonomic level and enumerated the cell numbers of each algal species for the epilimnion, metalimnion and hypolimnion of each study lake by microscopic enumeration. We also enumerated red autofluorescent algal cells on the CARD-FISH filter (see below) under blue excitation (470 nm, Filter 09, Carl Zeiss Microscopy GmbH, Jena, Germany) to estimate algal cell density (cells/mL).

2.4. Analyses for Planktonic Bacterial Density and Community Structure

To analyze the distribution of specific microbes related to SMM formation, the CARD-FISH analysis was performed for the water samples collected from each sampling depth. The aliquot of lake water samples fixed with paraformaldehyde solution was filtered through a white polycarbonate membrane filter (type GTTP, pore size 0.2 μm, Millipore, Billerica, MA, USA) and a brown polycarbonate membrane filter (type GTBP, pore size 0.2 μm, Millipore, Billerica, MA, USA) for the enumeration of CARD-FISH- and DAPI-stained bacterial cells, respectively. We applied the oligonucleotide probes (labeled with horseradish peroxidese) specific for eubacteria (EUB338), type I MOB (Mg84 + Mg705), *Synechococcus* (405_Syn) and archaea (ARCH915) for the filters at 46 °C for > 8h for hybridization (Table S4) because these microbial groups have been reported to influence the dissolved methane profile in lakes [15,36–38]. FITC-labeled tyramide solution and DAPI solution were then applied to the filter sections for the enumeration of cells on a fluorescence microscope (Zeiss Axio Lab.A1, Carl Zeiss Microscopy GmbH, Jena, Germany) under blue excitation (470 nm, Filter 10) and UV excitation (365 nm, Filter 01), respectively. We also enumerated the autofluorescent cyanobacterial cells under green excitation (545 nm, Filter 43) to estimate the density of cyanobacteria. At least 1000 cells were counted to estimate the cell density of the target bacteria (cells/mL).

To analyze the community structure of planktonic bacteria, 16S rRNA gene amplicon sequencing was performed for the filter samples collected at the epilimnion, metalimnion and hypolimnion of each lake. From the Sterivex filter samples, DNA was extracted using the PowerWater Sterivex DNA Isolation Kit (Qiagen, Hilden, Germany). From the DNA samples, sequencing libraries were prepared according to the Illumina 16S Metagenomic Sequencing Library Preparation protocol and then subjected to Illumina MiSeq sequencing. The resulting reads were processed with QIIME, an open-source bioinformatics platform for microbial community analysis [39], as follows. Quality-filtered sequences were clustered using the UCLUST algorithm with an identity threshold of 97%, to generate operational taxonomic units (OTUs). The representative sequences of the OTUs were subjected to chimera-checking to eliminate chimeric OTUs. The representatives were then used for taxonomic assignment of the OTUs via the Greengenes database, version 13_8. Greengenes annonation includes eukaryotic organelles (i.e., chloroplasts and mitochondria) as a taxonomic group in bacterial phyla. As we focused on bacterioplankton communities in the 16S rRNA gene analyses, we removed the chloroplast and mitochondria reads from the amplicon library for data analyses. The nucleotide sequences of the 16S rRNA gene obtained in the present study are available in the NCBI/SRA with the accession number of PRJNA599317.

2.5. ^1H–^{31}P NMR Analyses for Suspended Particles

For NMR analysis, we extracted the organic phosphorus fraction from the suspended particles collected on the 2 μm cartridge filter [31]. We separated the pleated filter paper from the polypropylene filter cartridge by using a cutter and tweezers. The separated filter papers were then placed into a Teflon vial with 50 mL Milli-Q water, and the vials were subsequently immersed for approximately 1 h in a water bath set at 60 °C for the extraction [40]. The extracts were then filtered through a GF/F glass fiber filter followed by membrane filter filtration (PVDF: nominal pore size 0.45 μm, Millex, Millipore, Billerica, MA, USA). The filtrates were frozen (−20 °C) and lyophilized to concentrate the extracts for NMR analysis. Just before the NMR analyses, the lyophilized extracts were re-dissolved in 0.9 mL of EDTA solution (20 mmol L^{-1}) with D$_2$O (0.1 mL, Fujifilm Wako Pure Chemical Corp., Osaka, Japan) and shaken at room temperature for 10 minutes. The solution was then filtered through a membrane filter (nominal pore size: 0.45 μm, Millex, Millipore, Billerica, MA, USA) and transferred into a 5 mm NMR tube. The ^1H–^{31}P NMR spectrum of the extracts was recorded at 500.2MHz (for ^1H) and 202.5 MHz (for ^{31}P) via a JNM-ECA 500 spectrometer (JEOL Ltd., Tokyo, Japan), equipped with a 5 mm auto-tune probe. These spectra were externally referenced to 4,4-dimethyl-4-silapentane-1-sulfonic acid (DSS) and D$_3$PO$_4$ (δ = 0 ppm), respectively. The datasets of ^1H–^{31}P HMBC spectra were acquired with 2048 and 256 data points and sweep widths of 9384 and 20,259 Hz for the x and y dimensions, respectively. The obtained spectra were processed with a 2 Hz line broadening using Delta, version 5.

2.6. Data Analyses

A generalized linear model (GLM) was developed to identify environmental variables that affect the dissolved CH_4 concentrations in the study lakes. We included physical (depth, water temperature, and percentage of surface irradiance), chemical (DIN, SRP, DOC, DOP, and TN/TP), and biological (Chl a) variables as the explanatory variables. Two variables (i.e., SRP and Chl a) were transformed as log_{10} $(x + 1)$ because the data were not normally distributed by visual examination. A GLM analysis with a gamma error distribution and an inverse link function was used to analyze the factors that influenced the response variable (dissolved CH_4). We built candidate models containing all possible combinations of the nine explanatory variables ($2^9 - 1 = 511$ models) and selected the best model based on the AIC value. The relative importance of an explanatory variable was evaluated by the sum of Akaike weights (w) over all models, including that variable [41]. To identify the indirect causal relationships of the environmental variables with SMM formation, we also constructed GLMs for the environmental variables that most strongly affected the dissolved CH_4 concentrations. When we constructed the GLM for DOP values, we transformed the response variable as DOP + 1 because the gamma regression analysis did not allow for the inclusion of variables with negative or zero values. The GLM analyses and model-selection procedures were performed using the MuMIn package, version 1.40.0 [42] and the MASS package, version 7.3-48 [43] of the statistical R software 3.3.3 (R Development Core Team 2017).

For the community analyses of planktonic bacteria, 16S rRNA gene sequence read counts were used to calculate the relative abundance of each OTU for eplimnetic, metalimnetic, and hypolimnetic bacterial communities, by assuming that the amplicon reads could be used as a surrogate for the abundance of each bacterial population (total number of amplicon reads = 8848–29,331 reads per sample). In this calculation, we removed the taxonomically unassigned OTUs from the analyses, as the percentage of their amplicon reads accounted only for <1.6% of the total reads. We also confirmed a significant positive correlation between the total number of bacterial read counts per unit of lake volume and the cell density of eubacteria obtained from the CARD-FISH analyses (EUB338, $r = 0.50$, $p = 0.015$). Finally, the taxonomic composition data were combined into the phylum level for the analyses. Non-metric multidimensional scaling (NMDS) analysis was performed to identify the relationships between bacterioplankton community composition and the environmental variables of the epilimnion, metalimnion, and hypolimnion of each study lake. To obtain an NMDS ordination with a low stress value at <0.2 (an indication of well-represented data by a two-dimensional representation [44]), the phyla with a relative abundance of <0.4% of the total read counts when all of the samples were combined were grouped into a category as "Others". For the environmental variables, the average values of 10 variables (i.e., depth, temperature, percentage of surface irradiance, CH_4, Chl a, DIN, SRP, DOC, DOP, TN/TP) were determined for each of epilimnion, metalimnion and hypolimnion, and were then incorporated in the analyses. The NMDS analysis (based on the Bray–Curtis dissimilarity) was performed by using the vegan 2.4-5 package [45] of the R software 3.3.3.

Finally, we examined the relationship between the degree of SMM development and lake environmental variables to identify the factors controlling the oxic CH_4 peak. According to our previous study [15], the SMM is herein defined as the layer of the local CH_4 maximum that forms below the surface water. The degree of SMM development for each study lake was quantified by peak SMM, namely the maximum CH_4 concentration in the SMM minus the atmospheric equilibrium CH_4 concentration at that depth. The peak SMM for the study lakes was regressed against the environmental variables responsible for SMM formation.

3. Results

3.1. SMM Formation in Lakes

The vertical profiles of dissolved CH_4 concentrations revealed that the SMM developed in aerobic layers in all of the nine study lakes (Figure 1). Further, peak SMM concentrations (range = 62.5–592.6 nM) showed the highest values in the vertical CH_4 profiles for all of the study lakes. The SMM peak tended to occur within the metalimnion, or in depths adjacent to the metalimnion where DO saturation levels were high (range = 85–131%). In addition, the peak depth and vertical profile of dissolved CH_4 (i.e., subsurface peak of CH_4) were similar to those of DO saturation, except in Lake Nojiri.

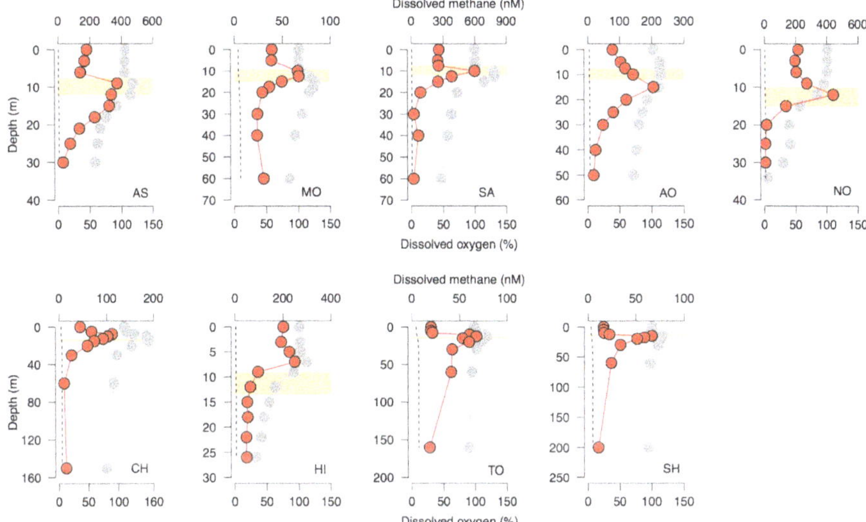

Figure 1. Vertical profiles of dissolved CH_4 concentrations in the nine study lakes of Japan during the July to September in 2016–2017 period. AS: Lake Ashinoko, MO: Lake Motosu, SA: Lake Saiko, AO: Lake Aoki, NO: Lake Nojiri, CH: Lake Chuzenji, HI: Lake Hibara, TO: Lake Toya, SH: Lake Shikotsu. Red and gray circles refer to the dissolved methane (nM) and oxygen (%) concentrations, respectively. Dotted lines denote the atmospheric equilibrium concentration of dissolved CH_4 in lake water (3.7–7.2 nM). Shaded areas denote the range of the thermocline.

Regarding the other environmental variables, water temperature and percentage of surface irradiance were highest at the lake surface and decreased with depth; this pattern did not correspond with the vertical profile of the dissolved CH_4 concentration for any of the study lakes (Figures S2 and S3). Although the overall pattern of Chl *a* and DOP profiles did not match the CH_4 profile, their peaks were located at depths similar to the CH_4 peaks (Figures S4 and S5). In most of the study lakes, DIN concentrations tended to be higher in the hypolimniion and lower in both the epilimnion and metalimnion, the latter of which exhibited high CH_4 concentrations (Figure S6). Although SRP did not show a clear overall trend, the peak depth and vertical profile of SRP were similar to those of CH_4 in Lake Motosu, Lake Saiko, Lake Toya and Lake Shikotsu (Figure S7). Similarly, the vertical patterns of DOC were variable, but its concentrations tended to be higher in the metalimnion or nearby layers (except in Lake Nojiri), as observed in the CH_4 profile (Figure S8). The observed values of lake TN/TP ratio (median = 151, range = 32–875) were high relative to the Redfield ratio (N:P = 16:1) for all of the study lakes (Figure S9). The vertical TN/TP profile tended to show a pattern opposite to that of the CH_4 profile—except in Lake Shikotsu, where a close association was observed for both variables.

The cross-lake comparison showed that the development of SMM varied among the study lakes (Figure 1). The highest value of peak SMM was observed in Lake Saiko (an oligotrophic lake), while small CH_4 peaks were observed in three ultraoligotrophic lakes (Lake Motosu, Lake Toya, and Lake Shikotsu).

3.2. Relationships between Dissolved CH_4 and Physicochemical Variables

GLM analyses were performed to identify the physicochemical variables affecting the water-column CH_4 concentrations in lakes. In the analyses, six environmental variables (i.e., depth, water temperature, percent of surface irradiance, DIN, DOP, and TN/TP ratio) were retained in the best model selected by AIC (Figure 2). The sum of Akaike weights (w) further identified that depth, DIN, DOP, and TN/TP ratio, were the most important variables explaining the variability of dissolved CH_4 concentrations in the lakes (Table S5).

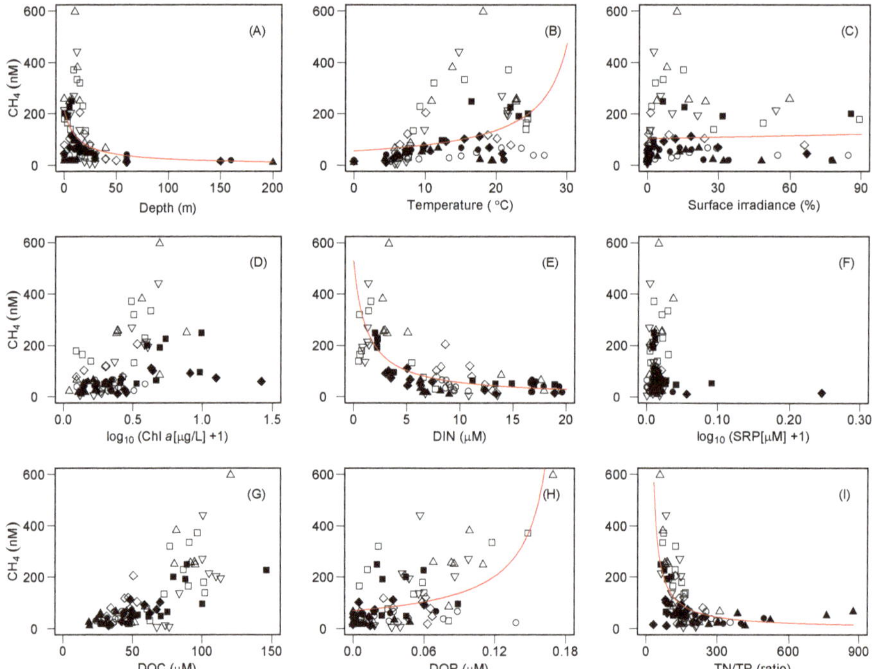

Figure 2. Relationships between environmental variables and dissolved CH_4 concentrations in the study lakes. Univariate regression lines are shown for the explanatory variables that were retained in the best generalized linear models (GLMs) with a gamma distribution of the response variable. □: Lake Ashinoko, ○: Lake Motosu, △: Lake Saiko, ◇: Lake Aoki, ▽: Lake Nojiri, ♦: Lake Chuzenji, ■: Lake Hibara, ●: Lake Toya, ▲: Lake Shikotsu.

We observed a strong negative nonlinear relationship between DIN and dissolved CH_4 concentrations when we combined the data from all nine study lakes (Figure 2 and Table S6). The CH_4 concentration increased rapidly with a decrease in the DIN concentration. Similar curvilinear patterns were observed for depth and the TN/TP ratio; the lake waters with shallower depths and a lower TN/TP ratio increased the water-column CH_4 concentrations. On the other hand, a positive effect of DOP on dissolved CH_4 concentrations was observed for the study lakes.

To explore the potential factors controlling the chemical variables that strongly affect dissolved CH_4, we further constructed GLMs for DIN, DOP, and TN/TP (Table S5). The results showed that DOC concentration was the best explanatory variable for all three influential variables (Figure 3 and Table S6). DOC exhibited negative nonlinear relationships with both DIN and the TN/TP ratio. In contrast, DOC showed a positive relationship with DOP concentrations.

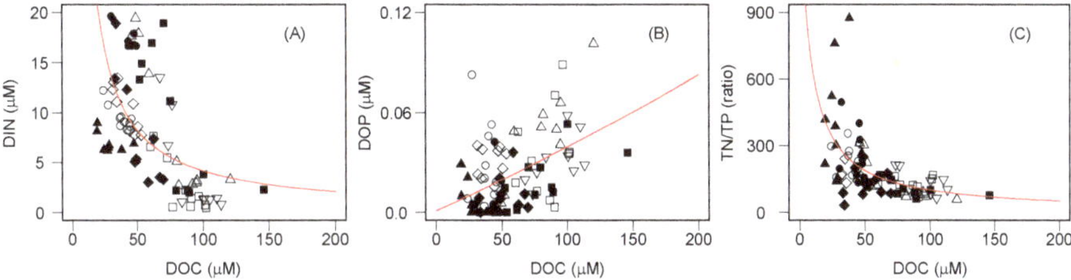

Figure 3. Relationships between dissolved organic carbon (DOC) and chemical variables in the study lakes. Univariate regression lines estimated by the best GLMs with a gamma distribution of the response variable (y) are shown. See the legend for Figure 2 for a description of the symbols.

3.3. Community Analyses of Planktonic Algae and Bacteria

Microscopic enumeration for eukaryotic algae identified six phytoplankton classes present in the study lakes (Table S7). Among these phytoplanktonic algal groups, Chlorophyceae or Bacillariophyceae predominated in the epilimnion of most of the study lakes except for Lake Toya, where Dinophyceae (53.6%) was the dominant group. Similarly, Chlorophyceae and/or Bacillariophyceae tended to predominate in the algal communities in the metalimnion, although Cryptophyceae and Dinophyceae were relatively abundant in the metalimnion of Lake Ashinoko. In the hypolimnion, Bacillariophyceae accounted for 54–97% of the relative abundance of algal communities, although Chlorophyceae was relatively abundant in the hypolimnion community of Lake Chuzenji (45.8%). There was no phytoplanktonic algal group that consistently exhibited a metalimnetic peak in terms of relative abundance for any of the study lakes.

The 16S rRNA gene amplicon sequencing analyses detected 6510 OTUs of bacteria (except unassigned OTUs), including 47 phyla, from the epilimnetic, metalimnetic, and hypolimnetic layers of all the study lakes. Actinobacteria, Verrucomicrobia, Proteobacteria, Cyanobacteria, Bacteroidetes and Armatimonadetes accounted for, respectively, 1.8–47.2%, 2.7–45.0%, 10.4–41.1%, 0.02–29.1%, 5.7–26.1%, and 0.1–5.8% of the total bacterial community in the epilimnion, metalimnion, and hypolimnion (Figure 4).

The NMDS analyses showed that the community composition of epilimnetic and metalimnetic microbes differed from that of the hypolimnetic community (Figure 4). The results revealed that the relative abundance of Proteobacteria, Verrucomicrobia, Actinobacteria, Bacteroidetes, and Cyanobacteria tended to be high in the epilimnion and metalimnion, whereas Planctomycetes, Acidobacteria, Chloroflexi, and Chlorobi were relatively abundant in the hypolimnion. We further observed the association of community structure and environmental variables. In epilimnetic and metalimnetic layers with high CH_4 and DOP concentrations, Cyanobacteria, and Verrucomicrobia tended to be relatively abundant.

The results of epifluorescent microscopic analyses (CARD-FISH, DAPI, and autofluorescent enumeration) showed that both Cyanobacteria and *Synechococcus* cell density showed significant positive relationships with water-column dissolved CH_4 concentrations (Figure 5). *Synechococcus* accounted for 21–96%, 64–96%, and 22–75% of the total cyanobacterial density in the epilimnion, metalimnion, and hypolimnion of the study lakes, respectively. In contrast to the patterns of Cyanobacteria and *Synechococcus*, phytoplankton biomass (Chl *a*), as well as the cell density of eubacteria (EUB338), type I MOB (Mg84 + Mg705), archaea (ARCH915), and algae, showed no significant relationships with dissolved CH_4 concentrations.

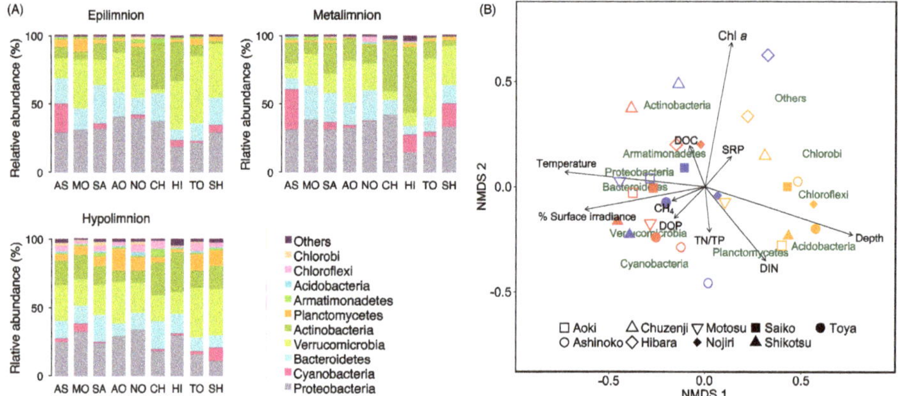

Figure 4. Taxonomic composition of bacterioplankton in the study lakes revealed by 16S rRNA amplicon sequencing analyses (**A**) and the result of non-metric multidimensional scaling (NMDS_ analysis for the relationship between lake physicochemical variables and the community composition of planktonic bacteria (**B**). See the legend for Figure 2 for the abbreviations of lake names appearing in panel (A). Red, blue and orange symbols refer to the bacterioplankton samples obtained from the epilinmion, metalimnion, and hypolinmion, respectively.

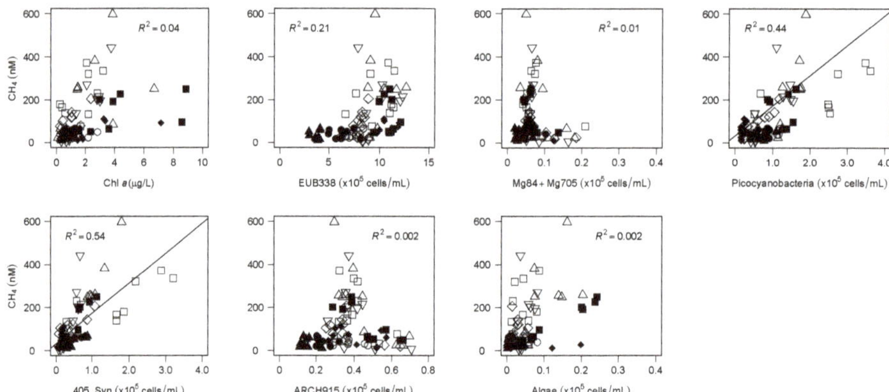

Figure 5. Relationships between the dissolved CH_4 concentration and the cell density of microbial communities in study lakes. EUB338, bacteria; Mg84 + Mg705, type-I methane-oxidizing bacteria; 405_Syn, *Synechococcus*; ARCH915, Archaea (see Table S4). GLM with a Gaussian error distribution and an identity link function was used for the variables, while regression lines shown when significant ($p <$ 0.05). See the legend for Figure 2 for a description of the symbols.

We then analyzed the vertical profile of *Synechococcus* density in relation to the CH_4 profiles, which showed that both profiles were closely associated with each other (Figure 6). In addition, a negative curvilinear pattern was found for the relationship of *Synechococcus* cell density with both depth and DIN concentration, as was observed in the patterns of dissolved CH_4 concentrations (Figure S10). In contrast, a positive nonlinear relationship between *Synechococcus* cell density and DOP concentrations was found.

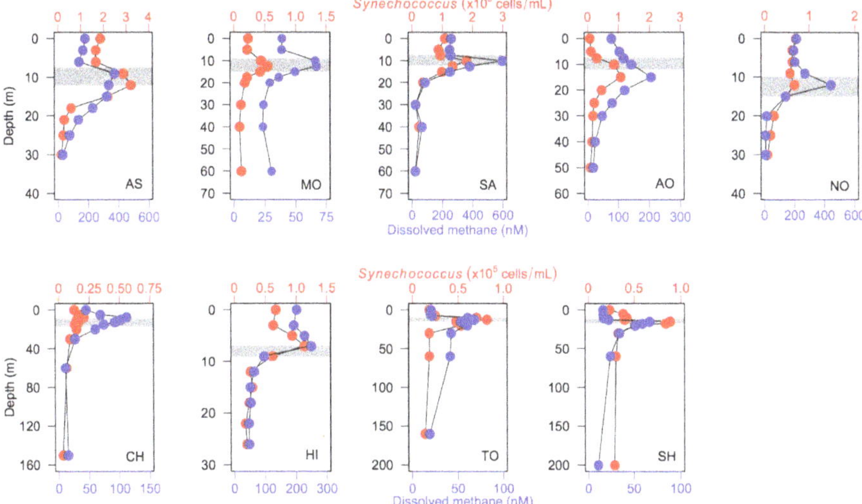

Figure 6. Vertical distribution of *Synechococcus* density (red circles) in relation to the dissolved CH$_4$ profile (blue circles) in the study lakes. Shaded areas denote the range of the thermocline.

3.4. Relationship between Peak SMM and Environmental Variables

A correlation analysis between the peak SMM and environmental variables revealed that the peak SMM was positively and linearly correlated with lake DOC, DOP, and *Synechococcus* cell density observed at the same depth (Figure 7 and Table S6). Lakes with higher concentrations of both DOC and DOP and higher *Synechococcus* cell density tended to develop a larger SMM peak. Moreover, a negative nonlinear relationship of peak SMM was observed for both DIN concentration and the TN/TP ratio, implying that lakes with lower DIN and TN/TP values experienced a rapid increase in peak SMM.

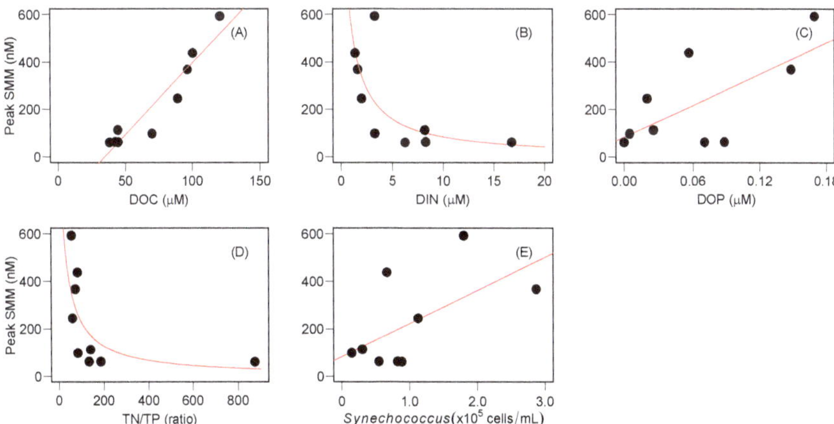

Figure 7. Relationships between lake environmental variables and the peak subsurface methane maximum (SMM) for the study lakes. For the relationships of both dissolved inorganic nitrogen (DIN) and TN/TP, the regression lines were estimated using generalized linear models (GLMs) with a gamma error distribution and an inverse link function for each variable, while the GLM with a Gaussian error distribution and an identity link function was used for the other variables. For explanatory variables, the values at the peak SMM depths were used for the analyses.

3.5. Identification of Phosphonates in Suspended Particles

Particulate suspended matter collected from the metalimnion was characterized by a two-dimensional ^{1}H–^{31}P NMR analysis. The spectral region of 20–31 ppm in ^{31}P NMR analysis is known to represent the chemical shifts of phosphonate compounds [25,46]. Our mass spectra at 1.2 ppm (^{1}H) and 25 ppm (^{31}P) revealed the presence of phosphonate in the suspended particles of the metalimnion in Lake Toya (Figure 8).

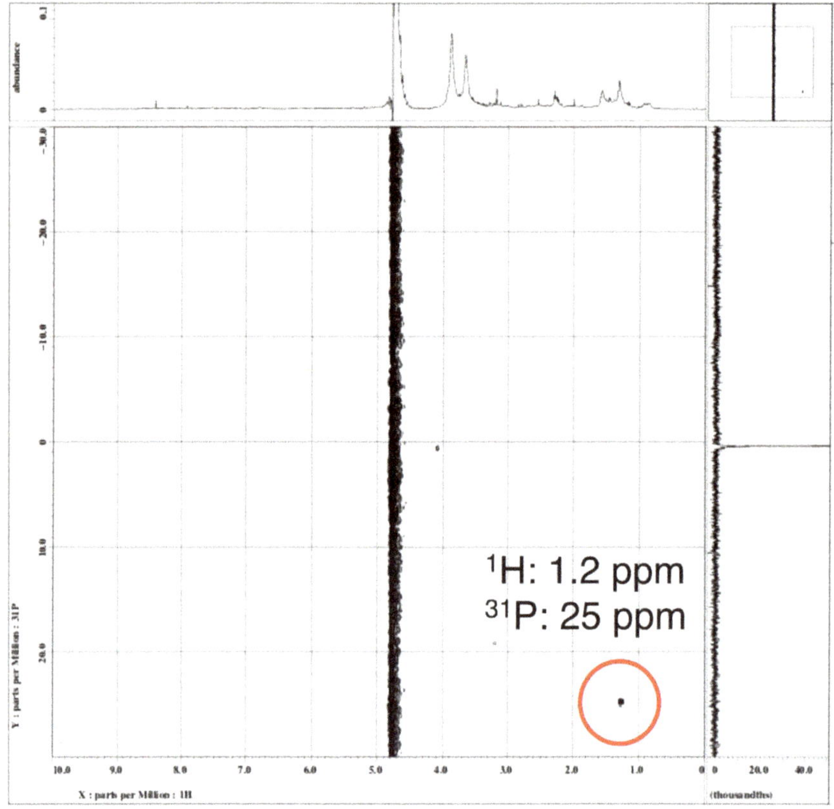

Figure 8. Two-dimensional (^{1}H-^{31}P) NMR microscopy for suspended particles collected from the metalimnion showed the presence of phosphonate compounds (black spot inside the red circles) in the metalimnion of Lake Toya.

4. Discussion

In the present study, we observed the formation of the CH_4 maximum in oxic subsurface layers (i.e., SMM) within or near the metalimnion in all of the study lakes (Figure 1). The seasonal development of metalimnetic CH_4 peaks during the stratification period was also reported by previous studies in deep freshwater lakes [8–10,15] as well as in pelagic marine ecosystems [16,17]. These results suggest that the development of the CH_4 peak in aerobic subsurface waters may be a common phenomenon in aquatic ecosystems.

SMM formation was unlikely to couple with the dissolution of atmospheric CH_4 because the observed CH_4 peak concentrations (67–597 nM, Figure 1) were one or two orders of magnitude higher than the dissolved CH_4 equilibrated with the atmosphere (range = 3.4–7.4 nM). Tributary inflow is known to contribute to SMM development in lakes [9]. However, we observed apparent SMM development even in lakes with no tributary streams (i.e., Lake Motosu and Lake Aoki). For the other lakes, moreover, the discharge-weighted average of the dissolved CH_4 concentrations of tributary streams (range = 8.8–88 nM) was generally lower than the peak CH_4 concentrations—except in Lake Ashinoko, where there is a small tributary (discharge < 0.02 m^3/s) with a high CH_4 concentration (Table S3). Another potential source of SMM formation might be the transportation of CH_4 from anoxic littoral and profundal sediments [20,21]. However, the vertical profile of dissolved Mn concentrations, a tracer of water mass transported from anoxic environments [9,15], showed a pattern unrelated to the CH_4 profiles (Figure S11). Moreover, the vertical gradient of CH_4 profiles suggests no substantial contribution of hypolimnetic CH_4 to the SMM of any of the study lakes (Figure 1). Therefore, we argue that the diffusion of anoxic CH_4 from anoxic littoral and profundal sediments was also unlikely to be the source of the CH_4 supersaturation observed in our study.

Recent studies have suggested that the microbial degradation of phosphonates (such as MPn and 2-AEP) can explain aerobic methane production in oligotrophic waters [12,13,15,16,24,25]. The marine and freshwater microbial community (e.g., *Trichodesmium*, *Pseudomonas*, *Synechococcus*, SAR11) can express the C-P lyase operon (*phn* genes) to utilize phosphonic compounds under phosphate-starved conditions [15,16,24,25]. As a result of the cleavage of C-P bonds of phosphonates, these microbes can produce CH_4. In fact, the batch-culture experiments of lake water amended with MPn confirmed in situ aerobic CH_4 production by planktonic bacteria in one of the study lakes (Lake Saiko) [15]. Considering such circumstantial evidences, as well as the fact that the DO saturation profile resembled the dissolved CH_4 pattern (Figure 1), we argue that photosynthesis-related biogeochemical processes mediated by planktonic microbes may be relevant to the development of SMM.

The relationships between dissolved CH_4 and limnological variables suggest that DOC controls the availability of DIN, DOP, and the TN/TP ratio, thereby influencing the water-column CH_4 concentration in the study lakes (Figures 2 and 3). In particular, the remarkable negative nonlinear relationships of DOC with DIN and the subsequent nonlinear effects of DIN on dissolved CH_4 concentrations were found. The nonlinear relationship between DOC and nitrate (NO_3) has recently been reported from a wide variety of aquatic environments along a hydrologic continuum from soils to streams and lakes to coastal and pelagic ocean ecosystems [47–50]. It has been hypothesized that NO_3 accumulation may occur in aquatic environments with low DOC concentrations due to an organic C limitation for assimilation and/or denitrification by heterotrophic microbes [50]. In contrast, NO_3 depletion may occur in aquatic environments with high DOC concentrations due to the sufficient supply of organic C for heterotrophs. These patterns imply that stoichiometric controls of DOC–NO_3 relationships over microbial activity may regulate the fate of N in aquatic ecosystems. In fact, NO_3 is the dominant species of DIN in the study lakes (median = 85%, range = 0–98%). Moreover, even when we separately analyzed individual lakes, a significant negative correlation of DOC and DIN was found for each of them (r = −0.69–−0.90, p < 0.05), except Lake Shikotsu (r = −0.44, p > 0.05). Furthermore, a strong nonlinear effect of DOC on the TN/TP ratio was also evident (Figure 2), suggesting that the availability of organic carbon controls N accumulation in the water column, thereby influencing the elemental stoichiometric balance of nitrogen and phosphorus in lakes.

Previous studies explained that aerobic CH_4 production from phosphonate decomposition by planktonic microbes occurs under P-starved environments because the organisms utilize phosphonates as an alternative P source [12,13,15,16,24]. However, the present study showed that higher CH_4 concentrations tended to occur in lower DIN concentrations; no clear pattern was found for the relationship between SRP and CH_4 (Figure 2). These patterns seemed to be intuitively paradoxical, as a lower N availability was likely to stimulate aerobic CH_4 production. However, the observed TN/TP ratio in the study lakes (median = 151, range = 32–875, Table S2) was much higher than the Redfield ratio (N/P = 16), suggesting that lake productivity may be more P-limited than N-limited in these lakes. Therefore, we predict that P-starved planktonic microbes utilize phosphonate compounds to form SMM in the oxic waters of the study lakes. The positive relationship between DOP, which may contain dissolved organic phosphonate compounds, and CH_4 concentrations, also supports this argument

(Figure 2). Moreover, the NMR analysis revealed the presence of particulate phosphonates in the metalimnion of Lake Toya, where SMM formation was observed (Figure 8). A previous study also detected 2-aminoethylphosphonate (2-AEP), a possible precursor of aerobic methane production [51], from the suspended particles in the epilimnion of Lake Saiko [31]. Therefore, we hypothesized that planktonic bacteria in deep unproductive lakes may store P as phosphonate molecules and utilize such intracellular P compounds via the enzymatic liberation of C-P bonds under P-limited environments [15]. Although we were unable to detect phosphonates in the suspended fractions from the other study lakes, this is probably due to the insensitivity of NMR; prolonged data acquisition and/or a larger volume of lake water filtration may be necessary to obtain the spectra. In addition, further development of quantitative analyses for the dissolved species of phosphonate compounds [52] is necessary to understand the phosphonate and CH_4 dynamics in lakes.

The 16S rRNA gene amplicon sequencing analyses and the CARD-FISH analyses revealed that cyanobacteria, especially *Synechococcus*, were related to dissolved CH_4 concentrations and SMM formation. The composition of cyanobacteria estimated from amplicon reads was associated with the gradient of the CH_4 concentration (Figure 4). Moreover, a close association of the vertical CH_4 profile with *Synechococcus* distribution was observed across the lakes (Figure 6), as found in a previous study [15]. The batch-culture experiments conducted in previous studies also showed that *Synechococcus* strains have the ability to produce CH_4 in oxic conditions as a result of phosphonate (such as MPn and 2-AEP) utilization [15,51] or unknown photosynthesis-related processes [26]. Although other heterotrophic bacteria, such as *Limnohabotans* and *Pseudomonas*, are also known to carry the C-P lyase (*phnJ*) gene [13,15], the strong correlation between the CH_4 concentration and *Synechococcus* cell density (Figure 5), as well as the association of the DO saturation profile with the CH_4 peak (Figure 1), suggested that *Synechococcus* (photosynthetic cyanobacteria) may be one of the potential drivers of SMM formation in deep freshwater lakes.

Synechococcus, a group of ubiquitous freshwater picocyanobacteria, often predominate in the planktonic communities in the epilimnion and metalimnion of unproductive lakes during the summer stratification period, where nutrient depletion occurs due to the prevention of vertical water mixing and nutrient exhaustion by competitors, such as eukaryotic algae [53–55]. Our previous studies also indicated that the development of nutrient-depleted environments by stratification during midsummer (i.e., low N and P conditions) may favor the growth of *Synechococcus* because their small cell size is advantageous for nutrient uptake under oligotrophic conditions due to efficient nutrient diffusion per unit of cell volume [15,55]. The present results add a further potential scenario for picocyanobacterial dominance in the metalimnion; that is, the reduced nutrient availability associated with the increase in DOC loading in lakes may promote the population growth of *Synechococcus*. This may be the reason for the cascading negative responses of biogeochemical elements from DOC to DIN, and from DIN to both *Synechococcus* and CH_4, that were observed in our study. In fact, a cross-lake comparison revealed that lakes with relatively high DOC and low DIN concentrations enhanced the degree of SMM development (Figure 7). Likewise, lakes with more abundant *Synechococcus* populations and a higher DOP pool promoted the development of SMM (Figure 7). Therefore, we conclude that the stoichiometric balance between DOC and DIN in the water column may regulate nutrient availability for bacterioplankton communities (e.g., *Synechococcus*) and may subsequently control CH_4 production and SMM formation in deep freshwater lakes (Figure 9). Although the variability of DOC and DIN concentrations could also be explained by the extracellular release of DOC and DIN uptake by phytoplankters [28,29], the relationship between phytoplankton biomass (Chl *a*) and DOC concentration, as well as that between Chl *a* and DIN, was unclear (Figure S12). Moreover, the vertical profiles of DOC and *Synechococcus* density were not closely associated with each other for the study lakes (Figure S13), except Lake Saiko and Lake Hibara. Therefore, we argue that phytoplankton (including *Synechococcus*) per se may not be a major factor controlling the DOC and DIN variability, even though further studies are necessary to clarify the causal relationships between microbial activities and lake biogeochemical conditions.

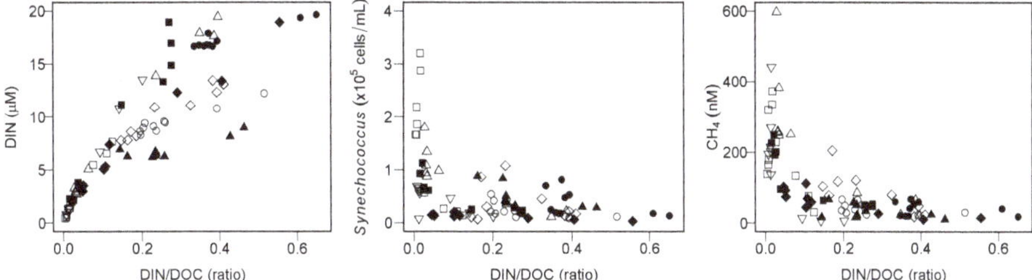

Figure 9. Effects of the stoichiometric balance between DIN and DOC on nitrogen availability, *Synechococcus* density and dissolved CH_4 concentration in the study lakes. See the legend for Figure 2 for a description of the symbols.

Although the present studies are conducted in monomictic or dimictic lakes where stratification occurs seasonally, enhanced SMM development might be predicted in meromictic lakes. This is because the lack of vertical mixing in such permanently stratified lakes may reduce the up-welling supply of nutrients from hypolimnion [56], which might result in a low DIN/DOC ratio in shallow water, as observed in the present study (Figure 9). If this is the case, the projected alteration in the lake mixing regime from dimictic to monomictic, or monomictic to meromictic, under climate change might increase the aerobic CH_4 production and emission from the nutrient-depleted oxic lake waters. Further studies are required to precisely predict the effect of lake-mixing regime on SMM formation.

5. Conclusions

The formation of SMM might have a significant impact on methane emissions from deep freshwater lakes because methanotrophs generally consume the majority of CH_4 and prevent fugitive methane emissions from deep profundal sediments or hypolimnetic zones in lakes [36,37]. In contrast, CH_4, produced by aerobic methanogenesis within the SMM of the metalimnetic layer, might easily leak into the atmosphere because of the proximity of CH_4 production sites to the lake surface. Therefore, the present study predicted that DOC loading triggers the cascading responses of biogeochemical processes, from N depletion to picocyanobacterial domination, which may promote CH_4 production and emission to the atmosphere from the SMM layer. Although previous studies also reported that DOC might influence the CH_4 emissions of freshwater ecosystems [57,58], these studies considered DOC as a direct substrate for anaerobic methanogenesis in sediments. In contrast, our study suggested that organic C exerts stoichiometric control over CH_4 production in the oxic layer of lakes. DOC concentrations are now expected to increase due to ongoing climate change, which may result in significant changes to the structure and functioning of lake ecosystems [48]. For example, increasing concentrations of DOC often change the light and thermal environments in water columns, thereby affecting the primary and secondary productivity of lake food webs, as well as the thermal stratification regime in lakes. An increased DOC supply can also provide energy subsidies to heterotrophic consumers and consequently influence the metabolic balance of lake ecosystems. Moreover, DOC can change the chemical environments of lake ecosystems, such as their pH and dissolved iron concentrations. We argue that, in addition to these phenomena, increased organic carbon loading under changing environments may promote aerobic CH_4 production and emission from the oxic layer of deep freshwater lakes.

Supplementary Materials: The following are available online at http://www.mdpi.com/2073-4441/12/2/402/s1, supplementary data accompany this paper with Figure S1: Geographic location of nine study lakes in Japan, Figure S2: Vertical profiles of dissolved CH_4 concentration (nM, red circles) and water temperature (°C, gray circles) in nine study lakes during the period from July to September in 2016–2017. AS: Lake Ashinoko, MO: Lake Motosu, SA: Lake Saiko, AO: Lake Aoki, NO: Lake Nojiri, CH: Lake Chuzenji, HI: Lake Hibara, TO: Lake Toya, SH: Lake Shikotsu. Shaded areas denote the range of the thermocline, Figure S3: Vertical profiles of dissolved CH_4 concentration (nM, red circles) and the percentage of surface irradiance (%, gray circles) in nine study lakes during the period from July to September in 2016–2017. See legend of Figure S2 for the abbreviation of lake names and the descriptions of shaded areas, Figure S4: Vertical profiles of dissolved CH_4 (nM, red circles) and chlorophyll *a* (µg/L, gray circles) concentrations in nine study lakes during the period from July to September in 2016–2017. See legend of Figure S2 for the abbreviation of lake names and the descriptions of shaded areas, Figure S5: Vertical profiles of dissolved CH_4 (nM, red circles) and DOP (µM, gray circles) concentrations in nine study lakes during the period from July to September in 2016–2017. See legend of Figure S2 for the abbreviation of lake names and the descriptions of shaded areas, Figure S6: Vertical profiles of dissolved CH_4 (nM, red circles) and DIN (µM, gray circles) concentrations in nine study lakes during the period from July to September in 2016–2017. See legend of Figure S2 for the abbreviation of lake names and the descriptions of shaded areas, Figure S7: Vertical profiles of dissolved CH_4 (nM, red circles) and SRP (µM, gray circles) concentrations in nine study lakes during the period from July to September in 2016–2017. See legend of Figure S2 for the abbreviation of lake names and the descriptions of shaded areas, Figure S8: Vertical profiles of dissolved CH_4 (nM, red circles) and DOC (µM, gray circles) concentrations in nine study lakes during the period from July to September in 2016–2017. See legend of Figure S2 for the abbreviation of lake names and the descriptions of shaded areas, Figure S9: Vertical profiles of dissolved CH_4 concentration (nM, red circles) and TN/TP (ratio, gray circles) concentrations in nine study lakes during the period from July to September in 2016–2017. See legend of Figure S2 for the abbreviation of lake names and the descriptions of shaded areas, Figure S10: Relationships between *Synechococcus* cell density (cells/mL) and lake physicochemical variables in study lakes. Univariate regression lines were shown for the explanatory variables that were retained in the best model of GLMs with gamma distribution of response variable. See the legend of Figure S10 for symbol description, Figure S11: Vertical profiles of dissolved CH_4 concentration (nM, red circles) and Mn (µM, gray circles) concentrations in nine study lakes during the period from July to September in 2016–2017. See legend of Figure S2 for the abbreviation of lake names and the descriptions of shaded areas, Figure S12: Relationships between phytoplankton biomass ($\log_{10}$ (Chl *a*), µg/L) and two variables (DIN and DOC, µM) in study lakes, Figure S13: Vertical distribution of *Synechococcus* density (red circles) in relation to the DOC profile (gray circles) in the study lakes. Shaded areas denote the range of the thermocline. See legend of Figure S2 for the abbreviation of lake names, Table S1: Watershed and lake morphological variables of the nine study lakes in Japan, Table S2: Water quality variables of the nine study lakes in Japan, Table S3: Water quality variables of tributary streams for study lakes, Table S4: The CARD-FISH probes used in the study, Table S5: Relative importance for limnological explanatory variables used in the generalized linear models (GLMs) with a gamma error distribution and an inverse link function for each response variable. Relative importance was evaluated by the sum of Akaike weights (*w*) for each model (CH_4, DIN, DOP and TN/TP), Table S6: Results of the generalized linear models (GLM) assessing the univariate relationship between environmental variables and dissolved CH_4 concentrations (as shown in Figure 2), the univariate relationship between DOC and chemical variables (Figure 3), and the univariate relationship between lake environmental variables and the peak subsurface maximum (SMM) in nine study lakes (Figure 7). GLMs were constructed with a gamma error distribution and an inverse link function for each variable ($1/y = ax + b$), except for the variables with an asterisk whose relationships were estimated with GLMs with a Gaussian error and an identity link function ($y = ax + b$), Table S7: Relative abundance (%) of phytoplanktonic algae in the epilimnion, metalimnion and hypolimnion of Table S6. Relative abundance (%) of phytoplanktonic algae in the epilimnion, metalimnion and hypolimnion of study lakes.

Author Contributions: T.I., Y.I., K.Y., D.I., T.K. and H.S. performed the field survey and laboratory analyses. S.K., T.I., Y.I. and K.Y. performed the CARD-FISH analyses. H.K. and T.K. performed the genome analyses and database research with S.K. and T.I. R.S. performed the ^{1}H–^{31}P NMR analyses. S.K., Y.I., K.Y., and T.I. performed the data analyses and presentation. S.K. and T.I. wrote the manuscript with the assistance of the other co-authors. All authors have read and agreed to the published version of the manuscript.

Funding: This research was funded by Grants-in-Aid (B) (16H02935) and by the Grant-in-Aid for Young Scientists (A) (23681003) from the Japan Society for the Promotion of Science, the Grant for Joint Research Program of the Institute of Low Temperature Science, Hokkaido University in 2016, and the Grant from the Nippon Life Insurance Foundation in 2015.

Acknowledgments: We thank the laboratory members for their support of our study. We also thank Denboh and the staff of Lake Toya Field Station at Hokkaido University and the staff of Shikotsuko Fishermen's Cooperative for their support during our field survey.

Conflicts of Interest: The authors declare no competing financial interests.

References

1. Kirschke, S.; Bousquet, P.; Ciais, P.; Saunois, M.; Canadell, J.G.; Dlugokencky, E.J.; Bergamaschi, P.; Bergmann, D.; Blake, D.R.; Bruhwiler, L.; et al. Three decades of global methane sources and sinks. *Nat. Geosci.* **2013**, *6*, 813–823. [CrossRef]
2. Ciais, P.; Sabine, C.; Bala, G.; Bopp, L.; Brovkin, V.; Canadell, J.; Chhabra, A.; DeFries, R.; Galloway, J.; Heimann, M.; et al. Carbon and other biogeochemical cycles. In *Climate Change 2013: The Physical Science Basis, Contribution of Working Group I to the Fifth Assessment Report of the Intergovernmental Panel on Climate Change*; Stocker, T.F., Qin, D., et al., Eds.; Cambridge University Press: Cambridge, UK, 2013; pp. 465–570.

3. Saunois, M.; Bousquet, P.; Poulter, B.; Peregon, A.; Ciais, P.; Canadell, J.G.; Dlugokencky, E.J.; Etiope, G.; Bastviken, D.; Houweling, S.; et al. The global methane budget 2000–2012. *Earth Syst. Sci. Data* **2016**, *8*, 697–751. [CrossRef]

4. Bastviken, D.; Cole, J.; Pace, M.; Tranvik, L. Methane emissions from lakes: Dependence of lake characteristics, two regional assessments, and a global estimate. *Glob. Biogeochem. Cycles* **2004**, *18*, 1–12. [CrossRef]

5. Bastviken, D.; Tranvik, L.J.; Downing, J.A.; Crill, P.M.; Enrich-Prast, A. Freshwater methane emissions offset the continental carbon sink. *Science* **2011**, *331*, 50. [CrossRef] [PubMed]

6. Thauer, R.K.; Kaster, A.K.; Seedorf, H.; Buckel, W.; Hedderich, R. Methanogenic archaea: Ecologically relevant differences in energy conservation. *Nat. Rev. Microbiol.* **2008**, *6*, 579–591. [CrossRef]

7. Conrad, R. The global methane cycle: Recent advances in understanding the microbial processes involved. *Environ. Microbiol. Rep.* **2009**, *1*, 285–292. [CrossRef]

8. Utsumi, M.; Nojiri, Y.; Nakamura, T.; Nozawa, T.; Otsuki, A.; Takamura, N.; Watanabe, M.; Seki, H. Dynamics of dissolved methane and methane oxidation in dimictic Lake Nojiri during winter. *Limnol. Oceanogr.* **1998**, *43*, 10–17. [CrossRef]

9. Murase, J.; Sakai, Y.; Sugimoto, A.; Okubo, K.; Sakamoto, M. Sources of dissolved methane in Lake Biwa. *Limnology* **2003**, *4*, 91–99. [CrossRef]

10. Grossart, H.P.; Frindte, K.; Dziallas, C.; Eckert, W.; Tang, K.W. Microbial methane production in oxygenated water column of an oligotrophic lake. *Proc. Natl. Acad. Sci. USA* **2011**, *108*, 19657–19661. [CrossRef]

11. Tang, K.W.; McGinnis, D.F.; Frindte, K.; Brüchert, V.; Grossart, H.P. Paradox reconsidered: Methane oversaturation in well-oxygenated lake waters. *Limnol. Oceanogr.* **2014**, *59*, 275–284. [CrossRef]

12. Yao, M.; Henny, C.; Maresca, J.A. Freshwater bacteria release methane as a by-product of phosphorus acquisition. *Appl. Environ. Microbiol.* **2016**, *82*, 6994–7003. [CrossRef] [PubMed]

13. Wang, Q.; Dore, J.E.; McDermott, T.R. Methylphosphonate metabolism by *Pseudomonas* sp. populations contributes to the methane oversaturation paradox in an oxic freshwater lake. *Environ. Microbiol.* **2017**, *19*, 2366–2378. [CrossRef] [PubMed]

14. Bižić-Ionescu, M.; Ionescu, D.; Günthel, M.; Tang, K.W.; Grossart, H.P. Oxic methane cycling: New evidence for methane formation in oxic lake water. In *Biogenesis of hydrocarbons: Handbook of Hydrocarbon and Lipid Microbiology*; Stams, A.J.M., Sousa, D.Z., Eds.; Springer International Publishing: Cham, Switzerland, 2018; pp. 379–400.

15. Khatun, S.; Iwata, T.; Kojima, H.; Fukui, M.; Aoki, T.; Mochizuki, S.; Naito, A.; Kobayashi, A.; Uzawa, R. Aerobic methane production by planktonic microbes in lakes. *Sci. Total Environ.* **2019**, *696*, 133916. [CrossRef]

16. Karl, D.M.; Beversdorf, L.; Björkman, K.M.; Church, M.J.; Martinez, A.; Delong, E.F. Aerobic production of methane in the sea. *Nat. Geosci.* **2008**, *1*, 473–478. [CrossRef]

17. Damm, E.; Helmke, E.; Thoms, S.; Schauer, U.; Nöthig, E.; Bakker, K.; Kiene, R.P. Methane production in aerobic oligotrophic surface water in the central Arctic Ocean. *Biogeosciences* **2010**, *7*, 1099–1108. [CrossRef]

18. Bogard, M.J.; Del Giorgio, P.A.; Boutet, L.; Chaves, M.C.G.; Prairie, Y.T.; Merante, A.; Derry, A.M. Oxic water column methanogenesis as a major component of aquatic CH_4 fluxes. *Nat. Commun.* **2014**, *5*, 5350. [CrossRef]

19. Tang, K.W.; McGinnis, D.F.; Ionescu, D.; Grossart, H.P. Methane production in oxic lake waters potentially increases aquatic methane flux to air. *Environ. Sci. Technol. Lett.* **2016**, *3*, 227–233. [CrossRef]

20. Encinas Fernández, J.; Peeters, F.; Hofmann, H. On the methane paradox: Transport from shallow water zones rather than in situ methanogenesis is the major source of CH_4 in the open surface water of lakes. *J. Geophys. Res. Biogeosciences* **2016**, *121*, 2717–2726. [CrossRef]

21. Donis, D.; Flury, S.; Stöckli, A.; Spangenberg, J.E.; Vachon, D.; McGinnis, D.F. Full-scale evaluation of methane production under oxic conditions in a mesotrophic lake. *Nat. Commun.* **2017**, *8*, 1661. [CrossRef]

22. DelSontro, T.; del Giorgio, P.A.; Prairie, Y.T. No longer a paradox: The interaction between physical transport and biological processes explains the spatial distribution of surface water methane within and across lakes. *Ecosystems* **2018**, *21*, 1073–1087. [CrossRef]

23. Schmale, O.; Wäge, J.; Mohrholz, V.; Wasmund, N.; Gräwe, U.; Rehder, G.; Labrenz, M.; Loick-Wilde, N. The contribution of zooplankton to methane supersaturation in the oxygenated upper waters of the central Baltic Sea. *Limnol. Oceanogr.* **2018**, *63*, 412–430. [CrossRef]

24. Carini, P.; White, A.E.; Campbell, E.O.; Giovannoni, S.J. Methane production by phosphate-starved SAR11 chemoheterotrophic marine bacteria. *Nat. Commun.* **2014**, *5*, 4346. [CrossRef] [PubMed]

25. Repeta, D.J.; Ferrón, S.; Sosa, O.A.; Johnson, C.G.; Repeta, L.D.; Acker, M.; Delong, E.F.; Karl, D.M. Marine methane paradox explained by bacterial degradation of dissolved organic matter. *Nat. Geosci.* **2016**, *9*, 884–887. [CrossRef]

26. Bižić, M.; Klintzsch, T.; Ionescu, D.; Hindiyeh, M.Y.; Günthel, M.; Muro-Pastor, A.M.; Eckert, W.; Urich, T.; Keppler, F.; Grossart, H.-P. Aquatic and terrestrial cyanobacteria produce methane. *Sci. Adv.* **2020**, *6*, eaax5343. [CrossRef] [PubMed]

27. Lenhart, K.; Klintzsch, T.; Langer, G.; Nehrke, G.; Bunge, M.; Schnell, S.; Keppler, F. Evidence for methane production by the marine algae *Emiliania huxleyi*. *Biogeosciences* **2016**, *13*, 3163–3174. [CrossRef]

28. Baines, S.B.; Pace, M.L. The production of dissolved organic matter by phytoplankton and its importance to bacteria: Patterns across marine and freshwater systems. *Limnol. Oceanogr.* **1991**, *36*, 1078–1090. [CrossRef]

29. Litchman, E.; de Tezanos Pinto, P.; Edwards, K.F.; Klausmeier, C.A.; Kremer, C.T.; Thomas, M.K. Global biogeochemical impacts of phytoplankton: A trait-based perspective. *J. Ecol.* **2015**, *103*, 1384–1396. [CrossRef]

30. Wetzel, R.G. The phosphorus cycle. In *Limnology lake and river ecosystems*, 3rd ed.; Elsevier Academic Press: San Diego, CA, USA, 2001; pp. 239–288.

31. Shinohara, R.; Iwata, T.; Ikarashi, Y.; Sano, T. Detection of 2-aminoethylphosphonic acid in suspended particles in an ultraoligotrophic lake: A two-dimensional nuclear magnetic resonance (2D-NMR) study. *Environ. Sci. Pollut. Res.* **2018**, *25*, 30739–30743. [CrossRef]

32. Solórzano, L. Determination of ammonia in natural waters by phenolhypochlorite method. *Limnol. Oceanogr.* **1969**, *14*, 799–801. [CrossRef]

33. Crumpton, W.G.; Isenhart, T.M.; Mitchell, P.D. Nitrate and organic N analyses with second-derivative spectroscopy. *Limnol. Oceanogr.* **1992**, *37*, 907–913. [CrossRef]

34. Bendschneider, K.; Robinson, R.J. A new spectrophotometric method for the determination of nitrite in sea water. *J. Mar. Res.* **1952**, *11*, 87–96.

35. Neal, C.; Neal, M.; Wickham, H. Phosphate measurement in natural waters: Two examples of analytical problems associated with silica interference using phosphomolybdic acid methodologies. *Sci. Total Environ.* **2000**, *251–252*, 511–522. [CrossRef]

36. Kojima, H.; Iwata, T.; Fukui, M. DNA-based analysis of planktonic methanotrophs in a stratified lake. *Freshw. Biol.* **2009**, *54*, 1501–1509. [CrossRef]

37. Tsutsumi, M.; Iwata, T.; Kojima, H.; Fukui, M. Spatiotemporal variations in an assemblage of closely related planktonic aerobic methanotrophs. *Freshw. Biol.* **2011**, *56*, 342–351. [CrossRef]

38. Borrel, G.; Jézéquel, D.; Biderre-Petit, C.; Morel-Desrosiers, N.; Morel, J.P.; Peyret, P.; Fonty, G.; Lehours, A.C. Production and consumption of methane in freshwater lake ecosystems. *Res. Microbiol.* **2011**, *162*, 832–847. [CrossRef] [PubMed]

39. Caporaso, J.G.; Kuczynski, J.; Stombaugh, J.; Bittinger, K.; Bushman, F.D.; Costello, E.K.; Fierer, N.; Pẽa, A.G.; Goodrich, J.K.; Gordon, J.I.; et al. QIIME allows analysis of high-throughput community sequencing data. *Nat. Methods* **2010**, *7*, 335–336. [CrossRef]

40. Metcalf, W.W.; Griffin, B.M.; Cicchillo, R.M.; Gao, J.; Janga, S.C.; Cooke, H.A.; Circello, B.T.; Evans, B.S.; Martens-Habbena, W.; Stahl, D.A.; et al. Synthesis of methylphosphonic acid by marine microbes: A source for methane in the aerobic ocean. *Science* **2012**, *337*, 1104–1107. [CrossRef]

41. Burnham, K.P.; Anderson, D.R. *Model Selection and Multi-Model Inference: A Practical Information—Theoretical Approach*, 2nd ed.; Springer-Verlag: New York, NY, USA, 2002; pp. 1–488.

42. Bartón, K. MuMIn: Model Selection and Model Averaging Base on Information Criteria. R Package Version 1.40.0. Available online: https://CRAN.R-project.org/package=MuMIn (accessed on 1 October 2017).

43. Ripley, B.; Venables, B.; Bates, D.; Hornik, K.; Gebhardt, A.; Firth, D. MASS: Support Functions and Datasets for Venables and Ripley's MASS. R Package Version 7.3-48. Available online: https://CRAN.R-project.org/package=MASS (accessed on 25 December 2017).

44. Clarke, K.R.; Warwick, R.M. *Change in Marine Communities: An Approach to Statistical Analysis and Interpretation*, 2nd ed.; PRIMER-E Ltd.: Plymouth, UK, 2001.

45. Oksanen, A.J.; Blanchet, F.G.; Kindt, R.; Legen, P.; Minchin, P.R.; Hara, R.B.O.; Simpson, G.L.; Solymos, P.; Stevens, M.H.H. Vegan: Community Ecology Package. R Package Version 2.4-5. Available online: https://CRAN.R-project.org/package=vegan (accessed on 1 December 2017).

46. Nanny, M.A.; Minear, R.A. Characterization of soluble unreactive phosphorus using ^{31}P nuclear magnetic resonance spectroscopy. *Mar. Geol.* **1997**, *139*, 77–94. [CrossRef]

47. Williams, C.J.; Frost, P.C.; Morales-Williams, A.M.; Larson, J.H.; Richardson, W.B.; Chiandet, A.S.; Xenopoulos, M.A. Human activities cause distinct dissolved organic matter composition across freshwater ecosystems. *Glob. Chang. Biol.* **2016**, *22*, 613–626. [CrossRef]

48. Corman, J.R.; Bertolet, B.L.; Casson, N.J.; Sebestyen, S.D.; Kolka, R.K.; Stanley, E.H. Nitrogen and phosphorus loads to temperate seepage lakes associated with allochthonous dissolved organic carbon loads. *Geophys. Res. Lett.* **2018**, *45*, 5481–5490. [CrossRef]

49. Maranger, R.; Jones, S.E.; Cotner, J.B. Stoichiometry of carbon, nitrogen, and phosphorus through the freshwater pipe. *Limnol. Oceanogr. Lett.* **2018**, *3*, 89–101. [CrossRef]

50. Taylor, P.G.; Townsend, A.R. Stoichiometric control of organic carbon-nitrate relationships from soils to the sea. *Nature* **2010**, *464*, 1178–1181. [CrossRef] [PubMed]

51. Gomez-Garcia, M.R.; Davison, M.; Blain-Hartnung, M.; Grossman, A.R.; Bhaya, D. Alternative pathways for phosphonate metabolism in thermophilic cyanobacteria from microbial mats. *ISME J.* **2011**, *5*, 141–149. [CrossRef] [PubMed]

52. Tsuji, K.; Maruo, M.; Obata, H. Determination of trace methylphosphonate in natural waters by ion chromatography. *Bunseki Kagaku* **2019**, *68*, 275–278. [CrossRef]

53. Tilman, D.; Kiesling, R.; Sterner, R.; Kilham, S.S.; Johnson, F.A. Green, bluegreen and diatom algae: Taxonomic differences in competitive ability for phosphorus, silicon and nitrogen. *Arch. für Hydrobiol.* **1986**, *106*, 473–485.

54. Agawin, N.S.R.; Duarte, C.M.; Agustí, S. Nutrient and temperature control of the contribution of picoplankton to phytoplankton biomass and production. *Limnol. Oceanogr.* **2000**, *45*, 591–600. [CrossRef]

55. Serizawa, H.; Amemiya, T.; Rossberg, A.G.; Itoh, K. Computer simulations of seasonal outbreak and diurnal vertical migration of cyanobacteria. *Limnology* **2008**, *9*, 185–194. [CrossRef]

56. Woolway, R.I.; Merchant, C.J. Worldwide alteration of lake mixing regimes in response to climate change. *Nat. Geosci.* **2019**, *12*, 271–276. [CrossRef]

57. Tranvik, L.J.; Downing, J.A.; Cotner, J.B.; Loiselle, S.A.; Striegl, R.G.; Ballatore, T.J.; Dillon, P.; Finlay, K.; Fortino, K.; Knoll, L.B.; et al. Lakes and reservoirs as regulators of carbon cycling and climate. *Limnol. Oceanogr.* **2009**, *54*, 2298–2314. [CrossRef]

58. Stanley, E.H.; Casson, N.J.; Christel, S.T.; Crawford, J.T.; Loken, L.C.; Oliver, S.K. The ecology of methane in streams and rivers: Patterns, controls, and global significance. *Ecol. Monogr.* **2016**, *86*, 146–171. [CrossRef]

Article

Protected Freshwater Ecosystem with Incessant Cyanobacterial Blooming Awaiting a Resolution

Nada Tokodi [1], Damjana Drobac Backović [1,*], Jelena Lujić [2], Ilija Šćekić [2], Snežana Simić [3], Nevena Đorđević [3], Tamara Dulić [4], Branko Miljanović [1], Nevena Kitanović [2], Zoran Marinović [2], Henna Savela [5], Jussi Meriluoto [1,4] and Zorica Svirčev [1,4]

[1] Department of Biology and Ecology, Faculty of Sciences, University of Novi Sad, Trg Dositeja Obradovića 3, 21000 Novi Sad, Serbia; nada.tokodi@dbe.uns.ac.rs (N.T.); branko.miljanovic@dbe.uns.ac.rs (B.M.); Jussi.Meriluoto@abo.fi (J.M.); zorica.svircev@dbe.uns.ac.rs (Z.S.)

[2] Department of Aquaculture, Szent István University, Páter Károly u. 1, 2100 Gödöllő, Hungary; jelena.lujic@mkk.szie.hu (J.L.); ilijas92@gmail.com (I.Š.); nevena.n.kitanovic@gmail.com (N.K.); zor.marinovic@gmail.com (Z.M.)

[3] Faculty of Science, Institute of Biology and Ecology, University of Kragujevac, Radoja Domanovića 12, 34 000 Kragujevac, Serbia; snezana.simic@pmf.kg.ac.rs (S.S.); nevena.djordjevic@pmf.kg.ac.rs (N.Đ.)

[4] Biochemistry, Faculty of Science and Engineering, Åbo Akademi University, Tykistökatu 6 A, 20520 Turku, Finland; waterlifea@gmail.com

[5] Finnish Environment Institute, Marine Research Centre, Agnes Sjöbergin katu 2, 00790 Helsinki, Finland; Henna.Savela@ymparisto.fi

* Correspondence: damjana.drobac@dbe.uns.ac.rs

Received: 27 November 2019; Accepted: 29 December 2019; Published: 31 December 2019

Abstract: For 50 years persistent cyanobacterial blooms have been observed in Lake Ludoš (Serbia), a wetland area of international significance listed as a Ramsar site. Cyanobacteria and cyanotoxins can affect many organisms, including valuable flora and fauna, such as rare and endangered bird species living or visiting the lake. The aim was to carry out monitoring, estimate the current status of the lake, and discuss potential resolutions. Results obtained showed: (a) the poor chemical state of the lake; (b) the presence of potentially toxic (genera *Dolichospermum, Microcystis, Planktothrix, Chroococcus, Oscillatoria, Woronichinia* and dominant species *Limnothrix redekei* and *Pseudanabaena limnetica*) and invasive cyanobacterial species *Raphidiopsis raciborskii*; (c) the detection of microcystin (MC) and saxitoxin (STX) coding genes in biomass samples; (d) the detection of several microcystin variants (MC-LR, MC-dmLR, MC-RR, MC-dmRR, MC-LF) in water samples; (e) histopathological alterations in fish liver, kidney and gills. The potential health risk to all organisms in the ecosystem and the ecosystem itself is thus still real and present. Although there is still no resolution in sight, urgent remediation measures are needed to alleviate the incessant cyanobacterial problem in Lake Ludoš to break this ecosystem out of the perpetual state of limbo in which it has been trapped for quite some time.

Keywords: cyanobacteria; blooms; microcystin; Lake Ludoš

---

1. Introduction

In the very north of Serbia there is an old and unusual lake, Lake Ludoš, with beautiful open water landscapes surrounded by reeds, wetlands and steppe. The environment is rich in many plant and animals species. European pond turtles, various amphibians, otters, moles, rabbits, foxes and roe deer have found their home there. What makes Lake Ludoš especially famous, and validates the name originating from the Hungarian word "lud" meaning goose, is that there are more than 200 bird species, including rare and endangered species, nesting or resting during their migration. Because of all this, Lake Ludoš is recognized and protected as a special nature reserve on the list of Ramsar sites.

One part of the lake is often visited by fishermen, but their catch mostly consists of the very resilient and adaptable Prussian carp (*Carassius gibelio* (Block, 1782)) which are quite small in size. This may be related to the fact that the water of Lake Ludoš has an intense green color throughout the year caused by cyanobacteria (e.g., [1]). Their extensive growth and blooming causes many problems in freshwater ecosystems, including this one. Cyanobacteria can produce cyanotoxins that affect other organisms, including valuable flora and fauna, especially aqueous organisms such as fish [2]. Cyanobacteria also present a threat to humans, such as fishermen, who may be exposed to cyanotoxins through contaminated food, inhalation and direct contact [3,4]. In addition to human health, the health of this important ecosystem is also jeopardized. Furthermore, this "disease" could also be transmissible, since there is a possibility that water birds visiting the lake during their migration path can disseminate toxic cyanobacteria [5].

Lake Ludoš is only one of many aquatic ecosystems in Serbia where cyanobacteria are present and blooming [6]. What sets this ecosystem apart from many others is that it has been known for perpetual blooming in the last 50 years. Previous research has shown that the lake is in a poor ecological state which leads to the question of whether the protection of this natural habitat in a bad ecological state is justified. The cyanobacterial problem, which can potentially affect every living being in the proximity of this ecosystem, has also been preserved as measures to improve the water quality have not been undertaken on the lake [5]. The problem of cyanobacteria in Lake Ludoš has been addressed during our previous research when potentially toxic cyanobacterial species, cyanotoxins in water, macrophytes and fish tissues were detected, as well as histological alterations and DNA damage in fish tissue (see [5]). Six years later further monitoring was carried out in order to estimate the current state of this ecosystem. Therefore, several investigative steps have been taken: monitoring of physical and chemical parameters of the lake; assessment of qualitative and quantitative analyses of cyanobacteria; the first survey of the cyanotoxin coding genes; determination of cyanotoxins in water and fish; analysis of histopathology of different fish organs; and discussion of potential health risks and resolutions.

Freshwater ecosystems throughout the world have similar problems in connection to cyanobacterial blooming and cyanotoxin production. Recent publication of a global geographical and historical overview of cyanotoxin distribution demonstrated the presence of well-known cyanotoxins in each continent (including 520 lakes) and their harmful consequences on human health [4]. Hence, this issue is of global concern. The present investigation aims to assist in making appropriate decisions and measures for the remediation of not only this, but many other old and rapidly aging and protected lakes.

2. Materials and Methods

2.1. Sampling Site and Sampling of Water and Fish

Lake Ludoš (Figure 1) is a one of the few preserved shallow lakes in the region. It has a maximum depth of 2.25 m, is 4.5 km long, and it represents a remnant of the Pannonian Sea. In most places the depth does not exceed 1 m, and it may be frozen for more than three months a year. It is located in the north part of Serbia, near the city of Subotica. The lake and the associated wetland ecosystem is highly valued due to the great biological diversity, and as such the area is classified among wetlands of international significance. The quality of the lake's water is of great importance for the preservation of the flora and fauna connected to this marshland ecosystem.

The lake is supplied with water from aquifers and Kereš River. However, in the northern part, Lake Ludoš receives water from the canal Palić-Ludoš which is the recipient of wastewaters from the Palić settlement. Water treatment of these wastewaters is still inadequate and the canal water is characterized by a high level of organic pollution, high concentrations of salt and very high nutrient concentrations. The inflow of untreated and partially purified waters in Lake Ludoš contributes to the deterioration of the water quality and the increase of the sludge quantity [7–11].

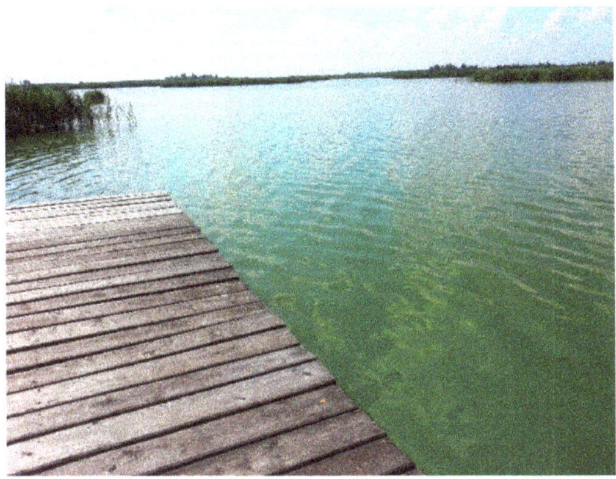

Figure 1. Pier view of the blooming Lake Ludoš in July 2018.

Water samples were collected from the surface water layer within the littoral zone (pier next to the visitor center) (46.103207 N, 19.821360 E) and from the center of the lake (46.102159 N, 19.821149 E) in March, May, July and September of 2018. Samples of Prussian carp were collected from the center of the lake before and after the summer (March and September 2018) with gillnets of various mesh sizes and a standard electrofishing device.

2.2. Analyses of Physical and Chemical Parameters

Multi-parameter WTW probes were used for carrying out in situ measurements in the field and the following physical and chemical parameters were determined: temperature, pH, conductivity, O_2 concentration and O_2 saturation. TSS (total suspended solids), TOC (total organic carbon), NO_3, detergents, COD (chemical oxygen demand) and BOD (biological oxygen demand) were measured in the laboratory conditions with a Pastel Ultraviolet (UV) Secomam.

2.3. Qualitative and Quantitative Analyses of Cyanobacteria

The phytoplankton samples for the cyanobacterial qualitative analysis were collected by sweeping a plankton net (netframe 25 cm ø, net mesh 23 μm). All samples were immediately preserved in a Lugol solution. Taxonomic identifications of cyanobacteria were made according to several taxonomic keys [12–15] and were done under a light microscope Motic BA310 using a Bresser (9MP) digital camera and Micro Cam Lab software. For the quantitative analysis of phytoplankton, the 15 L of water was collected by sweeping a plankton net at the depth of 0.3 m in March, while in May, July and September only 200 mL of water were collected directly from the lake as a result of high bloom density. The quantitative analysis was made by using the Utermöhl method [16] under a Motic AE 2000 inverted microscope. The phytoplankton individuals were sedimented and cyanobacteria quantified on the chamber (i.e., transects) with an inverted microscope at different magnifications depending on their size (100×, 400×) and expressed as the number of cells per mL.

2.4. Cyanotoxin Coding Gene Analyses

2.4.1. Samples—Reference Strains for Polymerase Chain Reaction (PCR) Analysis

Reference strains were obtained from Pasteur Culture Collection (PCC), National Institute for Environmental Studies Microbial Culture Collection (NIES), Australian National Algae Culture Collection (CS), and Finnish Environment Institute (SYKE). They consisted of:

- microcystin (MC) producers: PCC7820 (*Microcystis aeruginosa*), NIES-107 (*Microcystis wesenbergii*);

- cylindrospermopsin (CYN) producers: CS-505, CS-506 (*Cylindrospermopsis raciborskii*), SYKE-966 (*Anabaena lapponica*);
- saxitoxin (STX) producers: CS-337/01, CS-537/13 (*Dolichospermum circinale*);
- anatoxin-a (ATX) producer: ANA123 (*Dolichospermum circinale*).

2.4.2. DNA Extraction

Depending on the bloom density, 30–400 mL of water samples were filtered (pore size 2–3 μm), and filtride was freeze-dried. Approximately 10 mg of freeze-dried biomass was used for DNA extraction from reference strains. Genomic DNA from biomass of the reference strains and filtrides was extracted with the DNeasy Plant Mini Kit (QIAGEN, Hilden, Germany) according to manufacturer's instructions, with minimal modifications for the extraction from filtrides (double amount of Buffer AP1, RNase A and Buffer P3 was added to fully suspend the samples). During the initial steps of extraction, samples were homogenized using zirconia/silica disruption beads (0.5 mm) and by vortexing for 1 min. The quality was assessed spectrophotometrically (NanoDrop ND-1000, Thermo Scientific, Waltham, MA, USA), where A260/A280 ratio varied between 1.22 and 2.04.

2.4.3. Qualitative PCR

Qualitative PCR was run to analyze samples for the presence of MC (*mcyE*), CYN (*cyrJ*), STX (*sxtA, sxtG, sxtS*) and ATX (*anaC*) synthetase genes. PCR reaction mixtures were prepared in a total volume of 20 μL containing 1× Phire Reaction Buffer, 0.4 μL Phire II HotStart polymerase (Thermo Scientific), 0.2 mM deoxyribonucleotide triphosphates (dNTPs) (Thermo Scientific), 0.5 μL forward and reversed primers (Table 1), 2 μL of template and sterile deionized water. PCRs were run on a C1000 Touch Thermal Cycler (Bio-Rad, Helsinki, Finland) according to the following protocols: initial denaturation for 30 s at 98 °C; 40 cycles of 5 s at 98 °C, 5 s at 61 °C (HEPF, HEPR, sxtA1480_R stxA855_R), or 62 °C (cyrJ_F, cyrJ_R, sxtG432_F, sxtG928_R, sxtS205_F, sxtS566_R), or 52 °C (anaC-genF, anaC-genR) and 10 s at 72 °C; and a final extension of 1 min at 72 °C. To examine the potential inhibition of PCRs, an exogenous amplification control template was prepared containing 1 μL:1 μL (reference:sample). Following strains were used as a reference in the control template: PCC7820 for *mcyE*, CS-506 for *cyrJ*, CS-537/13 for *sxtA, sxtG, sxtS*, and ANA123 for *anaC*. Visualization of PCR products was performed on a 1.5% Top Vision agarose gel (Thermo Scientific) dyed with SYBR® Safe DNA gel stain. The observed bands were documented on Gel Doc™ XR (Bio-Rad) using Quantity One software (v. 4.6.9).

Table 1. List of primers used for qualitative polymerase chain reaction (PCR).

Gene	Primer	5′–3′ Sequence	Reference
mcyE	HEPF	TTTGGGGTTAACTTTTTTGGGCATAGTC	[17]
	HEPR	AATTCTTGAGGCTGTAAATCGGGTTT	
cyrJ	cyrJ_F	TTCTCTCCTTTCCCTATCTCTTTATC	[18]
	cyrJ_R	GCTACGGTGCTGTACCAAGGGGC	
sxtA	stxA855_F	GACTCGGCTTGTTGCTTCCCC	[19]
	sxtA1480_R	GCCAAACTCGCAACAGGAGAAGG	
sxtG	sxtG432_F	AATGGCAGATCGCAACCGCTAT	[19]
	sxtG928_R	ACATTCAACCCTGCCCATTCACT	
sxtS	sxtS205_F	GGAGTATTDGCGGGTGACTATGA	[20]
	sxtS566_R	GGTGGCTACTTGGTATAACTCGCA	
anaC	anaC-genF	TCTGGTATTCAGTCCCCTCTAT	[21]
	anaC-genR	CCCAATAGCCTGTCATCAA	

2.5. Cyanotoxin Analyses

2.5.1. Preparation of Water Samples for Liquid Chromatography–Tandem Mass Spectrometry (LC–MS/MS)

Depending on the bloom density, 30–100 mL of water samples were filtered (pore size 2–3 μm). Biomass on the filter was then freeze-dried. Afterwards, filtrides were placed in glass tubes and the toxin was extracted with 3 mL of 75% MeOH and ultrasonication. The extracts were centrifuged for 10 min at 10,000× g and two 1 mL aliquots of the supernatant were evaporated to dryness (50 °C nitrogen flow) in glass tubes. For MC analysis the sample was redissolved in 75% MeOH in 200 μL, and for CYN analysis the sample was redissolved in 200 μL H_2O. The samples were then filtered (0.2 μm GHP ACRODISC 13 Pall Life Sciences, Ann Arbot, MS, USA) into inserts and were ready for liquid chromatography–tandem mass spectrometry (LC–MS/MS) analysis.

The extracellular MC content was concentrated by solid-phase extraction (SPE) on Waters Oasis HLB (30 mg). The samples eluted with 5 mL 90% MeOH were placed into glass tubes, evaporated using nitrogen flow and redissolved in 200 μL of 75% MeOH. Subsequently, they were filtered (0.2 μm GHPACRODISC 13 Pall Life Sciences) into inserts and were ready for LC–MS/MS analysis.

2.5.2. Preparation of Fish Tissue Samples for LC–MS/MS

Prussian carp from Lake Ludoš was analysed for MCs in fish liver, gills, kidney, intestine, gonads (testis and ovaries), spleen, and muscle samples by LC–MS/MS. A total of 30 individuals (TL: 16.63 ± 2.24 cm, SL: 14.58 ± 1.68 cm, 18 female/12 male) were collected during two separate sampling surveys—spring (March) and autumn (September) of 2018. Different fish tissues were separately homogenized and freeze-dried. Samples of the same organ of all the individuals were pooled together. Before further preparation, several fish tissues were spiked in order to test the preparation method. Freeze-dried fish tissue samples (100 mg) were placed into glass tubes and 5 mL of 75% MeOH was added for extraction of cyanotoxins. Homogenization was performed on ice for 30 second, and the samples were then ultrasonicated in a bath sonicator for 15 min and further extracted with a probe sonicator. Samples were then centrifuged for 10 min at 10,000× g followed by an addition of 1 mL of hexane to 2 mL of the supernatants obtained. The hexane (lipid) layer was removed using glass pipettes and the remaining samples were evaporated (50 °C nitrogen flow) in glass tubes. Finally, samples were redissolved in 300 μL 75% MeOH and filtered (0.2 μm GHPACRODISC 13 Pall Life Sciences) into inserts. The fish tissue samples were then ready for LC-MS/MS analysis.

2.5.3. LC–MS/MS

Toxin analyses were performed by LC–MS/MS [22]. The analytical targets consisted of nine MC variants (MC-dmRR, MC-RR, MC-dmYR, MC-YR, MC-dmLR, MC-LR, MC-LY, MC-LW and MC-LF) and CYN.

2.6. Analyses of Fish Histology

Captured Prussian carp from Lake Ludoš used for cyanotoxin detection were also used for histological analyses. Liver, kidney, gill, intestine, spleen, gonad and muscle samples were dissected from each fish and fixed in 4% formaldehyde. Additionally, 6 individuals of common carp (*Cyprinus carpio* L.) (TL: 18.35 ± 6.24 cm, 3 females/3 males) were obtained from the Department of Aquaculture, Szent István University, Hungary. These fish were kept in a recirculation system (Sentimento Kft., Hungary), under a 12 h light/12 h dark cycle at 24 ± 0.2 °C and served as a control group in this study.

After fixation of at least three days, samples were processed by standard histological procedure. Gill and muscle samples were decalcified beforehand. For tissue processing, samples were dehydrated in graded series of ethanol, cleared in xylene and subsequently embedded in paraffin wax blocks. Three five-μm-thin sections per tissue per individual were cut and placed onto glass slides, and stained with haematoxylin and eosin (H&E) dyes. Sections were examined under a Nikon Eclipse 600 microscope and photographed using a QImaging Micro Publisher 3.0 digital camera.

3. Results

3.1. Physical and Chemical Parameters of Water Samples

The recent investigations of Lake Ludoš during 2018 corroborated the continuously poor chemical state of the lake. pH levels, saturation with O_2 as well as electrical conductivity were high during the investigated period (Table 2).

Table 2. Physical and chemical parameters of water from Lake Ludoš in 2018.

Physical and Chemical Parameters	March	May	July	September
temperature (°C), in situ	10.5	20.3	25.9	17.5
pH, in situ	8.3	8.9	8.9	8.6
concentration O_2, in situ (mg/L)	16.2	9.5	26.9	8.5
saturation O_2, in situ (%)	146.7	99.7	>300	90.5
conductivity, in situ (μS/cm)	873.5	915	874.5	967.5
TSS (mg/dm^3)	43	104.5	46.8	48.4
TOC (mg/dm^3)	8.45	10.4	9.5	13.2
NO_3 (mg/dm^3)	≤0.5	≤0.5	≤0.5	≤0.5
detergents (mg/dm^3)	2.05	2.5	2.1	3.1
COD (mgO$_2$/dm^3)	23.85	31.8	27.7	38
BOD (mgO$_2$/dm^3)	12	15	13.5	18.9

3.2. Presence of Cyanobacterial Species in Water Samples

Most dominant cyanobacterial species were *Limnothrix redekei* (Van Goor) Meffert and *Pseudanabaena limnetica* (Lemmermann) Komárek (Table 3, Figure 2). Furthermore, both *Microcystis* species, *Microcystis aeruginosa* (Kützing) Kützing and *Microcystis wesenbergii* (Komárek) Komárek, were numerous during the whole investigated period. At the end of the summer, an invasive *Raphidiopsis raciborskii* (Woloszynska) Aguilera, Berrendero Gómez, Kastovsky, Echenique and Salerno (basionym *Cylindrospermopsis raciborskii* (Woloszynska) Seenayya and Subba Raju) also started to occur. Usually, more cells per mL were found in the pier samples compared to the center samples; however, the same species were present in both sampling sites.

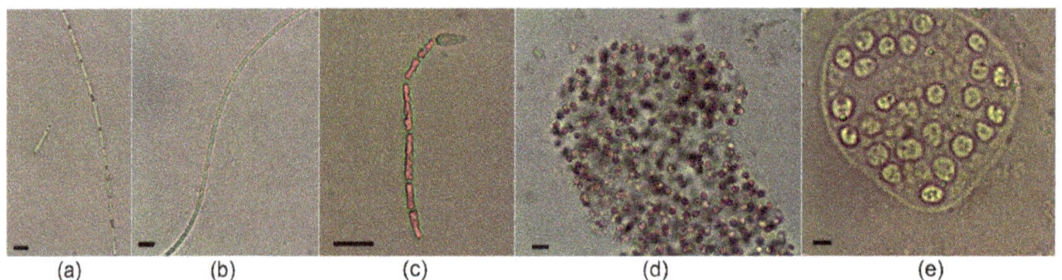

Figure 2. Dominant cyanobacterial species from Lake Ludoš in 2018: (**a**) *Limnothrix redekei*; (**b**) *Pseudanabaena limnetica*; (**c**) *Raphidiopsis raciborskii*; (**d**) *Microcystis aeruginosa*; (**e**) *Microcystis wesenbergii*; Scale bars: 10 μm.

Table 3. Qualitative and quantitative composition of cyanobacteria from Lake Ludoš in 2018.

Sampling Period	March		May		July		September	
	Center cells/mL	Pier cells/mL	Center cells/mL	Pier cells/mL	Center cells/mL	Pier cells/mL	Center cells/mL	Pier cells/mL
Cyanobacteria								
Chroococcus limneticus Lemmermann	13,980	15,100	13,210	17,650	31,000	53,700	68,100	82,300
Raphidiopsis raciborskii (Woloszynska) Aguilera, Berrendero Gómez, Kastovsky, Echenique and Salerno	–	–	–	–	–	–	48,000	65,000
Dolichospermum flos-aquae (Brébisson ex Bornet and Flahault)	–	–	–	–	86,500	58,300	21,000	187,300
Limnothrix redekei (Van Goor) Meffert	10,140,000	9,855,000	10,983,000	10,287,000	11,180,000	13,100,000	9,983,100	10,150,000
Microcystis aeruginosa (Kützing) Kützing *Microcystis wesenbergii* (Komárek) Komárek	205,410 104,670	320,210 200,930	198,310 163,750	285,400 210,820	185,600 321,700	232,160 485,300	185,300 1,504,100	213,100 1,756,100
Oscillatoria sp. Vaucher ex Gomont	–	–	+	+	+	+	+	+
Planktothrix agardhii (Gomont) Anagnostidis and Komárek	43,000	74,600	51,000	62,600	62,700	70,200	85,400	91,300
Pseudanabaena limnetica (Lemmermann) Komárek	12,785,000	14,521,000	13,956,000	15,281,000	15,813,000	17,323,000	13,685,000	15,321,000
Woronichinia sp. A. A. Elenkin	–	–	+	+	+	+	+	+
Σ	23,292,060	24,986,840	25,365,270	26,144,470	27,680,500	31,322,660	25,580,000	27,866,100

Legend: (+) present taxa with abundance less than 0.1% of total; (−) not present in sample.

3.3. Presence of Cyanotoxin Coding Genes in Biomass Samples

Biomass samples were tested for the presence of cyanotoxin coding genes, including MCs, STX, CYN and ATX (Table 4).

Table 4. The prevalence of *mcyE, sxtA, sxtG, sxtS, cyrJ* and *anaC* PCR products in Lake Ludoš.

Sampling Period	March		May		July		September	
PCR Products	Center	Pier	Center	Pier	Center	Pier	Center	Pier
mcyE	+	+	+	+	+	+	+	+
sxtA	/	/	/	/	−	−	−	−
sxtG	+	−	−	+	+	+	+	+
sxtS	−	−	−	−	+	+	+	+
cyrJ	−	−	−	−	−	−	−	−
anaC	−	−	−	−	−	−	−	−

Legend: (+) amplified; (−) not amplified; (/) not analysed.

MC coding gene *mcyE* (472 bp, Figure 3a) was amplified in all samples. STX coding genes *sxtG* (519 pb, Figure 3c) and *sxtS* (382 bp, Figure 3d) were amplified in a total of 6 and 4 samples, respectively. The *sxtG* gene was observed in all sampling seasons, while *sxtS* gene was observed only in samples collected in August and September. Saxitoxin coding gene *sxtA* (648 bp, Figure 3b), CYN coding gene *cyrJ*, and ATX coding gene *anaC* were not amplified in this study. Significant inhibition of the PCR reaction was not observed in exogenous amplification control templates.

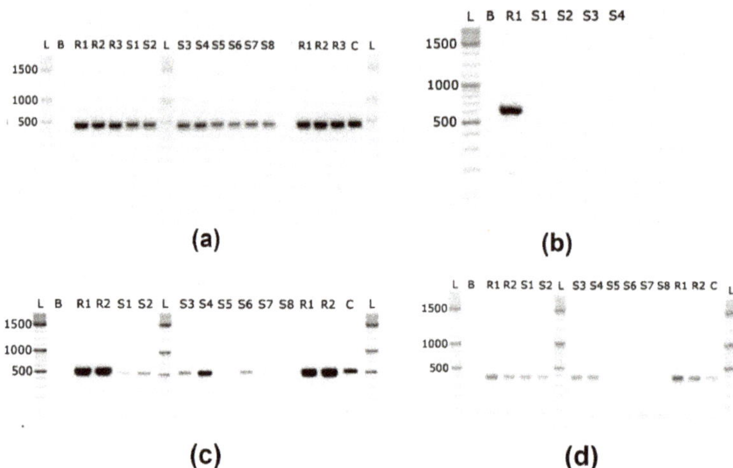

Figure 3. Visualization of PCR products on agarose gel: (**a**) *mcyE*; (**b**) *sxtA*; (**c**) *sxtG*; (**d**) *sxtS*. Legend: L—ladder; B—blank; R1, R2—reference strains; C—exogenous amplification control; S1—sample September—center; S2—sample September—pier; S3—sample July—center; S4—sample July—pier; S5—sample May—center; S6—sample May—pier; S7—sample March—center; S8—sample March—pier.

3.4. Presence of Cyanotoxins in Water and Fish Samples

Biomass and extracellular content of water samples were tested for the presence of cyanotoxins. Several MC variants were noted in the total content (Table 5) including most commonly occurring MC-LR and MC-RR, however, their concentrations were rather low during the whole investigated period. CYN was not detected during the investigated period in Lake Ludoš, even though there were some cyanobacterial species that could potentially produce this cyanotoxin.

Table 5. Presence and highest concentrations of microcystin (MC) variants in Lake Ludoš during 2018.

Microcystin Variants (µg/L)	March	May	July	September
MC-LR	–	0.1	0.022	0.285
dmMC-LR	0.023	–	–	–
MC-RR	–	–	–	0.006
dmMC-RR	0.003	–	0.002	–
MC-LF	–	–	–	+

Legend: (+) present in sample; (−) not present in sample.

Analyses of several fish tissue samples did not show a presence of investigated MC variants in spring nor autumn samples.

3.5. Histological Alterations in Fish Samples

During both investigated periods, spring (before) and autumn season (during and after the bloom), most affected fish organs were liver, kidneys and gills. Liver samples of individuals from the control group displayed typical organization of the hepatic parenchyma, with cord-like formations of hepatocytes interspaced with sinusoids and radially arranged around blood vessels (Figure 4a). Hepatocytes were polygonal, with a clearly visible cell membrane and large round nuclei with distinguishable nucleoli. In contrast, microscopic examination of fish from Lake Ludoš revealed severe alterations of liver histology which were observed in both March and September sampling groups. Livers of these fish showed changes in architectural structure, with less prominent cord-like organization and sinusoid capillaries no longer clearly distinguishable (Figure 4b). Loss of shape and rounding of hepatocytes was most characteristic in the March group, with large groups of cells displaying ball-or onion-like shape (Figure 4c). Altered hepatocytes typically had darker nuclei with condensed chromatin and no discernable nucleoli. Signs of kariopyknosis, predominantly in the September sampling group, were indicative of necrosis (Figure 4d). Most prominent alterations present in all examined samples were glycogen depletion and vacuolization of hepatocytes. Many cells had a completely clear cytoplasm, which suggest hypervacuolization (Figure 4e).

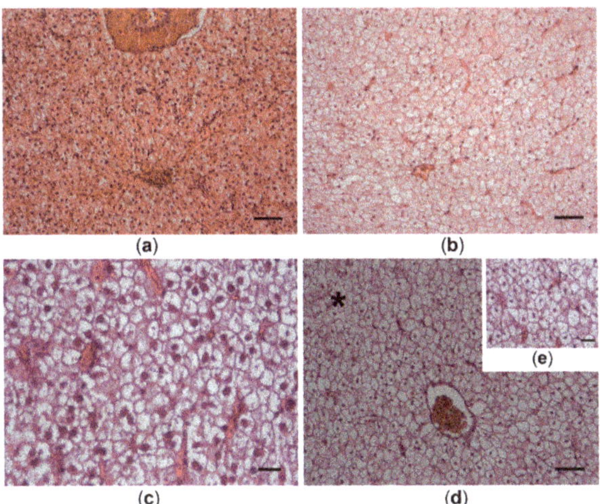

Figure 4. Histopathological alterations in the liver of Prussian carp *Carassius gibelio* from Lake Ludoš, 2018: (**a**) control fish; (**b**) loss of the cord-like parenchymal structure; (**c**) rounding of hepatocytes; (**d**) necrotic fields (asterisk) with hepatocytes displaying karyopyknosis; (**e**) hepatocytes displaying vacuolization and glycogen depletion. Haematoxylin and eosin (H&E) staining. A, B and D–50 µm; C and E–20 µm.

Renal corpuscles in the kidneys of the control individuals were round in shape and had relatively large glomeruli. The Bowman's capsule was continuous with thin intercapsular space. Both proximal and distal renal tubules had one layer of columnar epithelial cells with proximal segments having basal nuclei, and distal segments having central nuclei and less intensive cytoplasmic stain (Figure 5a). Even in controls, slight clogging of tubules and slight vacuolization was observed.

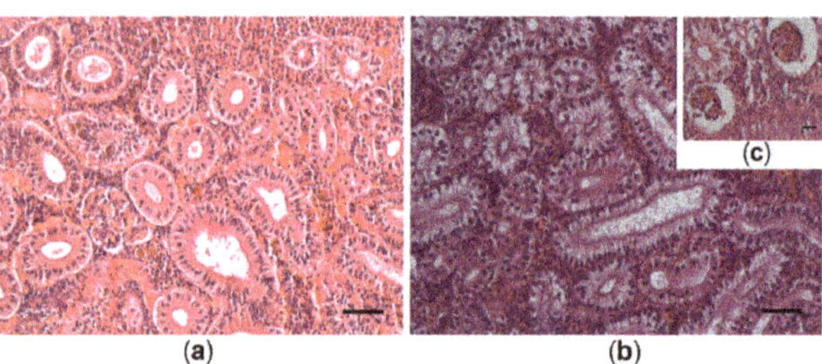

Figure 5. Histopathological alterations in the kidney of Prussian carp *Carassius gibelio* from Lake Ludoš, 2018: (**a**) control fish; (**b**) degeneration of tubules including vacuolization and separation of epithelial layer from the basal lamina; (**c**) reduction of glomeruli size and intense dilatation of Bowman's capsule. H&E staining. A and B–50 μm; C–20 μm.

Only individuals sampled in March had significant pathological alterations in the kidney. These included degeneration and loss of nephron formation, as well as interstitium structure. Renal corpuscles showed a reduction of glomeruli size, accompanied with dilatations of intercapsular space of the Bowman's capsule (Figure 5c). In tubules, epithelial cells showed intense vacuolization and in some cases the epithelial layer was separated from the basal lamina (Figure 5b). The number of tubules appeared clogged and in the process of necrosis. Cells in the necrotic area had pyknotic nuclei and displayed signs of cell membrane lysis, such as no discernable boundary between cells. In some fish, necrosis, hyalization of the interstitium and the presence of macrophage aggregates were evident. Certain alterations, such as vacuolization and tearing of the tubular epithelium, were also present in individuals from the September group; however, this was not as frequent and severe compared to the March group.

Gills of the control group showed no pathological changes (Figure 6a). Secondary lamellae regularly lined both sides of the primary lamellae (filament) and were covered with one layer of squamous epithelial cells. Contrarily, individuals from Lake Ludoš in both sampling periods (March and September) had noticeably altered gill structure. Several individuals displayed signs of hyperplasia, as well as hypertrophy of interlamellar cell mass, mainly epithelial and mucous cells. Such swelling of secondary lamellae and proliferation of interepithelial cells has led to a complete fusion of the secondary lamellae, especially noticeable in the September sampling group (Figure 6b). Other observed lesions included epithelial lifting and oedema, accompanied with hypertrophy of interepithelial chloride cells (Figure 6c). In some individuals, endothelial cells of the capillaries showed signs of telangiectasia (aneurysm), along with epithelial rupture and hemorrhage (Figure 6d).

Other examined organs of the bloom-exposed fish in the present study did not display histopathological alterations (not shown). Intestines of all groups showed normal histology, with villi regularly lining the lumen. A single layer of enterocytes with basal nuclei lined the surface of the vili, along with fewer goblet cells. Furthermore, sections of muscle tissue in all groups also showed no structural alterations. Spleen of all examined individuals displayed a normal structure, with aggregates of erythroid and lymphoid cells, surrounding blood vessels. Neither male nor female gonads had histopathological changes. Testes and ovaries had normal structural organization with germline cells present in different stages and numbers, which is dependent on the season and age of the fish.

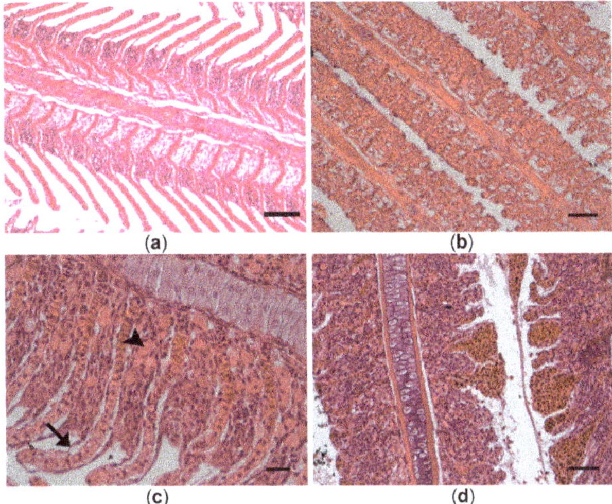

Figure 6. Histopathological alterations in the gills of Prussian carp *Carassius gibelio* from Lake Ludoš, 2018: (**a**) control fish; (**b**) fusion of the secondary lamellae; (**c**) intensive proliferations of chloride cells (arrowhead) and oedema (arrow); (**d**) telangiectasia. H&E staining. A and D–50 µm; B–100 µm; C–20 µm.

4. Discussion

4.1. Monitoring of the Water

4.1.1. Physical and Chemical Parameters in Lake Ludoš

The measurements of physical and chemical parameters showed several important findings:

- the lake water pH values were very high (almost 9), probably as a results of the activity of phytoplankton;
- O_2 saturation showed high values, and even supersaturation during July 2018, likely due to photosynthetic activity of phytoplankton;
- electrical conductivity was relatively high.

It is assumed that cyanobacteria are more resistant to solute increases compared to other phyla e.g., Chlorophyta [23]. Additionally, conductivity changes lead to decreased zooplankton—predators of phytoplankton, thus it is possible that a reduction in zooplankton would potentiate phytoplankton increase.

Regular monitoring showed similar findings in recent years (2013–2017) [7–11]. pH levels were between 8 and 9, O_2 saturation during the year was uneven with high values and supersaturation during summer and sometimes during autumn months, while high values of electric conductivity of the water seemed to be rising in the recent years. Additionally, COD was extremely high, similar to wastewaters, indicating a poor status of the lake and based on the measured parameters it is recommended that the lake should not to be used for any purpose [24]. BOD was uneven during the year, demonstrating the problem of instability of the system. The water of the Lake Ludoš is very rich in nitrogen compounds (mostly organically bound nitrogen), which led to increased biological production. Nitrates were uneven during the year and seemed to depend on the inflow from the Palić-Ludoš canal. Phosphates showed a similar trend as nitrates and their concentration seemed to be on the rise thus contributing to the deterioration of the water quality. Sediment analysis suggested rich deposits of nutrients which will contribute to the continual hypertrophic state of the lake [7–11].

The status of surface waters in terms of general quality can be shown by the Serbian Water Quality Index (SWQI). SWQI is based on 10 quality parameters (temperature, pH, conductivity, O_2 saturation, BOD for 5 days, suspended matter, total nitrogen oxides, orthophosphates, ammonia ions, coliform bacteria) that are aggregated into a composite indicator of the quality of surface waters, leading them to one index number [24]. SWQI of the water from Lake Ludoš through 2018 found it to be of either very low quality or low quality [25]. Similar findings were also noted for the period from 2013 to 2017 [7–11].

4.1.2. Cyanobacterial Community in Lake Ludoš

During 2018, the most dominant cyanobacterial species were *L. redekei* and *P. limnetica* (Table 3). Furthermore, both *Microcystis* species, *M. aeruginosa* and *M. wesenbergii*, were numerous during the whole investigated period. Same species were present in pier and center samples, although more cells were noted in the pier samples possibly due to lower depth, higher temperature, wave, and wind effects. In the recent years (2013–2017), several other cyanobacterial species were also frequently found: *Microcystis delicatissima*, *Oscillatoria agardhii* (current *Planktothrix agardhii*), *Oscillatoria putrida*, *Lyngbya limnetica* and *Anabaena spiroides* [7–11]. Furthermore, in our previous research during 2011 and 2012, similar cyanobacterial species were abundant: *L. redekei*, *P. limnetica*, *P. agardhii* and *Microcystis* spp. [5]. Most of the present species and genera are known as potential cyanotoxin producers.

Additionally, in the autumn of 2018, the invasive species *Raphidiopsis raciborskii* was also found. This invasive and potentially toxic species was first noted in Serbia in 2006 [26] and soon in Lake Ludoš as well [27], and since then it has been frequently found and blooming in the lake. The *R. raciborskii* presence is of particular concern due to its ability to expand its distribution rapidly (see [28]). Differences in toxin production are known among the strains: South American strains produce STX, Australian and Chinese strains produce CYN, and European and North American strains are considered to be non-toxic [28]. Furthermore, non-toxic and toxic strains can co-exist, and even co-occurring strains exhibit genomic variability [29]. Bearing in mind that this species is expanding its distribution in Europe, and in Serbia as well, it is necessary to establish whether this species is a greater threat than previously assumed, and how it succeeds in spreading so rapidly.

4.1.3. Presence of Cyanotoxin Genes and Cyanotoxins in Lake Ludoš

For the first time, Lake Ludoš was assessed for the presence of genes coding several frequently occurring cyanotoxins. During investigated months, the MC-coding gene *mcyE* was amplified in all the biomass samples. This is in accordance with the cyanobacterial species composition observed in the lake, and confirms the uniform distribution of MC-producing species throughout the year. Amplification of STX coding genes, *sxtG* and *sxtS* occurred in some of the samples. The *sxtG* gene was observed in all sampling seasons, while *sxtS* gene was observed only in samples collected in August and September. The STX *sxtA* coding gene was not amplified in this study. Since the specificity of *sxtA*, *sxtG* and *sxtS* primers has been previously validated by Savela et al. [19,20], the lack of targeted genes may be due to unexpected sequence dissimilarities between primers and target that can occur in natural populations. Large-scale gene mutations such as deletions and insertions resulting in non-functional gene clusters may also cause a lack of PCR amplification. Major deletions events in MC-coding gene cluster were previously observed in non-toxic *Planktothrix* strains [30], suggesting that similar events may occur in STX-producing strains. Similarly to this study, the lack of the some of the targeted STX genes was previously observed by Savela et al. (2017) in Polish lakes. As STXs were not measured directly, conclusions on correlation between gene presence/absence and toxin production cannot be made. However, since STXs have been detected in the lake Ludoš before [5], and a potential STX producer, *R. raciborskii*, was detected, additional studies are warranted to confirm potential production of STXs in the lake.

There are several publications that show MC presence in various water bodies in Serbia [6,31]. However, only a few publications demonstrate presence of cyanotoxins in Lake Ludoš. MCs were first noted in 2006 by Simeunović [32,33], thereafter the presence of MCs and low concentrations of STXs in the lake was observed during our previous investigation in 2011 and 2012 [5]. In 2018, analyses of water and biomass samples for the presence of MCs and CYN were performed by LC–MS/MS and several MC variants were detected in low concentrations. CYN was not detected, although some investigations have shown that some of the present cyanobacterial species potentially could produce this toxin (e.g., [34,35]).

4.2. Uptake of Cyanotoxins and Bloom Effects

Fish are directly influenced by water blooms, which can cause health impairments and mortality [36]. Primary route of exposure is by ingestion, either directly or through the food web, and indirectly via epithelial absorption [37]. Consequently, liver and gills are often the most affected organs during cyanobacterial blooms [38,39]. Furthermore, there is evidence of cyanotoxin accumulation in fish tissue, which ultimately endangers the health of animals and even humans that use them as a food source (e.g. [40,41]). Multiple other stress factors co-occuring with the blooms can additionally threaten fish, damaging tissues and impairing their development and survival [42].

A previous investigation of Lake Ludoš by the authors showed cyanotoxin uptake by macrophytes and fish. Accumulation of MC-RR was detected in the rhizome of *Phragmites australis*, cattail *Typha latifolia* and royal blue water lily *Nymphaea elegans* [5]. Furthermore, the same investigation demonstrated accumulation of two MC variants (MC-LR and MC-RR) in the tissue samples of the Prussian carp from Lake Ludoš. During 2011 and 2012, MCs were found in the intestine (MC-LR) and muscle samples (MC-RR), however, no MCs were found in the liver samples. In 2011, the fish gills were also positive for MC-RR, and in 2012, kidneys and gonads were found to accumulate MC-LR [5]. In 2011, histopathological changes were observed in liver, kidney, gills and intestine samples of the tested individuals. Furthermore, DNA damage was detected in Prussian carp as revealed via comet assay in 2012. Three tissues were selected and assessed: blood had the lowest level of DNA damage, while liver and gills showed more damage [5]. In another study from fishponds in Serbia, MC-RR accumulation in muscle tissues was recorded and histopathological changes were noted in liver, kidneys, gills, intestines and muscles of the *Cyprinus carpio* tissues [41]. Many histological changes in different fish species and tissues after exposure to cyanobacterial bloom or cyanotoxins were presented in a review by Svirčev et al. [2].

In addition to histological changes and accumulation, cyanotoxins can affect fish growth, development, reproduction, and survival. Embryos and larvae are especially sensitive to the effects of MCs, and exposure to these cyanotoxins in the early life stages can disrupt embryos, reduce survival and growth rate, and cause disorders such as: small head, curvature of the body and tail, enlarged and opaque yolk sachet, hepatobiliary abnormalities and cardiac disturbances [43–45]. Furthermore, cyanobacteria and their metabolites can have a wide range of negative effects on the adults of fish. They affect locomotor activity and fish behavior, impair ingestion rate and growth, disrupt heart function, ion regulation, cause changes in serum composition, trigger oxidative stress and disturb the reproduction of fish. Alongside direct harmful effects caused by cyanotoxins, cyanobacteria can adversely affect fish through the alteration of environmental conditions during blooming. Decreased dissolved oxygen, increase in ammonia concentration, changes in pH value and water temperature, or a combination of all of these factors can have detrimental changes on fish [46].

4.2.1. Cyanotoxin Accumulation in Fish Tissues from Lake Ludoš

Several fish tissues of *Carassius gibelio* were examined for the presence of MCs. As already stated, in our previous study accumulation of two microcystin variants (MC-LR and MC-RR) was noted in several fish tissues: muscle, intestines, kidneys, gonads and gills of *Carassius gibelio* [5], however, accumulation in tissues was not found during this investigation. Although spiked samples demonstrated that the method for the preparation of the samples is adequate, it is possible that cyanotoxins remained covalently bound to protein phosphatases in the tissue [47–50] resulting in negative results. Furthermore, it is possible that cyanotoxins were excreted and concentrations lowered by detoxification processes which vary between different species, organ, MC congener and metabolism [40,51–53], or even that concentrations in water were not high enough to be accumulated in high concentrations for detection.

4.2.2. Bloom Effects on Histopathology of Fish from Lake Ludoš

Histopathology has an important place as a biomarker of fish health status, as histological changes often occur in response to acute or chronic exposure of an organism to a pollutant or a hazardous chemical, as well as adverse conditions present in the water. Even though no accumulation of cyanotoxins were detected during this investigation, histological observation of tissues from Prussian carp taken from Lake Ludoš suggest serious health implications often attributed to water blooming and cyanobacteria-rich aquatic environments. In general, most affected organs during both the spring and autumn seasons were liver, kidneys and gills. Furthermore, presence of alterations in both sampling periods (spring and autumn) suggests that chemical state of the lake was poor and that harmful blooms probably occurred in the previous year.

Structural alterations of the liver observed in this study are similar to those found by other authors after the exposure of fish to cyanotoxins or extracts of cyanobacterial strains and blooms [54–60]. Most prominent changes were loss of parenchymal structure, rounding of cells, glycogen depletion, vacuolization and pyknosis, all of which were so far reported after exposure of fish to MCs [2]. These alterations, specifically vacuolization and increase of lipid content, were also observed in mammalian livers in reaction to MC [61–63]. Hepatotoxicity of MC might be attributed to its affinity to bind and inhibit eukaryotic serine-threonine protein phosphatases 1 and 2A (PP1 and PP2), enzymes that are important in maintaining cell homeostasis and tumor suppression signaling pathways [64–66]. Studies in mammalian models have shown that inhibition of PP1 and PP2 can cause cytoskeletal damage [67,68] and may lead to loss of cell-to-cell contact, rounding of hepatocytes, condensation of chromatin and nuclear pyknosis.

Kidneys may be good indicators of environmental stress since they receive the majority of postbranchial blood. Additionally, kidney tubule cells possess a transport mechanism similar to that of hepatocytes (the multispecific bile acid transport system) which is responsible for MC uptake into the cell [69]. A study undertaken by Fischer and Dietrich [56] has shown that due to this efficient uptake of toxins in carp, MC-induced kidney pathologies in carp develop rapidly and at lower toxin concentrations. Such a finding supports the results of this study, as the kidneys exhibited the strongest histopathological changes, particularly in the fish that were taken during the spring when the levels of cyanotoxins are lower. Histological changes observed in the present study supports previous research on the effects of MCs on fish [39,55,56]. Glomerulopathy, with dilatations of the Bowman's capsule, and vacuolization of tubules are the most common alterations observed after exposure of fish to MCs [70,71]. Necrosis and impairment of renal tubules can affect ion and water regulation in the kidney, thus damaging the survival capabilities of fish [72].

Fish gills constitute over 50% of the total surface area of the animal and are in direct contact with water, which makes them sensitive to pollutants and toxic chemicals. A number of histopathological alterations were detected in fish from Lake Ludoš, which can be associated with conditions during water blooms, mostly the presence of cyanotoxins [54,57,73]. Lifting of the lamellar epithelial cells caused by the fluid penetration and epithelial hyperplasia were the most common lesions. Along with epithelial hyperplasia, these are considered defensive mechanisms of gills, both of which reduce the uptake of xenobiotics [74,75]. Other persistent alterations, such as oedema and telangiectasia from the secondary lamellae could be attributed directly to MC toxicity. Gill ion pumps (Na^+ and K^+ ATPases) could be inhibited by MCs [76,77], leading to a decline in blood Na+ and Cl− concentration and ion exchange imbalance. This could result in swelling of secondary lamellae as well as proliferation of interepithelial chloride cells. Intensive hyperplasia decreases the space between lamellae and causes fusion, which increases the thickness of the water–blood barrier and decreases the oxygen uptake.

These lesions can cause capillary hemorrhage and significantly hinder gill functions, such as respiration, ion regulation, acid-base regulation and nitrogenous waste excretion [78]. The molecular actions of cyanotoxins in fish liver and gills involve increased generation of reactive oxygen species (ROS) which randomly attack all cell components, including proteins, lipids and nucleic acids [79,80]. An imbalance between the generation and removal of ROS in these tissues results in oxidative stress and extensive cellular tissue damage in these organs [81,82].

Previous research on fish from Lake Ludoš by the authors showed similar findings. The liver showed a loss of cordlike parenchymal structure; presence of onion-shaped hepatocytes with clear cytoplasm as well as pyknosis and fields of anucleated cells. Changes in the kidney were glomerulopathy with intense dilatation of Bowman's capsule as well as vacuolization of tubules and macrophage infiltration. Furthermore, histopathological alterations observed in the gills showed fusions of lamellae; oedema and epithelial lifting and intensive proliferations of chloride cells. Alterations were also found in intestines where intensive oedematous alteration in the lamina propria; desquamation of enterocytes; and hypertrophies of goblet cells was noted [5]. Although more than five years have passed since the previous research, problems found in fish tissues remain present and could be a result from the constant cyanobacterial blooming in this lake. The aforementioned alterations can severely impact the life quality of fish and consequently disturb the whole food chain.

It should also be noted that fishing at Lake Ludoš by local fisherman is frequent, and therefore *Carassius gibelio* is sometimes included in the human diet. It is necessary to monitor concentrations of cyanotoxins in water and fish to make sure that ingested concentrations of MCs are below tolerable daily intake of 0.04 μg MC-LR equiv./kg body weight/day [83], so that any health consequences could be prevented.

4.3. Potential Health Risks Caused by Cyanobacterial Blooms

4.3.1. Transfer to Protected Water Birds

Cyanotoxins can be transmitted through the food web to different consumers, such as various animals living in and around the lake, including birds. In the case of Lake Ludoš, this is particularly important since one of the main reasons for protection of this wetland is because it is a habitat for water birds. In addition to the contaminated food, birds can come into contact with cyanotoxins via direct contact with the blooming water. There are few papers dealing with the effect of cyanotoxins on birds. Early reports, such as that from Storm Lake in Iowa associated with *Anabaena flos–aquae* blooms, include estimated deaths of 5–7000 gulls, 560 ducks, 400 coots, 200 pheasants, 50 squirrels, 18 muskrats, 15 dogs, 4 cats, 2 hogs, 2 hawks, 1 skunk, 1 mink, plus "numerous" songbirds. It seems that neurotoxicity resulted in prostration and convulsions preceded death; milder cases displayed restlessness, weakness, dyspnoea and tonic spasms. Furthermore, 57 weak and partially paralyzed mallards were recovered following gastric lavage [84,85]. In Denmark in 1993, two grebes and a coot died during cyanobacterial bloom and *Anabaena lemmermannii* was found in the stomach contents all three birds together with low levels of anatoxin-a(S)-like compounds [86,87]. Death of 20 ducks in Shin–ike pond (Japan) were described after evident *M. aeruginosa* bloom and presence of MC was confirmed, indicating MC intoxication as a possible cause of death [88]. Cyanobacteria-related mortalities have been reported in three flamingo species, both wild and captive [89]. Four MC congeners and ATX found in cyanobacterial mats and stomach contents of dead lesser flamingos as well as faecal pellets collected from shorelines of Lake Bogoria (Kenya) [90]. Similar Lesser Flamingo poisonings have been reported from alkaline lakes in Tanzania, with toxic *Arthrospira fusiformis* implicated [91]. The presence of MCs in several organs was detected in the domestic duck (*Anas platyrhynchos*) and in the black-crowned night heron (*Nycticorax nycticoraxs*) (alongside fish and turtle) from Lake Taihu (China) during toxic *Microcystis* blooms [92]. After experimental exposure to *Microcystis* biomass containing MCs, histopathological changes were observed in the form of cloudy swelling of hepatocytes (shrunken nuclei with ring-like nucleoli, cristolysis within mitochondria and vacuoles with pseudomyelin structures), vacuolar dystrophy, steatosis, hyperplasia of lymphatic centers and vacuolar degeneration of the testicular germinative epithelium in male Japanese quails (*Coturnix coturnix japonica*) [93].

4.3.2. Transfer to Humans as a Result of Fish Consumption

Humans at the top of the food chain may also be endangered. Lake Ludoš is regularly frequented by the local fishermen, and year-long and day-long fishing permits are available for this protected freshwater ecosystem. Based on the available data it was found that fishermen with prolonged and cumulative exposure through contaminated water (direct contact, inhalation, oral) and food may suffer a type of poisoning with a long-term complications. Such an outcome was demonstrated through biochemical alterations of liver damage biomarkers in fishermen and children that consumed (in addition to water) ducks, fish, shrimps, snails, and other aquatic organisms grown in blooming lakes in China [94,95]. Serum analysis of 35 fishermen who worked and lived on fishing ships on the blooming Lake Chaohu (China) for over 5 years, drank the lake water and ate fish, shrimps, and snails, showed the presence of MC [94]. The values of serum enzymes were significantly higher in the children that were exposed to MCs for over 5 years through drinking MC contaminated water and food e.g., carp and duck in the Three Gorges Reservoir Region in China. Additionally, 0.9% of parents of high-exposed children self-reported a cancer diagnosis (9 of 994, including four with hepatic carcinoma) compared to 0.5% of parents of low-exposed children (1 of 183), and none of the parents of children in the unexposed group [94]. To confirm this, in south-west China where MC-LR was detected in water, fish, and ducks, people with abnormal indicators of renal function had a much higher mean level of MC-LR exposure than those with normal indicators [96]. The foregoing indicates that the blooming phenomenon should not be taken lightly, but rather should receive more scientific attention.

4.4. A Potential Resolution?

By some estimates, we lost 50% of the world's wetlands in the 20th century [97]. So far an effective and long-term solution to reduce cyanobacterial blooms worldwide has not been attained. Cyanobacterial blooms have been recorded in the wetlands of the Perth region (Australia), with *Microcystis aeruginosa* and *M. flos-aquae* being the most ubiquitous bloom-forming cyanobacteria. Furthermore, for the first time *Nodularia* blooms have been recorded in such low salinity waters in Australia, and hepatotoxins, microcystin and nodularin, were associated with the analysed blooms [98]. Lake Ludoš has been notoriously known for consistent cyanobacterial blooming. Wetland ecosystems need help in dealing with this issue.

Mitigation of the global expansion of cyanobacterial harmful blooms, together with a variety of traditional (e.g., nutrient load reduction) and experimental (e.g., artificial mixing and flushing, omnivorous fish removal) approaches has been presented in a review by Pearl et al. [99]. Virtually all mitigation strategies are influenced by climate change, which may require setting new nutrient input reduction targets and establishing nutrient-bloom thresholds for impacted waters. Physical-forcing mitigation techniques, such as flushing and artificial mixing, will need adjustments to deal with the ramifications of climate change. Current mitigation strategies need to be examined and the potential options for adapting and optimizing them in a world facing increasing human population pressure and climate change [99]. A notable approach on large-scale wetland restoration has been proposed in Ohio (USA). Through restoration of the Great Black Swamp, cyanobacterial blooming in Lake Erie should be mitigated [97]. The goal would be to restore rare and declining plant communities and species and to provide nutrient load reduction to Lake Erie, specifically total phosphorous. With the creation of the 20,000 of wetlands in "hot spots" of the former Great Black Swamp, removal of 18% of the 2617 metric tons/year of phosphorus loading by the Maumee River to Lake Erie would be expected [100]. This would lead to a cleaner lake and a more sustainable landscape in that region [97].

In order to reduce cyanobacterial population in Lake Ludoš, the potential efficiency of hydrogen peroxide treatment was examined in vitro. Although further research is needed, the initial laboratory results showed that this method may not be readily applicable, since the dense cyanobacterial population and the high load of organic matter (that consumes hydrogen peroxide) would require the use of harmfully high doses of hydrogen peroxide in order to fight the cyanobacteria [5]. Another recent study performed in Lake Ludoš demonstrated the use of H_2O_2 and the MC-degrading capacity of the enzyme MlrA. Results showed that the treatment decreased the abundance of the dominant cyanobacterial taxa and reduction of the intracellular concentration of MC by H_2O_2, but the reduction of the extracellular MC was not accomplished in combination with MlrA. Since H_2O_2 was found to induce the expression of *mcyB* and *mcyE* genes involved in MC biosynthesis, the use of H_2O_2 as a safe cyanobacteriocide still requires further investigation [101]. Several other authors have also suggested measures for water-quality improvement of this lake, including the identification of source water with lower nutrient content for maintaining the volume of the lake, sediment removal [102,103], and sediment phytoremediation [104]. Even if restoration of the endangered ecosystem were to be realized, continuous monitoring would still be preferable. Such system (South Florida Wetland Monitoring Network (SFWMN)) has been created in the Everglades (USA), with three real-time hydrologic, water-quality, and meteorological field stations. Besides research, data from these monitoring stations assist in a better understanding of wetlands dynamics and function [105].

Lake Ludoš, a significant ecosystem and a habitat for several endemic and relict plant species, is preserved at the moment, however, the problematic ecological state of the lake is also perpetuated, so justification for such action was questioned. Current research strongly supports the earlier findings that the ecological balance of this Ramsar site is impaired. By preserving the lake and its cyanobacterial problem the water birds and their habitat are not really protected and, quite the contrary, the lake and nearby ecosystems are put at risk. There is a likely probability that the birds visiting the contaminated lake during their migrations carry viable cyanobacterial cells on their feet, feathers, bills, gullets, and faecal material, thus contributing to the spreading of cyanobacteria and hence expanding the problem [5].

The future of Lake Ludoš is still unclear and a potential resolution is still not in sight. How long will this vital ecosystem stay in a state of limbo? Based on this research, and as an extension and confirmation of the previous investigations, it is possible to predict the continuation of cyanobacterial blooms in Lake Ludoš, degradation of protected habitat and negative effects on aquatic organisms. Food items derived from Lake Ludoš also present one path of human exposure to cyanotoxins, which together with the direct contact and ingestion of contaminated water, as well as inhalation, signifies a health risk. Therefore, it is necessary to continue the monitoring of this lake, and work on finding an effective treatment that will help this ecosystem, but also many others that suffer from the same problem. Recently, it was published that there are over 1000 recorded identifications of major cyanotoxins in more than 800 aquatic ecosystems from over 60 countries worldwide [4].

It is time to solve this problem but most if not all the options being considered are limited and/or inconsequential [97]. This paper should be seen as an invitation to scientists, engineers, competent authorities, policy makers and anyone else who can contribute to solving the problems of Lake Ludoš and finally ending the lake's perpetual state of limbo. Potential solutions should be comprehensive and holistic, and lead to a sustainable management of this marshland ecosystem and its services.

5. Conclusions

This investigation regarding the presence of cyanobacteria and cyanotoxins, together with the observations of effects on aquatic organisms, in Lake Ludoš during 2018 has resulted in the following findings:

- the poor chemical state of the lake based on the physical and chemical parameters;
- the presence of potentially toxic (genera *Dolichospermum*, *Microcystis*, *Planktothrix*, *Chroococcus*, *Oscillatoria*, *Woronichinia* and dominant species *L. redekei* and *Pseudanabaena limnetica*) and invasive cyanobacterial species *R. raciborskii*;
- the detection of MC and STX coding genes in biomass samples;

- the detection of several MC variants MC-LR, MC-dmLR, MC-RR, MC-dmRR, MC-LF) in low concentrations in water samples;
- histopathological alterations in fish liver, kidney and gills.

Additional emphasis has to be placed on the detection of MC and STX coding genes, which was performed for the first time in Lake Ludoš indicating these toxins as the main threat, as well as identification of demetylated forms of well-known toxins (MC-dmLR, MC-dmRR), and a more rare variant MC-LF in the water. Although present MC variants have a different toxicity, nonetheless, they contribute to possible adverse effects.

The results presented indicate that the potential threat to many organisms in the ecosystem, including birds and humans, is real and present. The persistent alarming condition of Lake Ludoš poses a great health risk and a "ticking (cyanobacterial) bomb" that could lead to the collapse of this special nature reserve. Urgent remediation measures are needed to alleviate the incessant cyanobacterial problem in Lake Ludoš and to break this ecosystem out of the perpetual state of limbo in which it has been trapped for quite some time.

Author Contributions: Funding acquisition, Z.S., J.M. and J.L.; conceptualization and design of the experiments, N.T., D.D.B., J.L. and Z.M.; investigation and field sampling, B.M. and I.Š.; phytoplankton analyses, S.S. and N.Đ.; histopatological analyses, N.K., J.L. and Z.M.; cyanotoxin analysis, N.T., D.D.B., T.D. and J.M.; cyanotoxin coding genes analysis, T.D. and H.S.; writing—original draft preparation, N.T.; writing—review and editing, N.T., Z.S. and J.M.; project administration, N.T. All authors have read and agreed to the published version of the manuscript.

Funding: This research was funded by the Ministry of Education, Science and Technological Development of the Serbian Government (project number: ON-176020, TR-31011, III-43002), Bilateral project Hungary-Serbia Invasive and blooming cyanobacteria in Serbian and Hungarian waters (451-03-02294/2015-09/3) and the Erasmus+ programme of the European Union (agreement number: 2017-1-FI01-KA107-034440).

Conflicts of Interest: The authors declare no conflict of interest.

References

1. Bury, N.R.; Eddy, F.B.; Codd, G.A. The effects of the cyanobacterium Microcystis aeruginosa, the cyanobacterial hepatotoxin hepatotoxin MC-LR, and ammonia on growth rate and ionic regulation of brown trout. *J. Fish Biol.* **1995**, *46*, 1042–1054.
2. Svirčev, Z.; Lujić, J.; Marinović, Z.; Drobac, D.; Tokodi, N.; Stojiljković, B.; Meriluoto, J. Toxicopathology induced by microcystins and nodularin: A histopathological review. *J. Environ. Sci. Health C* **2015**, *33*, 125–167. [CrossRef]
3. Svirčev, Z.; Drobac, D.; Tokodi, N.; Mijović, B.; Codd, G.A.; Meriluoto, J. Toxicology of microcystins with reference to cases of human intoxications and epidemiological investigations of exposures to cyanobacteria and cyanotoxins. *Arch. Toxicol.* **2017**, *91*, 621–650. [CrossRef]
4. Svirčev, Z.; Lalić, D.; Bojadžija Savić, G.; Tokodi, N.; Drobac Backović, D.; Chen, L.; Meriluoto, J.; Codd, G.A. Global geographical and historical overview of cyanotoxin distribution and cyanobacterial poisonings. *Arch. Toxicol.* **2019**, *93*, 2429–2481. [CrossRef]
5. Tokodi, N.; Drobac, D.; Meriluoto, J.; Lujić, J.; Marinović, Z.; Važić, T.; Nybom, S.; Simeunović, J.; Dulić, T.; Lazić, G.; et al. Cyanobacterial effects in Lake Ludoš, Serbia—Is preservation of a degraded aquatic ecosystem justified? *Sci. Total Environ.* **2018**, *635*, 1047–1062. [CrossRef]
6. Svirčev, Z.; Tokodi, N.; Drobac, D.; Codd, G.A. Cyanobacteria in aquatic ecosystems in Serbia: Effects on water quality, human health and biodiversity. *Syst. Biodivers.* **2014**, *12*, 261–270. [CrossRef]
7. Institute for Public Health, Subotica. *Monitoring Kvaliteta Vode Jezera Palić i Ludaš i Potoka Kereš u 2013 Godini*; Annual Report; Public Health Institute: Subotica, Serbia, 2014. Available online: http://www.subotica.rs/documents/zivotna_sredina/Monitoring/Voda/God/MH-2013-ovrsinskeVode.pdf (accessed on 30 December 2019).
8. Institute for Public Health, Subotica. *Monitoring Kvaliteta Vode Jezera Palić i Ludaš u 2014 Godini*; Annual Report; Public Health Institute: Subotica, Serbia, 2015. Available online: http://www.subotica.rs/documents/zivotna_sredina/Monitoring/Voda/God/MH-2014-ovrsinskeVode.pdf (accessed on 30 December 2019).

9. Institute for public health, Subotica. *Monitoring Kvaliteta Vode Jezera Palić, Ludaš i Kanala Palić-Ludaš u 2015 Godini*; Annual Report; Public Health Institute: Subotica, Serbia, 2016. Available online: http://www.subotica.rs/documents/zivotna_sredina/Monitoring/Voda/God/MH-2015-ovrsinskeVode.pdf (accessed on 30 December 2019).

10. Institute for Public Health, Subotica. *Monitoring Kvaliteta Vode Jezera Palić, Ludaš i Kanala Palič-Ludaš u 2016 Godini*; Annual Report; Public Health Institute: Subotica, Serbia, 2017. Available online: http://www.subotica.rs/documents/zivotna_sredina/Monitoring/Voda/God/MH-2016-ovrsinskeVode.pdf (accessed on 30 December 2019).

11. Institute for Public Health, Subotica. *Monitoring Kvaliteta Vode Jezera Palić, Ludaš i Kanala Palić-Ludaš u 2017 Godini*; Annual Report; Public Health Institute: Subotica, Serbia, 2018. Available online: http://www.subotica.rs/documents/zivotna_sredina/Monitoring/Voda/God/MH-2017-ovrsinskeVode.pdf (accessed on 30 December 2019).

12. Komárek, J.; Anagnostidis, K. Cyanoprokaryota 1. Teil: Chroococcaless. In *Süßwasserflora von Mitteleuropa*; Ettl, H., Gerloff, J., Heynig, H., Mollenhauer, D., Eds.; Spektrum Akademischer Verlag: Heidelberg/Berlin, Germany, 1998; pp. 1–548.

13. Komárek, J.; Anagnostidis, K. Cyanoprokaryota 2. Teil: Oscillatoriales. In *Süßwasserflora von Mitteleuropa*; Budel, B., Krienitz, L., Gärtner, G., Schagerl, M., Eds.; Spektrum Akademischer Verlag: Heidelberg/Berlin, Germany, 2005; pp. 1–759.

14. Komárek, J. Cyanoprokaryota 3. Teil: Heterocytous Genera. In *Süßwasserflora von Mitteleuropa*; Budel, B., Gärtner, G., Krienitz, L., Schagerl, M., Eds.; Springer Spektrum Verlag: Heidelberg/Berlin, Germany, 2013; pp. 1–1130.

15. Aguilera, A.; Berrendero, E.; Mez, G.; Kaštovský, J.; Echenique, R.; Graciela, L.; Salerno, G. The polyphasic analysis of two native *Raphidiopsis* isolates supports the unification of the genera *Raphidiopsis* and *Cylindrospermopsis* (Nostocales, Cyanobacteria). *Phycologia* **2018**, *57*, 130–146. [CrossRef]

16. Utermöhl, H. Zur vervolkmmung der quantitativen phytoplankton methodik. *Mitt. Int. Ver. Theor. Angew. Limnol.* **1958**, *9*, 1–38.

17. Dittman, E.; Mankiewicz-Boczek, J.; Gągała, I. SOP 6.2: PCR detection of microcystin and nodularin biosynthesis genes in the cyanobacterial orders Oscillatoriales, Chroococcales, Stigonematales, and Nostocales. In *Molecular Tools for the Detection and Quantification of Toxigenic Cyanobacteria*; Kurmayer, R., Sivonen, K., Wilmotte, A., Salmaso, N., Eds.; John Willey & Sons, Ltd.: Chichester, UK, 2017; pp. 175–178.

18. Mazmouz, R.; Chapuis-Hugon, F.; Mann, S.; Pichon, V.; Méjean, A.; Ploux, O. Biosynthesis of cylindrospermopsin and 7-epicylindrospermopsin in *Oscillatoria* sp. strain PCC 6506: Identification of the *cyr* gene cluster and toxin analysis. *Appl. Environ. Microb.* **2010**, *76*, 4943–4949. [CrossRef]

19. Savela, H.; Spoof, L.; Perälä, N.; Preede, M.; Lamminmäki, U.; Nybom, S.; Häggqvist, K.; Meriluoto, J.; Vehniäinen, M. Detection of cyanobacteria *sxt* genes and paralytic shellfish toxins in freshwater lakes and brakish waters on Åland Islands, Finland. *Harmful Algae* **2015**, *46*, 1–10. [CrossRef]

20. Savela, H.; Spoof, L.; Höysniemi, N.; Vehniäinen, M.; Mankiewicz-Boczek, J.; Jurczak, T.; Kokociński, M.; Meriluoto, J. First report of cyanobacterial paralytic shellfish toxin biosynthesis genes and paralytic shellfish toxin production in Polish freshwater lakes. *Adv. Oceanogr. Limnol.* **2017**, *8*, 61–70. [CrossRef]

21. Rantala-Ylinen, A.; Känä, S.; Wang, H.; Rouhiainen, L.; Wahlsten, M.; Rizzi, E.; Berg, K.; Gugger, M.; Sivonen, K. Anatoxin-a synthetase gene cluster of the cyanobacterium *Anabaena* sp. strain 37 and molecular methods to detect potential producers. *J. Appl. Environ. Microbiol.* **2011**, *77*, 7271–7278. [CrossRef]

22. Major, Y.; Kifle, D.; Spoof, L.; Meriluoto, J. Cyanobacteria and microcystins in Koka Reservoir (Ethiopia). *Environ. Sci. Pollut. Res.* **2018**, *25*, 26861–26873. [CrossRef]

23. Chakraborty, P.; Acharyya, T.; Raghunadh Babu, P.V.; Bandhyopadhyay, D. Impact of salinity and pH on phytoplankton community in a tropical freshwater system: An investigation with pigment analysis by HPLC. *J. Environ. Monit.* **2011**, *13*, 614–620. [CrossRef]

24. Službeni Glasnik Republike Srbije 50/12. Uredba o graničnim vrednostima zagađujućih materija u površinskim i podzemnim vodama i sedimentu i rokovima za njihovo dostizanje. Available online: https://www.paragraf.rs/propisi/uredba-granicnim-vrednostima-zagadjujucih-materija-vodama.html (accessed on 30 December 2019).

25. Public Health Institute. Available online: http://www.zjzs.org.rs/monitoring.php?obl=voda&id=929 (accessed on 25 November 2019).

26. Cvijan, M.; Fužinato, S. *Cylindrospermopsis raciborskii* (Cyanoprokaryota)-potential invasive and toxic species in Serbia. *Bot. Serbica* **2012**, *36*, 3–8.

27. Institute of Public Health. *Monitoring Kvaliteta Vode Jezera Palić i Ludaš i Potoka Kereš u 2012 Godini*; Annual report; Public Health Institute: Subotica, Serbia, 2013. Available online: http://www.subotica.rs/documents/ zivotna_sredina/Monitoring/Voda/God/MH-2012-ovrsinskeVode.pdf (accessed on 30 December 2019).

28. Burford, M.A.; Beardall, J.; Willis, A.; Orr, P.T.; Magalhaes, V.F.; Rangel, L.M.; Azevedo, S.M.F.O.E.; Neilan, B.A. Understanding the winning strategies used by the bloom forming cyanobacterium *Cylindrospermopsis raciborskii*. *Harmful Algae* **2016**, *54*, 44–53. [CrossRef]

29. Willis, A.; Woodhouse, J.N.; Ongley, S.E.; Jex, A.R.; Burford, M.A.; Neilan, B.A. Genome variation in nine co-occurring toxic *Cylindrospermopsis raciborskii* strains. *Harmful Algae* **2018**, *73*, 157–166. [CrossRef]

30. Christiansen, G.; Molitor, C.; Philmus, B.; Kurmayer, R. Nontoxic strains of cyanobacteria are the result of major gene deletion events induced by a transposable element. *Mol. Biol. Evol.* **2008**, *25*, 1695–1704. [CrossRef]

31. Svirčev, Z.; Tokodi, N.; Drobac, D. Review of 130 years of research on cyanobacteria in aquatic ecosystems in Serbia presented in a Serbian Cyanobacterial Database. *Adv. Oceanogr. Limnol.* **2017**, *8*, 153–160. [CrossRef]

32. Simeunović, J. Ekofiziološke Karakteristike Potencijalno Toksičnih i Toksičnih Vodenih Sojeva Cijanobakterija na Području Vojvodine. Ph.D. Thesis, University of Novi Sad, Novi Sad, Serbia, 2009.

33. Simeunović, J.; Svirčev, Z.; Karaman, M.; Knežević, P.; Melar, M. Cyanobacterial blooms and first observation of microcystin occurrences in freshwater ecosystems in Vojvodina region (Serbia). *Fresen. Environ. Bull.* **2010**, *19*, 198–207.

34. Messineo, V.; Melchiorre, S.; Di Corcia, A.; Gallo, P.; Bruno, M. Seasonal succession of *Cylindrospermopsis raciborskii* and *Aphanizomenon ovalisporum* blooms with cylindrospermopsin occurrence in the volcanic Lake Albano, Central Italy. *Environ. Toxicol.* **2010**, *25*, 18–27. [CrossRef]

35. Đorđević, N.B.; Matić, S.L.; Simić, S.B.; Stanić, S.M.; Mihailović, V.B.; Stanković, N.M.; Stanković, V.D.; Ćirić, A.R. Impact of the toxicity of *Cylindrospermopsis raciborskii* (Woloszynska) Seenayya & Subba Raju on laboratory rats in vivo. *Environ. Sci. Pollut. Res. Int.* **2017**, *24*, 14259–14272. [CrossRef]

36. Qiu, T.; Xie, P.; Ke, Z.; Li, L.; Guo, L. In situ studies on physiological and biochemical responses of four fishes with different trophic levels to toxic cyanobacterial blooms in a large Chinese lake. *Toxicon* **2007**, *50*, 365–376. [CrossRef]

37. Smith, J.L.; Haney, J.F. Food transfer, accumulation, and depuration of microcystins, a cyanobacterial toxin, in pumpkinseed sunfish (*Lepomis gibbosus*). *Toxicon* **2006**, *48*, 580–589. [CrossRef]

38. Marie, B.; Huet, H.; Marie, A.; Djediat, C.; Puiseux-Dao, S.; Catherine, A.; Trinchet, I.; Edery, M. Effects of a toxic cyanobacterial bloom (*Planktothrix agardhii*) on fish: Insights from histopathological and quantitative proteomic assessments following the oral exposure of medaka fish (*Oryzias latipes*). *Aquat. Toxicol.* **2012**, *114*, 39–48. [CrossRef]

39. Mitsoura, A.; Kagalou, I.; Papaioannou, N.; Berillis, P.; Mente, E.; Papadimitriou, T. The presence of microcystins in fish *Cyprinus carpio* tissues: A histopathological study. *Int. Aquat. Res.* **2013**, *5*, 8. [CrossRef]

40. Adamovsky, O.; Kopp, R.; Hilscherova, K.; Babica, P.; Palikova, M.; Paskova, V.; Navratil, S.; Blaha, L. Microcystin kinetics (bioaccumulation, elimination) and biochemical responses in common carp and silver carp exposed to toxic cyanobacterial blooms. *Environ. Toxicol. Chem.* **2007**, *26*, 2687–2693. [CrossRef]

41. Drobac, D.; Tokodi, N.; Lujić, J.; Marinović, Z.; Subakov-Simić, G.; Dulić, T.; Važić, T.; Nybom, S.; Meriluoto, J.; Codd, G.A.; et al. Cyanobacteria and cyanotoxins in fishponds and their effects on fish tissue. *Harmful Algae* **2016**, *55*, 66–76. [CrossRef]

42. Havens, K.E. Cyanobacteria blooms: Effects on aquatic ecosystems. *Adv. Exp. Med. Biol.* **2008**, *619*, 733–747.

43. Oberemm, A.; Becker, J.; Codd, G.; Steinberg, C. Effects of cyanobacterial toxins and aqueous crude extracts on the development of fish and amphibians. *Environ. Toxicol.* **1999**, *14*, 77–88. [CrossRef]

44. Liu, Y.; Song, L.; Li, X.; Liu, T. The toxic effects of microcystin-LR on embryo-larval and juvenile development of loach *Misguruns mizolepis* Gunthe. *Toxicon* **2002**, *40*, 395–399. [CrossRef]

45. Jacquet, C.; Thermes, V.; de Luze, A.; Puiseux-Dao, S.; Bernard, C.; Joly, J.-S.; Bourrat, F.; Edery, M. Effects of MC-LR on development of medaka fish embryos (*Oryzias latipes*). *Toxicon* **2004**, *43*, 141–147. [CrossRef]

46. Drobac, D. *Cyanotoxin Exposure and Human Health*; Andrejević Endowment: Belgrade, Serbia, 2018; p. 93.

47. MacKintosh, C.; Beattie, K.A.; Klumpp, S.; Cohen, P.; Codd, G.A. Cyanobacterial microcystin-LR is a potent and specific inhibitor of protein phosphatases 1 and 2A from both mammals and higher plants. *FEBS Lett.* **1990**, *264*, 187–192. [CrossRef]

48. Williams, D.E.; Craig, M.; Dawe, S.C.; Kent, M.L.; Andersen, R.J.; Holmes, C.F.B. C-14-Labeled microcystin-LR administered to Atlantic salmon via intraperitoneal injection provides in vivo evidence for covalent binding of microcystin-LR in salmon livers. *Toxicon* **1997**, *35*, 985–989. [CrossRef]

49. Williams, D.E.; Craig, M.; Dawe, S.C.; Kent, M.L.; Holmes, C.F.B.; Andersen, R.J. Evidence for a covalently bound form of microcystin-LR in salmon liver and Dungeness crab larvae. *Chem. Res. Toxicol.* **1997**, *10*, 463–469. [CrossRef]

50. Ibelings, B.W.; Bruning, K.; de Jonge, J.; Wolfstein, K.; Pires, L.D.M. Distribution of microcystins in a lake foodweb: No evidence for biomagnification. *Microb. Ecol.* **2005**, *49*, 487–500. [CrossRef]

51. Snyder, G.S.; Goodwin, A.E.; Freeman, D.W. Evidence that channel catfish, *Ictalurus punctatus* (Rafinesque), mortality is not linked to ingestion of the hepatotoxin microcystin-LR. *J. Fish Dis.* **2002**, *25*, 275–286. [CrossRef]

52. Xie, L.; Xie, P.; Ozawa, K.; Honma, T.; Yokoyama, A.; Park, H.-D. Dynamics of microcystins-LR and -RR in the phytoplanktivorous silver carp in a sub-chronic toxicity experiment. *Environ. Pollut.* **2004**, *127*, 431–439. [CrossRef]

53. Malbrouck, C.; Kestemont, P. Effects of microcystins on fish. *Environ. Toxicol. Chem.* **2006**, *25*, 72–86. [CrossRef]

54. Rodger, H.D.; Turnbull, T.; Edwards, C.; Codd, G.A. Cyanobacterial (blue-green algal) bloom associated pathology in brown trout, *Salmo trutta* L.; in Loch Leven, Scotland. *J. Fish Dis.* **1994**, *17*, 177–181. [CrossRef]

55. Carbis, C.R.; Rawlin, G.T.; Mitchell, G.F.; Anderson, J.W.; McCauley, I. The histopathology of carp, *Cyprinus carpio* L., exposed to microcystins by gavage, immersion and intraperitoneal administration. *J. Fish Dis.* **1996**, *19*, 199–207. [CrossRef]

56. Fischer, W.J.; Dietrich, D.R. Pathological and biochemical characterization of microcystin-induced hepatopancreas and kidney damage in carp (*Cyprinus carpio*). *Toxicol. Appl. Pharm.* **2000**, *164*, 73–81. [CrossRef]

57. Jiang, J.L.; Gu, X.Y.; Song, R.; Zhang, Q.; Geng, J.J.; Wang, X.Y.; Yang, L.Y. Time-dependent oxidative stress and histopathological alterations in *Cyprinus carpio* L. exposed to microcystin-LR. *Ecotoxology* **2011**, *20*, 1000–1009. [CrossRef]

58. Li, L.; Xie, P.; Li, S.; Qiu, T.; Guo, L. Sequential ultrastructural and biochemical changes induced in vivo by the hepatotoxic microcystins in liver of the phytoplanktivorous silver carp *Hypophthalmichthys molitrix*. *Comp. Biochem. Physiol. Part C Toxicol. Pharmacol.* **2007**, *146*, 357–367. [CrossRef]

59. Lecoz, N.; Malécot, M.; Quiblier, C.; Puiseux-Dao, S.; Bernard, C.; Crespeau, F.; Edery, M. Effects of cyanobacterial crude extracts from Planktothrix agardhii on embryo–larval development of medaka fish, *Oryzias latipes*. *Toxicon* **2008**, *51*, 262–269. [CrossRef]

60. Atencio, L.; Moreno, I.; Jos, A.; Pichardo, S.; Moyano, R.; Blanco, A.; Cameán, A.M. Dose-dependent antioxidant responses and pathological changes in tenca (*Tinca tinca*) after acute oral exposure to *Microcystis* under laboratory conditions. *Toxicon* **2008**, *52*, 1–12. [CrossRef]

61. Hooser, S.B.; Beasley, V.R.; Basgall, E.J.; Carmichael, W.W.; Haschek, W.M. Microcystin-LR-induced ultrastructural changes in rats. *Vet. Pathol.* **1990**, *27*, 9–15. [CrossRef]

62. Gupta, N.; Pant, S.C.; Vijayaraghavan, R.; Lakshmana Rao, P.V. Comparative toxicity evaluation of cyanobacterial cyclic peptide toxin microcystin variants (LR, RR, YR) in mice. *Toxicology* **2003**, *188*, 285–296. [CrossRef]

63. Sedan, D.; Laguens, M.; Copparoni, G.; Aranda, J.O.; Gianuzzi, L.; Marra, C.A.; Andrinolo, D. Hepatic and intestine alterations in mice after prolonged exposure to low oral doses of microcystin-LR. *Toxicon* **2015**, *104*, 26–33. [CrossRef]

64. Honkanen, R.E.; Zwiller, J.E.M.R.; Moore, R.E.; Daily, S.L.; Khatra, B.S.; Dukelow, M.; Boynton, A.L. Characterization of microcystin-LR, a potent inhibitor of type 1 and type 2A protein phosphatases. *J. Biol. Chem.* **1990**, *265*, 19401–19404.

65. Runnegar, M.; Berndt, N.; Kaplowitz, N. Microcystin uptake and inhibition of protein phosphatases: Effects of chemoprotectants and self-inhibition in relation to known hepatic transporters. *Toxicol. Appl. Pharm.* **1995**, *134*, 264–272. [CrossRef] [PubMed]

66. Xu, L.; Zhou, B.; Lam, P.K.S.; Chen, J.; Zhang, Y.; Harada, K.I. In vivo protein phosphatase 2A inhibition and glutathione reduction by MC-LR in grass carp (*Ctenopharyngodon idellus*). In Proceedings of the Ninth International Conference on Harmful Algal Blooms, Hobart, Australia, 7–11 February 2000; pp. 7–11.

67. Falconer, I.R.; Yeung, D.S. Cytoskeletal changes in hepatocytes induced by Microcystis toxins and their relation to hyperphosphorylation of cell proteins. *Chem-Biol. Interact.* **1992**, *81*, 181–196. [CrossRef]

68. Hiraga, A.; Tamura, S. Protein phosphatase 2A is associated in an inactive state with microtubules through 2A1-specific interaction with tubulin. *Biochem. J.* **2000**, *346*, 433–439. [CrossRef] [PubMed]

69. Runnegar, M.T.; Gerdes, R.G.; Falconer, I.R. The uptake of the cyanobacterial hepatotoxin microcystin by isolated rat hepatocytes. *Toxicon* **1991**, *29*, 43–51. [CrossRef]

70. Molina, R.; Moreno, I.; Pichardo, S.; Jos, A.; Moyano, R.; Monterde, J.G.; Cameán, A. Acid and alkaline phosphatase activities and pathological changes induced in Tilapia fish (*Oreochromis* sp.) exposed subchronically to microcystins from toxic cyanobacterial blooms under laboratory conditions. *Toxicon* **2005**, *46*, 725–735. [CrossRef]

71. Li, L.; Xie, P.; Lei, H.; Zhang, X. Renal accumulation and effects of intraperitoneal injection of extracted microcystins in omnivorous crucian carp (*Carassius auratus*). *Toxicon* **2013**, *70*, 62–69. [CrossRef]

72. Khoshnood, Z.; Jamili, S.; Khodabandeh, S. Histopathological effects of atrazine on gills of Caspian kutum *Rutilus frisii kutum* fingerlings. *Dis. Aquat. Organ.* **2015**, *113*, 227–234. [CrossRef]

73. Carbis, C.R.; Rawlin, G.T.; Grant, P.; Mitchell, G.F.; Anderson, J.W.; McCauley, I. A study of feral carp, Cyprinus carpio L.; exposed to Microcystis aeruginosa at Lake Mokoan, Australia, and possible implications for fish health. *J. Fish Dis.* **1997**, *20*, 81–91. [CrossRef]

74. Lujić, J.; Matavulj, M.; Poleksić, V.; Rašković, B.; Marinović, Z.; Kostić, D.; Miljanović, B. Gill reaction to pollutants from the Tamiš River in three freshwater fish species, *Esox lucius* L. 1758, *Sander lucioperca* L. 1758 and *Silurus glanis* L. 1758, A comparative study. *J. Vet. Med. Ser. C Anat. Histol. Embryol.* **2015**, *44*, 128–137. [CrossRef]

75. Agamy, E. Impact of laboratory exposure to light Arabian crude oil, dispersed oil and dispersant on the gills of the juvenile brown spotted grouper (*Epinephelus chlorostigma*): A histopathological study. *Mar. Environ. Res.* **2013**, *86*, 46–55. [CrossRef]

76. Gaete, V.; Canelo, E.; Lagos, N.; Zambrano, F. Inhibitory effects of Microcystis aeruginosa toxin on ion pumps of the gill of freshwater fish. *Toxicon* **1994**, *32*, 121–127. [CrossRef]

77. Bury, N.; Flik, G.; Eddy, F.; Codd, G. The effects of cyanobacteria and the cyanobacterial toxin microcystin-LR on Ca^{2+} transport and Na^+/K^+-ATPase in tilapia gills. *J. Exp. Biol.* **1996**, *199*, 1319–1326.

78. Jiraungkoorskul, W.; Upatham, E.S.; Kruatrachue, M.; Sahaphong, S.; Vichasri-Grams, S.; Pokethitiyook, P. Histopathological effects of Roundup, a glyphosate herbicide, on Nile tilapia (*Oreochromis niloticus*). *Sci. Asia* **2002**, *28*, 121–127. [CrossRef]

79. Cazenave, J.; Bistoni, M.A.; Pesce, S.F.; Wunderlin, D.A. Differential detoxification and antioxidant response in diverse organs of Corydoras paleatus experimentally exposed to microcystin-RR. *Aquat. Toxicol.* **2006**, *76*, 1–12. [CrossRef]

80. Prieto, A.I.; Pichardo, S.; Jos, A.; Moreno, I.; Cameán, A.M. Time-dependent oxidative stress responses after acute exposure to toxic cyanobacterial cells containing microcystins in tilapia fish (*Oreochromis niloticus*) under laboratory conditions. *Aquat. Toxicol.* **2007**, *84*, 337–345. [CrossRef]

81. Amado, L.L.; Monserrat, J.M. Oxidative stress generation by microcystins in aquatic animals: Why and how. *Environ. Int.* **2010**, *36*, 226–235. [CrossRef]

82. Hellou, J.; Ross, N.W.; Moon, T.W. Glutathione, glutathione S-transferase, and glutathione conjugates, complementary markers of oxidative stress in aquatic biota. *Environ. Sci. Pollut. R.* **2012**, *19*, 2007–2023. [CrossRef]

83. Chorus, I.; Falconer, I.R.; Salas, H.J.; Bartram, J. Health risks caused by freshwater cyanobacteria in recreational waters. *J. Toxicol. Environ. Heal. B* **2000**, *3*, 323–347.

84. Firkins, G.S. Toxic algae poisoning. *Iowa State Coll. Vet.* **1953**, *15*, 151–153.

85. Rose, E.F. Toxic algae in Iowa lakes. *Proc. Iowa Acad. Sci.* **1953**, *60*, 738–745.

86. Henriksen, P.; Carmichael, W.W.; An, J.; Moestrup, O. Detection of an anatoxin-a(s)-like anticholinesterase in natural blooms and cultures of cyanobacteria/ blue–green algae from Danish lakes and in the stomach contents of poisoned birds. *Toxicon* **1997**, *35*, 901–913. [CrossRef]

87. Onodera, H.; Oshima, Y.; Henriksen, P.; Yasumoto, T. Confirmation of anatoxin-a(s), in the cyanobacterium *Anabaena lemmermannii*, as the cause of bird kills in Danish lakes. *Toxicon* **1997**, *35*, 1645–1648. [CrossRef]

88. Matsunaga, H.; Harada, K.I.; Senma, M.; Ito, Y.; Yasuda, N.; Ushida, S.; Kimura, Y. Possible cause of unnatural mass death of wild birds in a pond in Nishinomiya, Japan: Sudden appearance of toxic cyanobacteria. *Nat. Toxins.* **1999**, *7*, 81–84. [CrossRef]

89. Codd, G.A.; Metcalf, J.S.; Morrison, L.F.; Krienitz, L.; Ballot, A.; Pflugmacher, S.; Wiegand, C.; Kotut, K. Susceptibility of flamingos to cyanobacterial toxins via feeding. *Vet. Rec.* **2003**, *152*, 722–723.

90. Krienitz, L.; Ballot, A.; Kotut, K.; Wiegand, C.; Pütz, S.; Metcalf, J.S.; Codd, G.A.; Pflugmacher, S. Contribution of hot spring cyanobacteria to the mysterious deaths of Lesser Flamingos at Lake Bogoria, Kenya. *FEMS Microbiol. Ecol.* **2003**, *43*, 141–148. [CrossRef]

91. Lugomela, C.; Pratap, H.B.; Mgaya, Y.D. Cyanobacteria blooms–a possible cause of mass mortality of Lesser Flamingos in Lake Manyara and Lake Big Momela, Tanzania. *Harmful Algae* **2006**, *5*, 534–541. [CrossRef]

92. Chen, J.; Zhang, D.; Xie, P.; Wang, Q.; Ma, Z. Simultaneous determination of microcystin contaminations in various vertebrates (fish, turtle, duck and water bird) from a large eutrophic Chinese lake, Lake Taihu, with toxic Microcystis blooms. *Sci. Total Environ.* **2009**, *407*, 3317–3322. [CrossRef]

93. Skočovská, B.; Hilscherova, K.; Babica, P.; Adamovský, O.; Bandouchová, H.; Horaková, J.; Knotková, Z.; Maršálek, B.; Pašková, V.; Pikula, J. Effects of cyanobacterial biomass on the Japanese quail. *Toxicon* **2007**, *49*, 793–803. [CrossRef]

94. Chen, J.; Xie, P.; Li, L.; Xu, J. First identification of the hepatotoxic microcystins in the serum of a chronically exposed human population together with indication of hepatocellular damage. *Toxicol. Sci.* **2009**, *108*, 81–89. [CrossRef]

95. Li, Y.; Chen, J.A.; Zhao, Q.; Pu, C.; Qiu, Z.; Zhang, R.; Weiqun, S. A cross-sectional investigation of chronic exposure to microcystin in relationship to childhood liver damage in the Three Gorges Reservoir Region, China. *Environ. Health Perspect.* **2011**, *119*, 1483–1488. [CrossRef]

96. Lin, H.; Liu, W.; Zeng, H.; Pu, C.; Zhang, R.; Qiu, Z.; Chen, J.A.; Wang, L.; Tan, Y.; Zheng, C.; et al. Determination of environmental exposure to microcystin and aflatoxin as a risk for renal function based on 5493 rural people in southwest China. *Environ. Sci. Technol.* **2016**, *50*, 5346–5356. [CrossRef] [PubMed]

97. Mitsch, W.J. Solving Lake Erie's harmful algal blooms by restoring the Great Black Swamp in Ohio. *Ecol. Eng.* **2017**, *108*, 406–413. [CrossRef]

98. John, J.; Kemp, A. Cyanobacterial blooms in the wetlands of the Perth region, taxonomy and distribution: An overview. *J. R. Soc. West. Aust.* **2006**, *89*, 51–56.

99. Pearl, H.W.; Gardner, W.S.; Havens, K.E.; Joyner, A.R.; McCarthy, M.J.; Newell, S.E.; Qin, B.; Scott, J.T. Mitigating cyanobacterial harmful algal blooms in aquatic ecosystems impacted by climate change and anthropogenic nutrients. *Harmful Algae* **2016**, *54*, 213–222. [CrossRef]

100. Scavia, D.; Allan, J.D.; Arend, K.K.; Bartell, S.; Beletsky, D.; Bosch, N.S.; Brandt, S.B.; Briland, R.D.; Daloglu, I.; DePinto, J.V.; et al. Assessing and addressing the reeutrophication of Lake Erie, Central basin hypoxia. *J. Gt. Lakes Res.* **2014**, *40*, 226–246. [CrossRef]

101. Dziga, D.; Tokodi, N.; Backović, D.D.; Kokociński, M.; Antosiak, A.; Puchalski, J.; Strzałka, W.; Madej, M.; Meriluoto, J.; Svirčev, Z. The effect of a combined hydrogen peroxide-MlrA treatment on the phytoplankton community and microcystin concentrations in a mesocosm experiment in Lake Ludoš. *Toxins* **2019**, *11*, 725. [CrossRef]

102. Dulić, S. Fitoplankton Kao Pokazatelj Eutrofizacije Ludaškog Jezera. Ph.D. Thesis, University of Novi Sad, Novi Sad, Serbia, 2002.

103. Seleši, Ð. *Voda Ludaškog Jezera*; Javno preduzeće Palić-Ludaš: Subotica, Serbia, 2006.

104. Radić, D.; Grujaničić, V.; Petričević, V.; Lalević, B.; Rudić, Z.; Božić, M. Macrophytes as remediation technology in improving Ludas lake sediment. *Fras. Environ. Bull.* **2013**, *22*, 1787–1791.

105. Zhang, L.; Thomas, S.; Mitsch, W.J. Design of real-time and long-term hydrologic and water quality wetland monitoring stations in South Florida, USA. *Ecol. Eng.* **2017**, *108*, 446–455. [CrossRef]

 *water*

Article

An Ecological Function Approach to Managing Harmful Cyanobacteria in Three Oregon Lakes: Beyond Water Quality Advisories and Total Maximum Daily Loads (TMDLs)

Eric S. Hall [1],*, Robert K. Hall [2], Joan L. Aron [3], Sherman Swanson [4], Michael J. Philbin [5], Robin J. Schafer [6], Tammy Jones-Lepp [7], Daniel T. Heggem [8], John Lin [8], Eric Wilson [9] and Howard Kahan [2]

[1] USEPA Office of Research and Development, NERL, Systems Exposure Division (SED), Ecological and Human Community Analysis Branch, Research Triangle Park, NC 27709, USA

[2] USEPA Region IX, WTR2, 75 Hawthorne St., San Francisco, CA 94105, USA; hall.robertk@epa.gov (R.K.H.); kahan.howard@epa.gov (H.K.)

[3] Aron Environmental Consulting, 5457 Marsh Hawk Way, Columbia, MD 21045, USA; joanaron@ymail.com

[4] Ecology, Evolution and Conservation Biology, University of Nevada, 1664 N. Virginia St., Reno, NV 89557, USA; sswanson@cabnr.unr.edu

[5] U.S. Dept. of the Interior Bureau of Land Management, Montana/Dakotas State Office, 5001 Southgate Drive, Billings, MT 59101, USA; mphilbin@blm.gov

[6] University of Puerto Rico, Río Piedras Campus, 14 Ave. Universidad, Ste. 1401, San Juan, PR 00925-2534, USA; robin.schafer@upr.edu

[7] USEPA Office of Research and Development, NERL, Exposure Methods and Measurement Division (EMMD), Environmental Chemistry Branch, Las Vegas, NV 89119, USA; tjoneslepp@gmail.com

[8] USEPA Office of Research and Development, NERL, Systems Exposure Division (SED), Ecosystem Integrity Branch, Las Vegas, NV 89119, USA; motoheggem2@gmail.com (D.T.H.); lin.john@epa.gov (J.L.)

[9] Gulf Coast STORET, LLC, 11110 Roundtable Dr., Tomball, TX 77375, USA; ericwilson@gulfcoaststoret.com

* Correspondence: hall.erics@epa.gov; Tel.: +1-919-541-3147

Received: 4 April 2019; Accepted: 24 May 2019; Published: 29 May 2019

Abstract: The Oregon Department of Environmental Quality (ODEQ) uses Total Maximum Daily Load (TMDL) calculations, and the associated regulatory process, to manage harmful cyanobacterial blooms (CyanoHABs) attributable to non-point source (NPS) pollution. TMDLs are based on response (lagging) indicators (e.g., measurable quantities of NPS (nutrients: nitrogen {N} and phosphorus {P}), and/or sediment), and highlight the negative outcomes (symptoms) of impaired water quality. These response indicators belatedly address water quality issues, if the cause is impaired riparian functions. Riparian functions assist in decreasing the impacts of droughts and floods (through sequestration of nutrients and excess sediment), allow water to remain on the land surface, improve aquatic habitats, improve water quality, and provide a focus for monitoring and adaptive management. To manage water quality, the focus must be on the drivers (leading indicators) of the causative mechanisms, such as loss of ecological functions. Success in NPS pollution control, and maintaining healthy aquatic habitats, often depends on land management/land use approaches, which facilitate the natural recovery of stream and wetland riparian functions. Focusing on the drivers of ecosystem functions (e.g., vegetation, hydrology, soil, and landform), instead of individual mandated response indicators, using the proper functioning condition (PFC) approach, as a best management practice (BMP), in conjunction with other tools and management strategies, can lead to pro-active policies and approaches, which support positive change in an ecosystem or watershed, and in water quality improvement.

Keywords: cyanobacteria; ecological function; ecosystems; harmful cyanobacterial bloom (CyanoHAB); proper functioning condition (PFC); total maximum daily load (TMDL); non-point source (NPS); point source (PS); Oregon Department of Environmental Quality (ODEQ); best management practice (BMP)

1. Introduction

Watersheds are complex ecosystems [1]. Nature is not static, but adjusts and adapts to natural climatic and anthropogenic stresses [2,3]. In aquatic environments, not all water pollution is from an external input. Pollution can come from the materials stored in riparian areas, and wetlands, due to their attributes, physical processes, and functions [4]. When an abundance of nutrients, increased warmth, higher salinity, and light are available, harmful cyanobacterial blooms (CyanoHABs) can occur [5]. Regulating water pollution is a key U.S. Clean Water Act (CWA) tool [6]. Point source (PS) pollution control strategies were very effective during the early decades of the United States Environmental Protection Agency (USEPA), when success was attributed to regulatory control structures capable of "breaking" the pathway of point source (PS) contamination to an endpoint cohort (i.e., humans, fish, etc.), and attaining water quality standards [6].

As the emphasis for pollution control moved from point sources to non-point sources, success has been sporadic [7], because a singular solution, or control structure(s), are not useful approaches for stemming non-point source (NPS) pollution. The Water Resources Development Act of 2007 (Available online: https://www.govinfo.gov/content/pkg/PLAW-110publ114/html/PLAW-110publ114.htm; accessed: 28 May 2019), specifies that federal water resources investments shall, "reflect national priorities, protect the environment, maximize sustainable economic development, avoid excessive use of floodplains and flood-prone areas, and protect and restore the functions of natural systems". For PSs, the CWA provided a framework (Sections 303(d), 305(b), 319) for identifying best management practices (BMPs) needed to reach required water quality standards. Once a waterbody is designated as impaired, per CWA Section 303(d), a total maximum daily load (TMDL) source assessment, load allocation, and in most states an implementation plan document is required. Creating a non-point source TMDL is often challenging, because a plan requires feasible, and accurate, accounting of allowable non-point source pollutant loadings and regulatory control methods. Leading indicators measure the performance of current and future ecological conditions that drive pollution [8,9]. The lack of success in addressing NPS pollution is partly because water quality, and aquatic organism measures, are typically lagging indicators [8–10].

A lagging indicator may eventually respond (e.g., excess sediment, nitrogen (N), and phosphorus (P)), but not soon enough to guide decisions needed to ensure progress [10–12]. Land cover and land use, in non-riparian watershed areas, can drive erosion or release and transport pollutants, into an expanded network of ditches and drains. Watershed vegetation is critical for slowing down hydraulic forces, which are magnified through peak hydrology, and efficient channel forms, and greatly influences the movement of eutrophication nutrients, such as nitrogen and phosphorus. Drivers (leading indicators) of physical function identify early interventions to prevent the loss of assimilation processes, and water quality deterioration. The objective of this study was to show that interdisciplinary, qualitative assessments of riparian and watershed function, and biophysical alterations at a local scale, such as the proper functioning condition (PFC) approach, can assist resource managers in prioritizing adaptive management practices (i.e., objective implementation, monitoring, etc.) [13–15].

1.1. Ecosystem Function

Bernhardt et al. [16], noted that most of the billions of dollars [17,18] spent to enhance water quality and improve in-stream habitat lacks the appropriate data and information to evaluate the ecological effectiveness of restoration activities. Therefore, it is essential to monitor vegetation, hydrology, soils, landform, and water quality parameters, as part of an ecological restoration project, to collect data/information essential for State and Tribal environmental managers to prioritize areas of focus and concern [19]. Ecosystem function is defined as a state of resiliency, which allows a riparian wetland system to remain intact during a 25 to 30-year flood event, and sustains an ecosystem's ability to produce ecosystem values related to physical and biological attributes [20,21]. Water quality and biological communities are affected if a stream's riparian function is impaired. A stream's shape evolves over time, in response to flow and sediment loads [22]. Increased volume and intensity of runoff, or improper watershed or riparian management, often leads to accelerated stream erosion, which may result in streams becoming straighter, steeper, and deeper through incision, or wider and shallower, through bank erosion and aggradation [23]. These reduce pool habitat, and fish cover, and destroy riparian vegetation [14,20].

Proper functioning condition (PFC) refers to how well the physical processes within a stream and wetland riparian area can sustain a state of resiliency [14,15,20,21]. This resiliency allows an area to provide desired and valued ecosystem services (e.g., fish habitat, livestock, and/or wildlife forage, water purification, carbon storage, and nutrient cycling) over time [21]. Functional ecosystems are resilient to disturbances (e.g., floods), in contrast to nonfunctional ecosystems, which fail to buffer against surges in flow from upland and upstream inputs. This is because ecosystems are interconnected communities of vegetation, hydrology, soils, landform, and micro-organisms linked by physical and chemical interactions [14]. For example, a change in vegetation impacts a lentic (still, fresh water) or lotic (rapidly-moving, fresh water) ecosystem's physical functions (assimilative capacity and bank stability). If vegetation deteriorates, and the system is not functioning properly, water quality will degrade. When the ecosystem is functioning, and moving in a positive direction, water quality will improve [9,10,20]. Therefore, it is important for a riparian system to have not only vegetation, but it must have the right kind of obligate stabilizing vegetation, appropriate for the setting [24], to protect banks [15,25,26].

1.2. Harmful Cyanobacterial Blooms

While cyanobacteria are one of the most primitive and pervasive forms of life in the aquatic environment, the over stimulation, and proliferation in eutrophying fresh waters, of cyanobacterial species capable of producing powerful toxins, can be detrimental to water bodies. Some of the negative impacts are: economic—from the loss of income due to the loss of recreational use, and loss of fishing resources; human health—from exposure to cyanobacterial toxins via drinking water, and; ecological—from hypoxia and fish kills, cattle and other domestic animal poisonings, from lakes and ponds that are afflicted with CyanoHABs [27]. Health-threatening cyanobacterial toxins are caused by a variety of species of cyanobacteria, with a lot of attention being focused on *Microcystis* species, which release the group of toxins known as microcystins [28]. Besides *Microcystis*, other species of CyanoHABs are rising to predominance in both the US, and globally, for example, the filamentous cyanobacteria *Dolichospermum planctonicum* (*syn: Anabaena plactonica*, [29]) and *Aphanizomenon flos-aquae* [30–32].

1.3. Oregon Department of Environmental Quality's Use of TMDLs to Manage CyanoHABs

The northwestern U.S. state of Oregon's Department of Environmental Quality (ODEQ) uses the TMDL process to manage CyanoHABs. A TMDL calculates the maximum amount of a pollutant that a waterbody can receive and still safely meet water quality standards. Harmful cyanobacteria in Oregon lakes are more strongly correlated with variation in total phosphorus, than whether the waterbody has the right amount of nitrogen relative to the amount of phosphorus [33]. TMDLs were designed to reduce sediment, nitrogen, and phosphorus. Because of the special nitrogen fixing properties of some species of cyanobacteria (e.g., *Dolichospermum planctonicum* (*syn: Anabaena plactonica*) and *Aphanizomenon* [29,32]), the reduction of phosphorus is the most effective means of reducing CyanoHABs [33–35]. TMDL levels were set individually for each targeted waterbody using information from the Atlas of Oregon Lakes [35], and benchmark water quality parameters established by USEPA. ODEQ combined its TMDL based strategy, with a solid surveillance and monitoring program for CyanoHABs [36], water quality and drinking water quality standards, waste water permitting, and some targeted educational outreach, including health advisories for CyanoHABs (Figure 1).

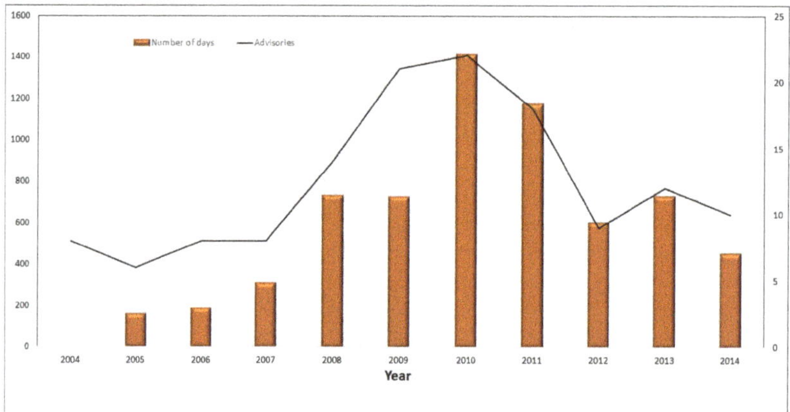

Figure 1. Oregon Department of Environmental Quality (ODEQ)'s number of harmful cyanobacterial blooms (CyanoHAB) waterbody advisories (line), and number of cumulative days per year (bars) under an advisory. (Available online: http://www.oregon.gov/oha/ph/HealthyEnvironments/Recreation/HarmfulAlgaeBlooms/Pages/index.aspx; accessed: 28 May 2019). <u>Note</u>: There were no CyanoHAB advisory days in Oregon for 2004.

As seen in Figure 1, ODEQ has had difficulty in reducing the number of advisories, and the number of waterbody days under an advisory, to below 2005 levels (when warnings were initiated). As a result, ODEQ used its regulatory authority to develop an early warning system to post health advisories, informing the public of the danger of increased exposure to CyanoHABs, and to "break" the exposure pathway (prevent contact with CyanoHABs in water and/or wildlife exposed to CyanoHABs). ODEQ and the USEPA marked the early warning system as a sign of success in solving immediate health problems. However, these existing protocols miss an opportunity to implement a long-term, sustainable solution to reduce the number of advisories. A sustainable solution is outlined for the Oregon Tenmile Lakes TMDL [37], which states that "improving watershed stream and wetland riparian functions was important in curtailing sediment and nutrient loads". This study recommends using the ODEQ TMDL approach, in conjunction with, time-series photography, CyanoHAB advisories, and the PFC methodology, as a preferred (best) practice, instead of using TMDLs without complementary approaches. This study illustrates the use of the proper functioning condition (PFC) methodology in assessing the state of the ecosystem of three Oregon lakes, and in developing a resource management plan that can enhance those ecosystems.

2. Methods

Time-series, remote-sensing data from Google Earth was used to evaluate the impact of different land use policies on the region, due to the consistency of coverage within the study area (the Tenmile Lakes (natural lake), Lemolo Lake (impoundment), and Diamond Lake (natural lake)). ODEQ has a well-developed surveillance and monitoring program [36], in coordination with educational outreach, waste water permitting, water quality and drinking water quality standards, and health advisories for CyanoHABs, linked with its TMDL approach. This was augmented with a proper functioning condition (PFC) stream and wetland riparian area assessment protocol [15,20], conforming to the United States Department of Agriculture's (USDA's) United States Forest Service (USFS) process.

PFC is a qualitative assessment process, which is implemented by a multidisciplinary/interdisciplinary team (ID team) [14,15,20,38]. The PFC assessment framework is a consistent, qualitative, science-based approach for considering stream and wetland hydrologic, vegetative, and geomorphic attributes and processes, at a point in time [15]. PFC is also an appropriate starting point for determining, prioritizing, and collecting information about riparian resources, developing monitoring needs [37], and providing context for quantitative data. An ID team must understand stream dynamics and its potential natural condition (PNC: i.e., the highest attainable ecological status of a riparian area without consideration of economic, political, or social constraints), and use their professional experience and judgment to accurately complete a qualitative assessment [38]. Use of quantitative data and techniques, i.e., field measurements [25,39] or remote sensing [20], is encouraged for individual or team calibration, or where opinions may differ [15].

PFC is used to describe both "an assessment process, and a defined, on-the-ground, condition of a riparian-wetland area". A PFC rating, having 17 attributes for a lotic ecosystem [15], and 20 attributes for a lentic ecosystem [20], relates how well the physical stream/riparian processes are functioning in comparison to its PNC. By focusing on physical functioning (hydrology, vegetation, soil, and landform attributes) the PFC protocol is designed to yield information about the biology of the plants and animals dependent on the riparian-wetland area [14,20].

After a PFC assessment is completed, it provides a rating for the "riparian-wetland area" of either "proper functioning condition", "functional-at-risk", "apparent trend", or "non-functional". Stream and wetland riparian areas in "proper functioning condition" (PFC) status sustain a state of resiliency appropriate for local ecological potential, and provide ecosystem services (e.g., wildlife and fish habitats, diminished flood impacts, and improved water quality). PFC refers to how well stream and wetland riparian physical processes can sustain a state of resiliency after a natural or anthropogenic disturbance [14,15,20,21]. A resilient, properly functioning, ecosystem can assimilate stressors and produce values related to both physical and biological attributes.

The "functional-at-risk" rating refers to a riparian area that is functioning, but with an existing soil, water, or vegetation attribute making it susceptible to degradation. The "apparent trend" rating is an assessment of the direction of change (e.g., upward or downward) in condition, either towards or away from the PNC or the PFC [14,20]. The apparent trend is determined by comparing the present condition with previous photos, trend studies, inventories, other documentation, or personal knowledge. The "non-functional" rating indicates the stream and wetland riparian area are in a degraded state.

2.1. Study Area

Streams differ in their potential to produce habitats, biota, and water quality for beneficial uses (i.e., swimmable, fishable, drinkable, etc.). Water quality standards (WQS) determine the allowable concentration of pollutant loads within the waterbody. The question is how to reduce pollutant loads, when many streams are themselves the source of sediment and/or nutrients, and this points to the importance of having a water quality management plan, as was developed for the Tenmile Lakes watershed [40], and Diamond Lake and Lemolo Lake [41], to mitigate those pollutant sources. The three lakes in Oregon that were studied are shown in Figure 2, and their general characteristics are provided in Table 1.

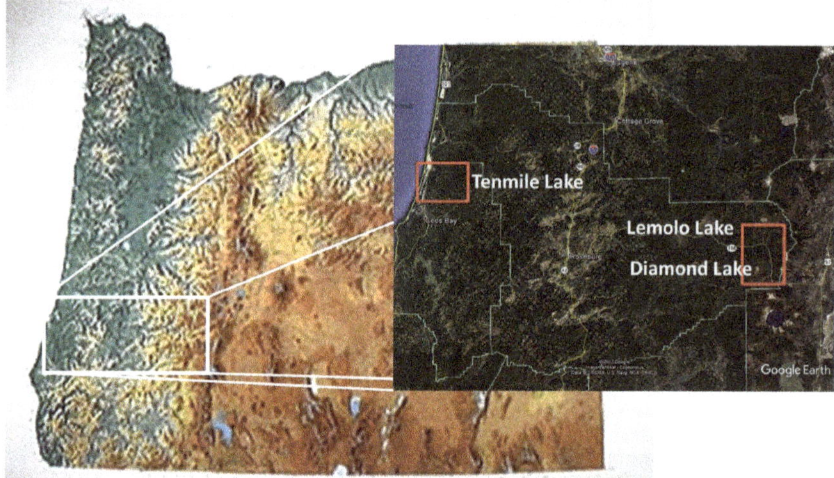

Figure 2. Map of the Lemolo Lake, Diamond Lake, and the Tenmile Lakes study areas. The larger image is a Landsat image from 15 July 2014. The inset image is from Google Earth.

Table 1. General characteristics of Diamond Lake (North Umpqua Watershed), Lemolo Lake (North Umpqua Watershed), and Tenmile Lakes (Tenmile Creek Watershed) in Oregon. Source: Reference [42].

Lake	Area (Acres)	Maximum Depth (Feet)	Average Depth (Feet)	Shoreline Length (Miles)
Diamond	3214	52	24	9
Lemolo	450	100	30	7.9
Tenmile	1627	22	10	22.9

2.1.1. Oregon Southwestern Coast and Range: Diamond Lake and Lemolo Lake

The North Umpqua River is a tributary to the Umpqua River, which headwaters are in the low-relief province of the Cascade Mountains of southwestern Oregon (Figure 2). This portion of the Cascades is underlain by highly permeable Pliocene and Quaternary lava flows, which have low rates of surface-water runoff and sediment transport [43]. Forest management prior to 1970 predominantly consisted of extensive "clear cutting" [44]. After 1970, the forest management approach changed to smaller patchwork "clear cuts" (Figure 3). The United States Forest Service (USFS) ecosystem function assessment indicates that stream channel stability in the managed areas of the North Umpqua River watershed did not significantly differ from those in unmanaged drainages [45]. The USFS determined these streams were relatively stable, and in proper functioning condition [45]. Around 2005, the forest management strategy was changed to "forest thinning", as seen around Lemolo Lake (Figure 3A,B).

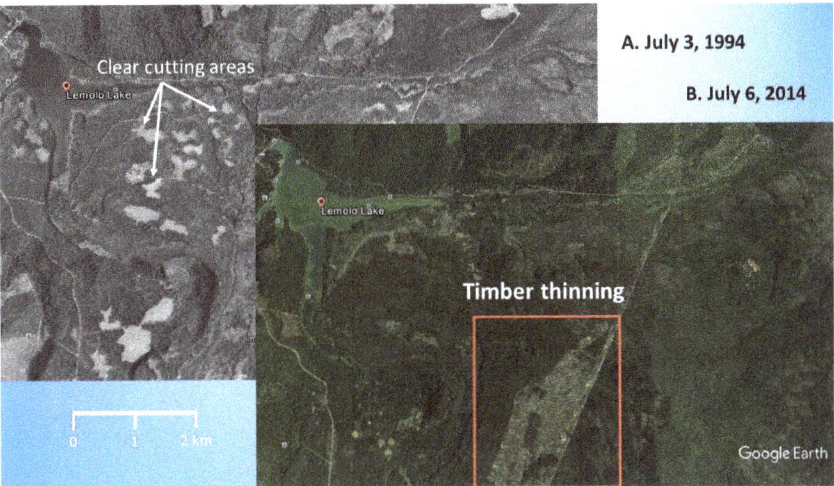

Figure 3. United States Department of Agriculture (USDA) black and white, and color aerial images, of the Lemolo Lake area, Oregon. Images were taken: (**A**) 3 July 1994 (black and white), and (**B**) 6 July 2014 (color). Bright patches in image A indicate areas of clear cutting. The red boundary in image B is an area currently being thinned. Source: Images obtained from Google Earth.

Phosphorus is associated with the decomposition of the volcanic rock of the Cascade Mountains, which have been "clear cut" (Figure 3A) [42]. Remote sensing analysis indicated there was a change in forest management from "clear cutting" (Figure 3A), prior to 2005, to "forest thinning" (Figure 3B).

Diamond Lake (Figure 4), a nitrogen-limited eutrophic lake, peaking at 0.5 mg/L total nitrogen (N), and nearly 0.2 mg/L ammonium-N in 2007 [41], is an important source of water and nutrients to the North Umpqua River [46]. Diamond Lake is listed as a "water-quality limited" waterbody (pH, CyanoHABs). Diamond Lake is naturally high in phosphorus, as high as 0.060 mg/L [41]. CyanoHABs have occurred in Diamond Lake since 2001 [46], resulting in closures of recreational water facilities. Increased CyanoHABs have been attributed to increased populations of the invasive, zooplankton-eating, Tui Chub fish [47]. Tui Chub feed on zooplankton that would otherwise control phytoplankton and CyanoHABs. The Final Environmental Impact Statement (FEIS) for the Diamond Lake Restoration Project advocated reducing Tui Chub fish biomass in Diamond Lake (natural lake) as the preferred restoration alternative (Available online: http://www.dfw.state.or.us/fish/local_fisheries/ diamond_lake/restoration-update.asp; accessed: 28 May 2019). However, dam releases from Diamond Lake, beginning in 2006 resulted in nutrients impacting the summer trophic status of the downstream portion of the North Umpqua River, and Lemolo Lake (impoundment) (Figure 5).

Figure 4. Satellite image of Diamond Lake, Oregon. Image taken 6 July 2014. Source: Image obtained from Google Earth.

Figure 5. USDA color aerial images of Lemolo Lake, Oregon. Images were taken: (**A**) 11 August 2012 and, (**B**) July 2014. Bright green areas within Lemolo Lake (image B) indicates an extensive cyanobacterial bloom occurring at the time the image was taken. Source: Images obtained from Google Earth.

Lemolo Lake, which is the upper-most impoundment on the North Umpqua River, is downstream from Diamond Lake (Figure 5). Hydrodynamic modeling [46,48] of the reservoir indicated an unusual mixing pattern, created by two stream inlets, (Diamond Lake and North Umpqua River), with greatly differing summer temperatures and nutrient loads (i.e., warmer waters from Diamond Lake, and colder phosphorus-enriched water from the North Umpqua River [48]). As seen in Figure 6, the number of days Lemolo Lake was under a CyanoHAB health advisory increased from 2006 to 2008, during and after the implementation of the USFS Diamond Lake restoration plan to eradicate the Tui Chub in 2006 [47].

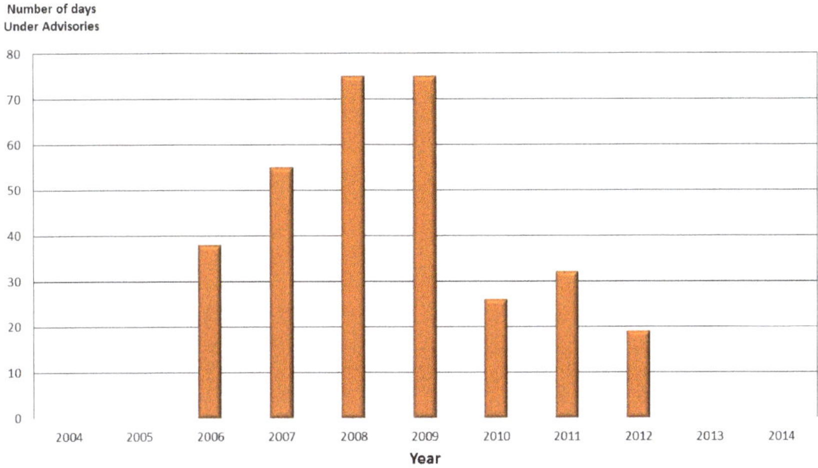

Figure 6. Oregon Department of Environmental Quality's (ODEQ) harmful cyanobacterial bloom (CyanoHAB) advisories for Lemolo Lake, and the number of days under an advisory. Current (post-2014) data has not been posted on the website: (Available online: http://www.oregon.gov/oha/ph/ HealthyEnvironments/Recreation/HarmfulAlgaeBlooms/Pages/index.aspx; accessed on 28 May 2019). <u>Note</u>: There were no CyanoHAB advisory days for Lemolo Lake in 2004, 2005, 2013, and 2014.

2.1.2. The Tenmile Lakes

The cause of the CyanoHABs in the Tenmile Lakes is similar to Lemolo Lake. The dominant species of cyanobacteria in the Tenmile Lakes are *Microcystis aeruginosa, Aphanizomenon flos-aquae*, and *Dolichospermum planctonicum (syn: Anabaena planctonica)* [29]. The Tenmile Lakes, consisting of North Tenmile and Tenmile Lake, are located on the south-central Oregon Coast (Figure 2). The lakes are highly productive coastal fisheries. However, salmonid populations have been declining [40]. Land use in the Tenmile Lakes watershed is predominantly urban along the shoreline of the lakes, agriculture (hay and grazing) in the flood plains of low gradient stream reaches, and logging in the upper watershed. The urban water management approach for the Tenmile Lakes is outlined in [49], which is accomplished through developing improvements to existing urban runoff control structures, and through the restriction of higher-density urban development. The Tenmile Lakes serve as the primary drinking water supply for lakeshore residents. Since the mid-1980s, the drinking water supply has been infused with high populations of cyanobacteria, *Microcystis aeruginosa*, and listed on the Oregon 303(d) list of impaired surface waters [40]. In 1997, Tenmile Lake was temporarily closed as a source of potable water. From 1997 to 2006, the cyanobacteria and toxin levels triggered five health advisories (Figure 7) related to lake water consumption (drinking water) and/or recreational contact with lake waters [40]. Oregon has continued to maintain a CyanoHAB Surveillance Program, and issues CyanoHAB health advisory guidelines on a regular basis [36].

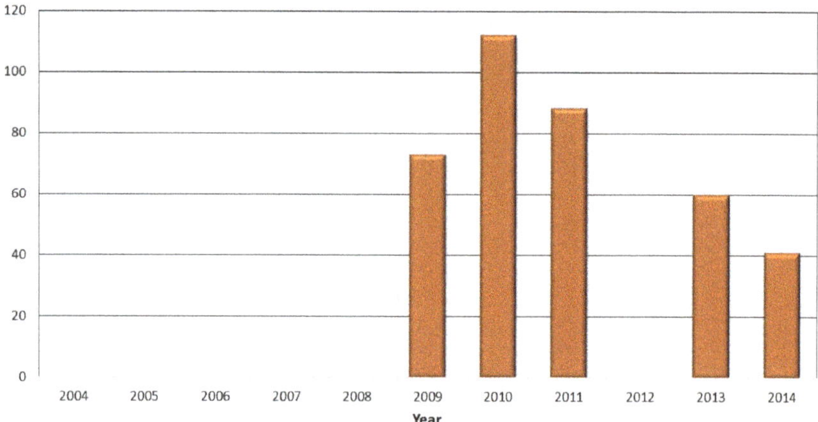

Figure 7. Oregon Department of Environmental Quality's (ODEQ) harmful cyanobacterial bloom (CyanoHAB) advisories for Tenmile Lakes and the number of days under an advisory. <u>Note</u>: There were no CyanoHAB advisory days for Tenmile Lakes in 2004, 2005, 2006, 2007, 2008, and 2012.

ODEQ [40] points out that water quality factors affecting weed and cyanobacterial growth occur in the presence of excessive nutrients and sediment. ODEQ [40] describes sediment lake core samples as having increased sediment accrual rates (above natural conditions). The continued water quality and fisheries problems in the Tenmile Lakes have prompted resource management agencies and the City of Lakeside to address the issues.

ODEQ maintains a list of total maximum daily loads (TMDLs) in Oregon approved by The United States Environmental Protection Agency ((EPA)—Available online: https://www.oregon.gov/deq/wq/tmdls/Pages/TMDLs-Approved-by-EPA.aspx; accessed: 28 May 2019). Sediment load is also linked to CyanoHABs, as phosphorus has an affinity for binding with fine particulate matter. Therefore, reducing nitrogen and phosphorus (i.e., sediment) delivery to the Tenmile Lakes is essential. The Tenmile Lakes sediment core samples [42] indicate increased sediment accrual rates began shortly after homesteading began in the area in the mid-to-late 1800s. ODEQ [42] describes sediment accumulation rate increases, with logging and clearing of farm and ranch land, in the 1800s. Logging, railroad, and road construction, occurred from 1910–1920, prior to riparian and water protection measures [42]. From the 1920s into the early 1940s, there was continued wetland-to-agriculture conversion, and timber harvesting. Post-World War II, 1945–1955, saw a surplus of wood products, resulting in a slowdown of timber harvesting. During this period, the ecosystem had a chance to recover. From 1955 to 1999, favorable economics increased timber harvesting activity, and accelerated urban and residential lakeshore development around the Tenmile Lakes, with the drainage of wetlands [42]. A plausible outcome of this increased engineered activity would be increased runoff and sediment load from the logged areas and roads. Adding to this scenario, as seen in Figures 8 and 9, Roberts Creek was moved to the south side of the valley from its original position in the center (<u>Note</u>: the dark spectral signature in center of the valley). Stream channel instability, and incision, could have occurred from altering the channel on the downslope side (south) of the flood plain to capture flood irrigation tail-water. The internal (sediment) loading of nutrients in the three lakes in the study area is greater than the nutrient loading received from watershed inputs [48]. Meeting the required internal loading levels led to the decision to eliminate 90–100% of the Tui Chub population, along with increased monitoring of the trout population to determine if a reduction in the trout population was warranted. The internal loading associated with the Tui Chub was estimated as being four times that of external loading.

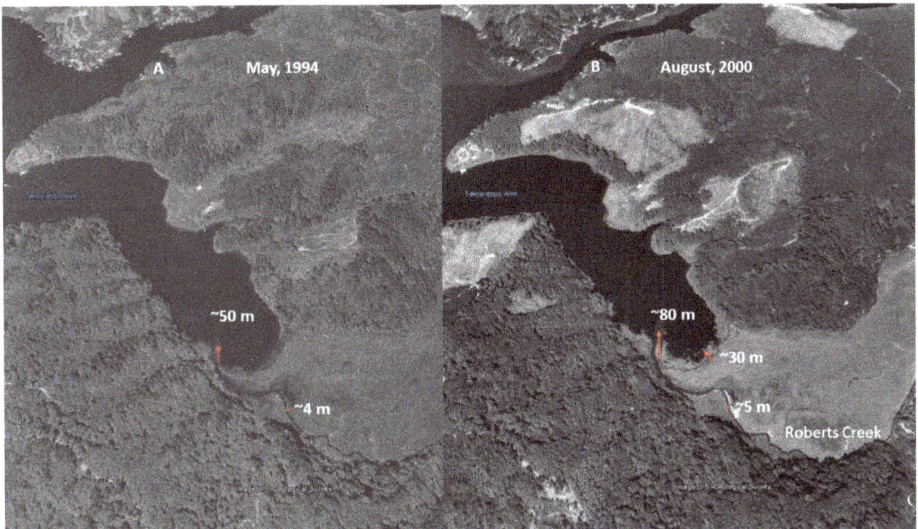

Figure 8. US Geological Survey aerial images of Tenmile Lake, Templeton Arm, and Roberts Creek. (**A**): The May 1994, black and white aerial image shows the Roberts Creek channel is incised, and the delta is expanding. (**B**): The August 2000, black and white aerial image shows the Roberts Creek channel has expanded (<u>Note</u>: the white area indicates stream bank slumping), and the delta has extended into the lake approximately 30 m farther than the May 1994 image. It also shows increased clear cuts on the slopes in three areas north of (above) the lake. Source: Images obtained from Google Earth. <u>Note</u>: Red arrows show the extent of delta growth.

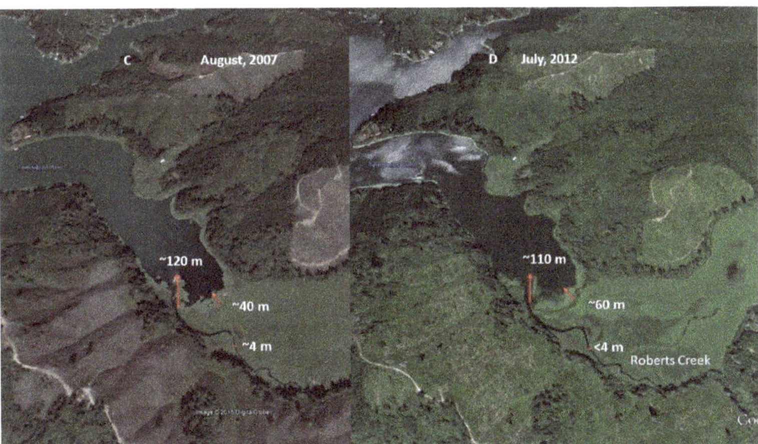

Figure 9. US Geological Survey color aerial images of Tenmile Lake, Templeton Arm, and Roberts Creek. (**C**): The August 2007, aerial image shows the Roberts Creek channel is incised, and the delta is expanding. (**D**): The July 2012, aerial image shows the Roberts Creek channel is unchanged, and the delta has been reduced by approximately 10 m from the August 2007 image. The August 2007 image also shows increased clear cuts on the slopes in three areas south of (below) the lake. Source: Images obtained from Google Earth. <u>Note</u>: Red arrows show the extent of delta growth.

Flood irrigation of the meadow required building a diversion dam, and a channel, to flood the upslope side of the field (Figures 8 and 9—north side), consequently incising Roberts Creek, for the irrigation tail-water to run into (Figures 8 and 9—south side). Lowering the base of the channel caused incision, with a nick point or nick zone, to migrate up the river and tributaries, thus disconnecting the stream from its flood plain. Channelization, and the incising of Roberts Creek, funnels the flow within the channel, resulting in an increased peak flow, causing further incision and bank erosion. Incision, with decreased floodplain access, greatly diminishes sediment and nutrient capture during floods, and greatly diminishes nutrient uptake by riparian vegetation, and the denitrification process [10].

The increased sediment load and dominance of invasive fish species (e.g., Tui Chub, bass) in the Tenmile Lakes have vastly altered the historic condition of the coastal lake ecosystem [42]. These factors indicate changes in the functional condition (i.e., physical processes) of the lakes and watershed, resulting in decreased water quality, loss of salmonid habitat, altered aquatic food chain dynamics, and associated increased CyanoHABs (Figures 8 and 9).

The nutrient report [50], and the TMDL study [37], discusses the importance of restoring and improving stream and wetland riparian functions. Nevertheless, there is no mention of improvement in ecosystem functions. To improve stream and wetland riparian functions, the focus should be on restoring the hydrological connection of the wetlands, to filter and store sediment, prior to entering the lake. The proper ecological, hydrological, and geomorphological functioning of lake and river catchments is important, and including this in the total catchment management activities to protect water quality and reduce CyanoHABs is essential.

3. Results: Ecosystem Function and Best Management Practices (BMPs)

Maintaining viable, economically sustainable, forest or agricultural land, is a key element in any federal, state, or local water quality protection program. Implementing a forest, or grazing management plan, and measures to minimize water quality impairment due to forest harvesting and associated activities, is the responsibility of the owners of the land [51]. Therefore, land management measures to reduce upland (i.e., forest, agricultural) NPS runoff should focus on restoration of ecosystem physical functions [9]. As noted in Reference [12], determining the physical processes driving upland and stream and wetland riparian ecological functions enables stakeholders to identify sources and severity of impairment to predict, and avoid, a decline in water quality.

Ecosystem function is notably distinct from other water quality management tools [51]. As seen in Figure 9D, it is difficult to determine which best management practices (BMPs) were applied to the logged area (e.g., replanting, road removal). Because of this, other land managers (private, public) require extra steps to determine if the BMPs used could be replicated on their site. To improve BMP sharing, the USEPA maintains a database that tracks restoration projects using the Clean Water Act Section 319 (CWA 319) NPS project tracking system called the Grants Reporting and Tracking System (GRTS—Available online: https://iaspub.epa.gov/apex/grts/f?p=grts:95; accessed: 28 May 2019). GRTS is the central oversight, management, and data-sharing tool for the Clean Water Act's Section 319 Program. All data in GRTS is entered by state, territory, or tribal agencies receiving Section 319 grants. Unfortunately, for the Tenmile Lakes, the GRTS CWA 319 tracking system did not have information on the BMPs used.

After implementation of the Tui Chub eradication program, the number of days Lemolo Lake was under an advisory were increased in 2007, peaking in 2008 and 2009. In 2010, the number of days under a health advisory was reduced by almost two-thirds (Figure 6). When Diamond Lake was drained, to lower the volume of the lake, and apply less Rotenone pesticide, to eradicate the Tui Chub [45], it had a downstream impact on Lemolo Lake [46,47]. Increased nutrient load into Lemolo Lake overwhelmed the aquatic ecosystem, causing eutrophication, and creating a large area of CyanoHAB (Figure 5), originating from the phosphorus-rich Upper Umpqua River.

In Diamond Lake, during the second year of the Tui Chub eradication program (2007), dramatic improvements were noted in water clarity and aquatic community populations [52]. In 2009, the cyanobacterial level had further reduced to a point that Diamond Lake could meet its water quality standards [52]. As seen in Figure 5, Lemolo Lake would take a few more years to meet that goal.

Ecosystem resiliency is the ability of the local ecology to respond in a positive manner to an acute and/or chronic stressor. Maintaining healthy aquatic and riparian habitats depends on adaptive "management" strategies, allowing for, or facilitating, natural recovery of lentic and lotic functions. In the case of Diamond Lake, the Oregon Department of Fish and Wildlife (ODFW) and the USFS implemented a management plan to eradicate the Tui Chub and save the lake. For the rest of the North Umpqua watershed, changes in the USFS management plan, from 1955 to the present, had the impact of restoring upland, and riparian ecological functions, in the other parts of the watershed. By 1998 [45], lotic ecosystems were close to, or in, proper functioning condition (PFC). As seen in Figure 6, the net result is that the number of days under a CyanoHAB health advisory were reduced to zero by 2013. The added impact was the decrease in sediment, and phosphorus, into the North Umpqua River and Lemolo Lake, as the logging scars healed [44]. As seen in Figure 3B, Lemolo Lake was eutrophic, and had CyanoHABs in the late summer, but CyanoHABs were not reaching a level where there was harm to humans, and therefore requiring CyanoHAB health advisories.

As displayed in a May 1994 black and white (BW) aerial image (Figure 8A), the physical condition of Roberts Creek had an incised stream channel, with an increasing delta, into the Templeton Arm of Tenmile Lake. Delta building indicates two potential sources of sediment: stream channel incision and subsequent channel evolution [14], and other potential land management impacts (e.g., grazing, timber harvesting, and road construction in the harvested areas). In an August 2000 image (Figure 8B), the delta extended approximately 30 m further than in May 1994. The stream channel had widened, and matted CyanoHABs were present, and extending out from the shore. The increase in the delta from 1994 to 2000 matched the sediment increase in the Tenmile Lake core during the same period [37].

As shown in the August 2007 USGS color aerial images of Roberts Creek (Figure 9C), the delta extended approximately 70 m further into the lake than in 1994. In Figure 8B, logging occurred along the north side, and west end, of the Templeton Arm of Tenmile Lake. In 2007 (Figure 9C), logging extended along the length of the Templeton Arm, and into the lower Roberts Creek catchment. The spatial area of the CyanoHAB was larger, and extended further along the shoreline than in August 2000 (Table 2). The July 2012 color aerial image (Figure 9D) shows that the Roberts Creek channel was unchanged, and the delta had decreased in length, by approximately 10 m, from the August 2007 image (Table 2). However, CyanoHABs were much more intensive, and occupy large portions of the shoreline. By 2012, the logged areas were revegetating, and showing some recovery of upland vegetation.

Table 2. Measured distance (m) of United States Geological Survey (USGS) aerial images, using Google Earth, for the Roberts Creek delta, approximate stream channel width, and extent of cyanobacterial growth into Tenmile Lake.

Feature	May 1994	August 2000	August 2007	July 2012
Delta	50 m	80 m	120 m	110 m
Stream Channel Width	4 m	5 m	4 m	<4 m
Cyanobacterial Extent	N/A	30 m	41 m	57 m

The aerial image, time-series analysis of the upper Roberts Creek shows a reach undergoing a management change (Figure 10). As seen in Figure 10, the potential of this narrow meandering stream is to have hydric soils, stabilizing herbaceous riparian vegetation (sedges, rushes), and to be connected to a broad floodplain. The presence of woody plants indicates that the soils are more oxygenated for this setting, which points to a degraded riparian ecosystem. In 2005, the stream channel was incised, wide, and with woody vegetation (Figure 10A). The floodplain, now a terrace, experienced incursion of upland woody plants and/or willows (Figure 10A). This indicates that the floodplain has been drying out, allowing for the migration of upland plants into the former floodplain. As seen in Figure 10B, the rancher implemented a management plan (building fences) to protect the stream. It appears the rancher removed the woody plants to promote the appropriate herbaceous plant community for hay and forage production.

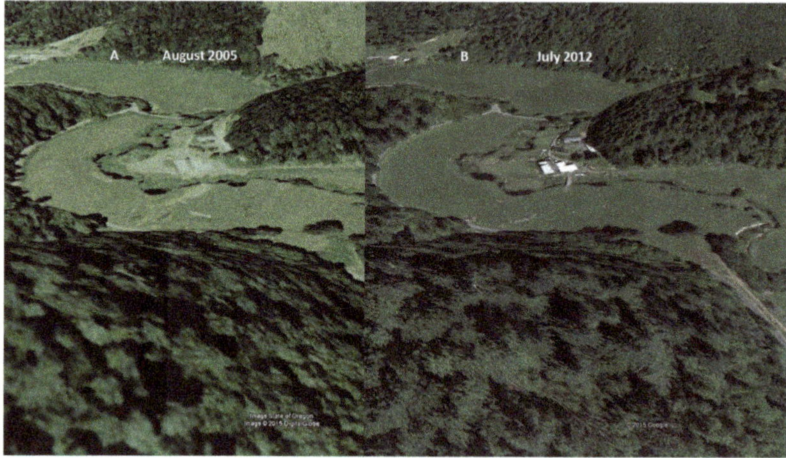

Figure 10. (**A**): August 2005 color image of the middle Roberts Creek watershed. Image shows that channel incision has migrated up into the sub-watersheds. The stream riparian area shown on the ranchland was prior to implementation of the management plan. The sub-watershed is dominated by woody plants. (**B**): USDA July 2012 color image after implementation of a natural resources management plan (fencing) for the riparian area. Some woody material has been removed. Source: State of Oregon, DigitalGlobe (2015).

4. Discussion

In aquatic environments, not all water pollution is from an external input, which makes attribution of pollution to its source/cause difficult, and managing aquatic ecosystems, and their pollution, problematic. The state of Oregon provides guidance for: classification and profiling of wetland and riparian locations [53], wetland monitoring and assessment [54], and a wetland assessment protocol [55], which provides considerable assistance to those tasked with managing aquatic ecosystems in the state of Oregon. Lewis and Wurtsbaugh [33] indicate reducing phosphorus is essential in curtailing the formation of CyanoHABs. Because of the unique relationship between phosphorus and sediment [56], the best method for reducing phosphorus loads is to reduce sediment from entering the lentic or lotic ecosystem. Sediment is a major pollutant across the United States [7], and frequently surges, in conjunction with functional ecosystem decline. The internal (sediment) loading of nutrients in the three lakes in the study area is greater than the nutrient loading received from watershed inputs [44], and was mitigated by the reduction of the Tui Chub population.

Vegetation provides roughness that slows water velocity. Dissipated energy is less likely, at any one spot, to exceed the critical shear stress of stream bank soils and stream substrate and cause erosion [57]. Similarly, when water slows, it deepens and spreads across accessible floodplain areas, where it slows further from friction with floodplain vegetation, and no longer has the velocity and turbulence to keep particles suspended. Reduced erosion and sediment deposition have a direct impact on water quality [10]. With the loss of ecological functions, stream and wetland riparian areas, and the aquatic environment, converts from a sink to a source of water pollution [58]. Ecological function potential is based on a concept of dynamic equilibrium in an ecosystem, that corresponds to measures of the physical setting [22,59–64]. Leading indicators are capable of measuring alterations to stream and wetland riparian functions [10], which provide essential ecosystem services [9].

A management response to CyanoHABs has three levels: Immediate—breaking the pathway of exposure by providing clean drinking water—this cannot be sustained indefinitely due to limited funding and uncontrollable water demand; Short-Term—breaking the pathway of exposure by initiating a warning system—this publicizes information on (lagging) indicators (CyanoHABs caused by excess nutrients) after they are detected; Long-Term—managing ecological functions to reduce transport of excess nutrients, thus preventing potential CyanoHABs. The immediate and short-term responses are unsustainable. However, keeping the flood channel narrow inhibits the processes of "self-healing" needed to restore sediment deposition and nutrient sequestration, and the assimilation processes that are important in improving water quality [10].

The goal of a successful resource management plan is to assist an ecosystem to respond in ways that enhance natural remediation [26,51,65,66], or quicken its pace to a particular level of functionality, or desired condition [9,12,15]. The first step is to assess the potential functioning condition, then determine the primary sources of nutrients. If NPSs are predominant, then it is essential to improve watershed, stream, and wetland riparian functions as part of any watershed management plan for long-term water quality improvement [11,12]. Properly functioning streams and wetland riparian ecosystems provide a steadying influence on water quality and aquatic habitat attributes. Restoring riparian functions will result in slowing the nutrient spiral, with flooding and floodplain deposition, and allow nutrient uptake, aquifer recharge, and reconstruction of quality habitat, and complex niches/food webs that interrelate riparian and aquatic ecosystems [10,67].

This research project illustrates that success in pollution control and in maintaining healthy aquatic habitats often depends on managing land to facilitate natural recovery of riparian functions [15]. Using best management practices throughout the watershed by using a combination of tools and resources, such as: time-series aerial photos (to display changes in overall ecosystem state, and in specific details, due to land use changes), TMDLs (to monitor the amount of nutrients, sediment, and CyanoHABs), CyanoHAB advisories (to inform the public and reduce their exposure to toxins), PFC (to characterize how well physical processes in a stream and wetland riparian area can sustain resiliency), and by addressing upland sources of accelerated nutrient supply to the stream, through appropriate forestry management approaches and road network design, can be applied to positively affect ecological function.

Author Contributions: R.K.H., D.T.H., and J.L., designed the project. J.L.A., and R.J.S., performed data analysis. E.S.H., and R.K.H., wrote and edited the manuscript. S.S., M.J.P., T.J.-L., E.W., and H.K., reviewed the manuscript. E.S.H., had the manuscript internally peer-reviewed, performed the final edits, and approved the final manuscript.

Funding: This research was funded by the United States Environmental Protection Agency, through its Office of Research and Development, under contract EP-15-Z-000082 to Dr. Joan L. Aron, and contract EP-14-Z-000031 to Dr. Robin J. Schafer.

Acknowledgments: The authors would like to thank Patti Tyler, USEPA Region 8, David Guiliano, USEPA Region 9, Adam Jorge, USEPA, Yongping Yuan, USEPA, Nilla Barros, USEPA, and Joyce Swanson for their critical review of this manuscript. Although this paper has been subjected to USEPA review and is approved for publication, it may not necessarily reflect official Agency policy. Mention of trade names and commercial products does not constitute endorsement or recommendation for use.

Conflicts of Interest: The authors declare no conflict of interest.

References

1. Yuechu, L.; Zhongxuan, Z.; Yi, S.; Hongbing, D.; Ping, Z.; Gang, W. Assessment methods of watershed ecosystem health. *Acta Ecol. Sin.* **2003**, *8*, 1606–1614.
2. Lichtenthaler, H.K. The stress concept in plants: An introduction. *Ann. N. Y. Acad. Sci.* **1998**, *851*, 187–198. [CrossRef]
3. Marazzi, L.; Gaiser, E.E.; Eppinga, M.B.; Sah, J.P.; Zhai, L.; Castañeda-Moya, E.; Angelini, C. Why Do We Need to Document and Conserve Foundation Species in Freshwater Wetlands? *Water* **2019**, *11*, 265. [CrossRef]
4. Weller, D.E.; Jordon, T.E.; Cornell, D.L. Heuristic models for material discharge from landscapes with riparian buffers. *Ecol. Appl.* **1998**, *8*, 1156–1169. [CrossRef]
5. World Health Organization (WHO). *Cyanobacterial Toxins: Microcystin-LR in Drinking Water*; World Health Organization: Geneva, Switzerland, 2003.

6. Federal Water Pollution Control Act (FWPCA, 1972). Public Law 92-500, 86 Stat. 816 (Amended 1977 and 1987, Referred to as the Clean Water Act, Codified at 33 U.S.C. 1251–1387, 1988). Available online: https://www.epa.gov/sites/production/files/2017-08/documents/federal-water-pollution-control-act-508full.pdf (accessed on 28 May 2019).

7. U.S. Environmental Protection Agency (USEPA). *National Lakes Assessment: A Collaborative Survey of the Nation's Lakes*; EPA 841-R-09-001; U.S. Environmental Protection Agency, Office of Water and Office of Research and Development: Washington, DC, USA, 2009. Available online: https://www.epa.gov/sites/production/files/2013-11/documents/nla_newlowres_fullrpt.pdf (accessed on 28 May 2019).

8. Swanson, S.R.; Hall, R.K.; Heggem, D.T.; Lin, J.; Kozlowski, D.F.; Gibson, R.J. Leading or Lagging Indicator for Water Quality Management, Abstracts with Programs. In Proceedings of the 8th National Monitoring Conference, Portland, OR, USA, 30 April–4 May 2012.

9. Aron, J.L.; Hall, R.K.; Philbin, M.J.; Schafer, R.J. Using watershed function as the leading indicator for water quality. *Water Policy* **2013**, *15*, 850–858. [CrossRef]

10. Swanson, S.; Kozlowski, D.; Hall, R.; Heggem, D.; Lin, J. Riparian Proper Functioning Condition Assessment to Improve Watershed Management for Water quality. *J. Soil Water Conserv.* **2017**, *72*, 168–182. [CrossRef]

11. Kozlowski, D.; Swanson, S.; Hall, R.; Heggem, D. *Linking Changes in Management and Riparian Physical Functionality to Water Quality and Aquatic Habitat: A Case Study of Maggie Creek, NV*; EPA/600/R-13/133; U.S. Environmental Protection Agency: Washington, DC, USA, 2013. Available online: https://cfpub.epa.gov/si/si_public_record_report.cfm?Lab=NERL&dirEntryId=257579 (accessed on 28 May 2019).

12. Hall, R.K.; Guiliano, D.; Swanson, S.; Philbin, M.J.; Lin, J.; Aron, J.L.; Schafer, R.J.; Heggem, D.T. An Ecological Function and Services Approach to Total Maximum Daily Load (TMDL) Prioritization. *J. Environ. Monit.* **2014**. [CrossRef]

13. Prichard, D.; Berg, F.; Hagenbuck, W.; Krapf, R.; Leinard, R.; Leonard, S.; Manning, M.; Noble, C.; Staats, J. *Riparian Area Management: A User Guide to Assessing Proper Functioning Condition and the Supporting Science for Lentic Areas*; Technical Reference 1737-16; Revised 2003; U.S. Department of the Interior (DOI), Bureau of Land Management (BLM): Washington, DC, USA, 1999. Available online: https://www.blm.gov/or/programs/nrst/files/Final%20TR%201737-16%20.pdf (accessed on 28 May 2019).

14. Prichard, D.; Anderson, J.; Correll, C.; Fogg, J.; Gebhardt, K.; Krapf, R.; Leonard, S.; Mitchell, B.; Staats, J. *Riparian Area Management, A User Guide to Assessing Proper Functioning Condition and the Supporting Science for Lotic Areas*; Technical Reference TR1737-15; U.S. Department of the Interior (DOI), Bureau of Land Management (BLM): Washington, DC, USA, 2015. Available online: https://www.blm.gov/or/programs/nrst/files/Final%20TR%201737-15.pdf (accessed on 28 May 2019).

15. Dickard, M.; Gonzales, M.; Elmore, W.; Leonard, S.; Smith, D.; Smith, S.; Staats, J.; Summers, P.; Weixelman, D.; Wyman, S. *Riparian Area Management—Proper Functioning Condition Assessment for Lotic Areas*, 2nd ed.; BLM Technical Reference 1737–15: Washington, DC, USA, 2015; Available online: http://www.remarkableriparian.org/pdfs/pubs/TR_1737-15.pdf (accessed on 28 May 2019).

16. Bernhardt, E.S.; Palmer, M.A.; Allan, J.D.; Alexander, G.; Barnas, K.; Brooks, S.; Carr, J.; Clayton, S.; Dahm, C.; Follstad-Shah, J.; et al. River Restoration Efforts. *Science* **2005**, *308*, 636–637. [CrossRef] [PubMed]

17. Vogan, C.R. Pollution Abatement and Control Expenditures, 1972–94. *Surv. Curr. Bus.* **1996**, 48–67. Available online: https://apps.bea.gov/scb/pdf/national/niparel/1996/0996eed.pdf (accessed on 28 May 2019).

18. Van Houtven, G.L.; Brunnermeier, S.B.; Buckley, M.C. *A Retrospective Assessment of the Costs of the Clean Water Act: 1972 to 1997*; U.S. Environmental Protection Agency Office of Water Office of Policy, Economics, and Innovation: Washington, DC, USA, 2000.

19. Norton, D.J.; Wickham, J.D.; Wade, T.G.; Kunert, K.; Thomas, J.V.; Zeph, P. A Method for Comparative Analysis of Recovery Potential in Impaired Waters Restoration Planning. *Environ. Manag.* **2009**, *44*, 356–368. [CrossRef]

20. Prichard, D.; Clemmer, P.; Gorges, M.; Meyer, G.; Shumac, K.; Wyman, S.; Miller, M. *Riparian Area Management: Using Aerial Photographs to Assess Proper Functioning Condition of Riparian-Wetland Areas*; Technical Reference 1737–12, Revised 1999; U.S. Department of the Interior (DOI), Bureau of Land Management (BLM): Washington, DC, USA, 1996. Available online: https://permanent.access.gpo.gov/lps111904/Final%20TR%201737-12.pdf (accessed on 28 May 2019).

21. Kondolf, G.M. The Espace de Liberte and restoration of fluvial process: When can the river restore itself and when must we intervene? In *River Conservation and Management*; Boon, P., Raven, P., Eds.; John Wiley and Sons: Chichester, UK, 2012; pp. 225–242.

22. Rosgen, D. *Applied River Morphology*; Wildland Hydrology: Pagosa Springs, CO, USA, 1996.

23. Schumm, S.A.; Harvey, M.D.; Watson, C.C. *Incised Channels: Morphology, Dynamics and Control*; Water Resources Publications: Littleton, CO, USA, 1984.

24. Weixelman, D.; Zamudio, D.; Zamudio, K. *Central Nevada Riparian Field Guide*; R4-ECOL-96-01; USDA Forest Service, Intermountain Region: Washington, DC, USA, 1996.

25. Winward, A.H. *Monitoring the Vegetation Resources in Riparian Areas*; Gen. Tech. Rep. RMRS-GTR-47; Department of Agriculture, Forest Service, Rocky Mountain Research Station: Fort Collins, CO, USA, 2002.

26. Burton, T.A.; Smith, S.J.; Cowley, E.R. *Riparian Area Management: Multiple Indicator Monitoring (MIM) of Stream Channels and Streamside Vegetation*; Technical Reference 1737-23; U.S. Department of the Interior (DOI), Bureau of Land Management (BLM), National Operations Center (NOC): Denver, CO, USA, 2011. Available online: https://www.fs.usda.gov/Internet/FSE_DOCUMENTS/fseprd558332.pdf (accessed on 28 May 2019).

27. Aráoz, R.; Molgó, J.; Tandeau de Marsac, N. Neurotoxic cyanobacterial toxins. *Toxicon* **2010**, *56*, 813–828. [CrossRef]

28. Kouzuma, A.; Watanabe, K. Exploring the potential of algae/bacteria interactions. *Curr. Opin. Biotechnol.* **2015**, *33*, 125–129. [CrossRef]

29. Wacklin, P.; Hoffmann, L.; Komárek, J. Nomenclatural validation of the genetically revised cyanobacterial genus Dolichospermum (Ralfs ex Bornet et Flahault) comb. nova. *Fottea* **2009**, *9*, 59–64. [CrossRef]

30. Lehman, P.W.; Kurobe, T.; Lesmeister, S.; Baxa, D.; Tung, A.; Teh, S.J. Impacts of the 2014 severe drought on the Microcystis bloom in San Francisco Estuary. *Harmful Algae* **2017**, *63*, 94–108. [CrossRef] [PubMed]

31. Sabart, M.; Crenn, K.; Perrière, F.; Abila, A.; Leremboure, M.; Colombet, J.; Jousse, C.; Latour, D. Co-occurrence of microcystin and anatoxin-a in the freshwater lake Aydat (France): Analytical and molecular approaches during a three-year survey. *Harmful Algae* **2015**, *48*, 12–20. [CrossRef] [PubMed]

32. Chorus, I.; Bartram, J. *Toxic Cyanobacteria in Water: A Guide to Their Public Health Consequences, Monitoring and Management*; World Health Organization: Geneva, Switzerland, 1999; ISBN 0-419-23930-8.

33. Lewis, W.M.; Wurtsbaugh, W.A. Control of Lacustrine Phytoplankton by Nutrients: Erosion of the Phosphorus Paradigm. *Int. Rev. Hydrobiol.* **2008**, *93*, 446–465. [CrossRef]

34. Schindler, D.W. Recent Advances in the Understanding and Management of Eutrophication. *Limnol. Oceanogr.* **2006**, *51*, 356–363. [CrossRef]

35. Johnson, D.; Petersen, R.; Lycan, D.; Sweet, J.; Newhaus, M.; Schaedel, A. *Atlas of Oregon Lakes*; OSU Press: Corvallis, OR, USA, 1985.

36. Oregon Health Authority (OHA). Public Health Division, Center for Health Protection, 2018, Oregon Harmful Algae Bloom Surveillance (HABS) Program, Recreational Use Public Health Advisory Guidelines. Cyanobacterial Blooms in Freshwater Bodies. September 2018. Available online: https://www.oregon.gov/oha/PH/HEALTHYENVIRONMENTS/RECREATION/HARMFULALGAEBLOOMS/Documents/Advisory%20Guidelines%20for%20Harmful%20Cyanobacteria%20Blooms%20in%20Recreational%20Waters.pdf (accessed on 28 May 2019).

37. ODEQ. Tenmile Lakes Watershed Total Maximum Daily Load. 2007. Available online: https://www.oregon.gov/deq/FilterDocs/scTenmiletmdl.pdf (accessed on 28 May 2019).

38. Prichard, D.; Bridges, C.; Krapf, R.; Leonard, S.; Hagenbuck, W. *Riparian Area Management: Process for Assessing Proper Functioning Condition for Lentic Riparian-Wetland Areas*; Technical Reference 1737-11; Revised 1998; U.S. Department of the Interior (DOI), Bureau of Land Management (BLM): Washington, DC, USA, 1994.

39. U.S. Environmental Protection Agency (USEPA). Wadeable Stream Assessment. A Collaborative Survey of Nation's Streams. EPA 841-B-06-002. December 2006. Available online: https://www.epa.gov/sites/production/files/2014-10/documents/2007_5_16_streamsurvey_wsa_assessment_may2007.pdf (accessed on 28 May 2019).

40. USGS. Water Quality and Algal Conditions in the North Umpqua River, Oregon, 1995–2007, and their Response to Diamond Lake Restoration. Open-File Report 2014–1098. 2014. Available online: https://pubs.usgs.gov/of/2014/1098/pdf/ofr2014-1098.pdf (accessed on 28 May 2019).

41. Diamond Lake Restoration EIS. Surface Water—Lakes: Lake Ecology and Water Quality. Available online: https://www.fs.usda.gov/Internet/FSE_DOCUMENTS/stelprdb5336132.pdf (accessed on 28 May 2019).

42. Oregon DEQ Harmful Algal Bloom Strategy, Appendix C. June 2011. Available online: https://www.oregon.gov/deq/FilterDocs/habsAppendixC.pdf (accessed on 28 May 2019).

43. Jefferson, A.; Grant, G.E.; Lewis, S.L.; Lancaster, S.T. Coevolution of hydrology and topography on a basalt landscape in the Oregon Cascade Range, USA: Earth Surface Processes and Landforms. *Earth Surf. Processes Landf. J. Br. Geomorphol. Res. Gr.* **2010**, *35*, 803–816. [CrossRef]

44. U.S. Forest Service (USFS). Vegetation Management Plan for The Unroaded Recreation Area West of Diamond Lake (Management Area 1), and the Diamond Lake and Lemolo Lake Recreation Composite Areas (Management Area 2), plus Watershed Analysis Iteration Updating recommendations made in the 1998 Diamond Lake/Lemolo Lake Watershed Analysis. 2008. Available online: https://www.fs.usda.gov/Internet/FSE_DOCUMENTS/stelprdb5335962.pdf (accessed on 28 May 2019).

45. U.S. Forest Service (USFS). *Lemolo and Diamond Lakes Watershed Analysis*; U.S. Department of Agriculture (USDA), U.S. Forest Service (USFS): Washington, DC, USA, 1998. Available online: https://www.fs.usda.gov/Internet/FSE_DOCUMENTS/stelprdb5335959.pdf (accessed on 28 May 2019).

46. ODEQ. Umpqua Basin Total Maximum Daily Load: Chapter 6, Diamond Lake and Lake Creek Aquatic Weeds, Dissolved Oxygen and Ph. 2006. Available online: http://www.oregon.gov/deq/FilterDocs/umpchpt6dialake.pdf (accessed on 28 May 2019).

47. Eilers, J.M.; Loomis, D.; St. Amand, A.; Vogel, A.; Jackson, L.; Kann, J.; Eilers, B.; Truemper, H.; Cornett, J.; Sweets, R. Biological effects of repeated fish introductions in a formerly fishless lake: Diamond Lake, Oregon, USA. *Fundam. Appl. Limnol. Arch. Für Hydrobiol.* **2007**, *169*, 265–277. [CrossRef]

48. ODEQ. Umpqua Basin Total Maximum Daily Load: Chapter 1. 2006. Available online: https://www.oregon.gov/deq/FilterDocs/umpchpt1overview.pdf (accessed on 28 May 2019).

49. Ward, T.A.; Tate, K.W.; Atwill, E.R.; Lile, D.F.; Lancaster, D.L.; McDougald, N.; Barry, S.; Ingram, R.S.; George, H.A.; Jensen, W.; et al. A comparison of three visual assessments for riparian and stream health. *J. Soil Water Conserv.* **2003**, *58*, 83–88.

50. Eilers, J.; Vache, K.; Kann, J. *Tenmile Lakes Nutrient Study—Phase II Report*; E&S Environmental Chemistry Inc.: Corvallis, OR, USA, 2002.

51. U.S. Environmental Protection Agency (USEPA). *National Management Measures to Control Nonpoint Source Pollution from Forestry*; EPA 841-B-05-001; U.S. Environmental Protection Agency (USEPA): Washington, DC, USA, 2005. Available online: https://www.epa.gov/nps/nonpoint-source-pollution-technical-guidance-and-tools (accessed on 28 May 2019).

52. Eilers, J.; Truemper, H. Diamond Lake Recovery—Again, Lakeline, Summer. 2010, pp. 23–26. Available online: http://www.dfw.state.or.us/fish/local_fisheries/diamond_lake/docs/eilers.pdf (accessed on 28 May 2019).

53. Oregon Department of State Lands. Manual for the Oregon Rapid Wetland Assessment Protocol (ORWAP), Paul, R. Adamus, Kathy Verble. November 2016. Available online: https://www.oregon.gov/dsl/WW/Documents/ORWAP_3_1_Manual_Nov_2016.pdf (accessed on 28 May 2019).

54. Mitra, A.; Flynn, K.J. Promotion of harmful algal blooms by zooplankton predatory activity. *Biol. Lett.* **2006**, *2*, 194–197. [CrossRef] [PubMed]

55. ODEQ. Tenmile Lakes Watershed Quality Management Plan (WQMP). February 2007. Available online: https://www.oregon.gov/deq/FilterDocs/scTenmilewqmp.pdf (accessed on 28 May 2019).

56. Søndergaard, M.; Jensen, J.S.; Jeppesen, E. Role of sediment and internal loading of phosphorus in shallow lakes. *Hydrobiologia* **2003**, *506*, 135–145. [CrossRef]

57. Katz, S.B.; Segura, C.; Warren, D.R. The influence of channel bed disturbance on benthic Chlorophyll *a*: A high resolution perspective. *Geomorphology* **2017**. [CrossRef]

58. Wentz, D.A.; Brigham, M.E.; Chasar, L.C.; Lutz, M.A.; Krabbenhoft, D.P. *Mercury in the Nation's Streams—Levels, Trends, and Implications*; U.S. Geological Survey Circular: Reston, VA, USA, 2014.

59. Leopold, L.B.; Wolman, M.G.; Miller, J.P. *Fluvial Processes in Geomorphology*; Dover Publications, Inc.: San Francisco, CA, USA, 1964; p. 522. ISBN 0-486-68588-8.

60. Rosgen, D.L. A classification of natural rivers. *Catena* **1994**, *22*, 169–199. [CrossRef]

61. Rosgen, D. *Watershed Assessment of River Stability and Sediment Supply (WARSSS)*; Wildland Hydrology: Ft. Collins, CO, USA, 2006; ISBN-13: 978-0-9791308-0-9.

62. Brierley, G.J.; Fryirs, K. River styles, a geomorphic approach to catchment characterization: Implications for river rehabilitation in Bega catchment, New South Wales, Australia. *Environ. Manag.* **2000**, *25*, 661–679. [CrossRef] [PubMed]

63. Kondolf, G.M.; Montgomery, D.R.; Piegay, H.; Schmitt, L. Geomorphic Classification of Rivers and Streams. In *Tools in Fluvial Geomorphology*; Kondolf, G.M., Piegay, H., Eds.; Wiley: West Sussex, UK, 2003; Chapter 7; pp. 171–204.

64. Brierley, G.; Fryirs, K.; Cook, N.; Outhet, D.; Raine, A.; Parsons, L.; Healey, M. Geomorphology in action: Linking policy with on-the-ground actions through applications of the River Styles framework. *Appl. Geogr.* **2011**, *31*, 1132–1143. [CrossRef]
65. Magadza, C.H.D. Kariba Reservoir: Experience and lessons learned. *Lakes Reserv. Res. Manag.* **2006**, *11*, 271–286. [CrossRef]
66. Stone, D.; William, B. Addressing Public Health Risks for Cyanobacteria in Recreational Freshwaters: The Oregon and Vermont Framework. *Integr. Environ. Assess. Manag.* **2007**, *3*, 137–143. [CrossRef]
67. Oregon Division of State Lands. Guidebook for Hydrogeomorphic (HGM)—Based Assessment of Oregon Wetland and Riparian Sites: Statewide Classification and Profiles; Paul R. Adamus. February 2001. Available online: https://www.oregon.gov/dsl/WW/Documents/hydro_guide_class.pdf (accessed on 28 May 2019).

Article

Effect of Phenyl-Acyl Compounds on the Growth, Morphology, and Toxin Production of *Microcystis aeruginosa* Kützing

Natalia Herrera [1,*], Maria Teresa Florez [2], Juan Pablo Velasquez [1] and Fernando Echeverri [1,*]

[1] Organic Chemistry Natural Products Group, Institute of Chemistry, Universidad de Antioquia, Calle 67 No. 53-108, 050010 Medellín, Colombia; juanp.velasquez@udea.edu.co

[2] Environmental Management and Modeling Research Group (GAIA), Universidad de Antioquia, Calle 67 No. 53-108, 050010 Medellín, Colombia; mariateresa.florez@gmail.com

* Correspondence: nahelo241980@gmail.com (N.H.); fernando.echeverri@udea.edu.co (F.E.); Tel.: +57-4219-6595 (F.E.)

Received: 20 November 2018; Accepted: 16 January 2019; Published: 30 January 2019

Abstract: The proliferation of cyanobacteria and, consequently, the production of cyanotoxins is a serious public health concern; for their control, several alternatives have been proposed, including physical, chemical, and biological methods. In the search for new alternatives and a greater understanding of the biochemical process involved in the blooms' formation, we report here the effect of eight phenyl-acyl compounds in the growth of *Microcystis aeruginosa* Kützing (assesed as cell density/count and Chl *a* fluorescence concentration) morphology, and production of the toxin microcystin-LR (MC-LR). Caffeic acid and eugenol decreased the growth of *M. aeruginosa* Kützing and the levels of Chl *a*. However, 3,5-dimethoxybenzoic acid and syringic acid caused the opposite effect in the growth; 2′and 4′only affected the Chl *a*. A reduction in the concentration of the MC-LR toxin was detected after treatment with syringic acid, caffeic acid, and eugenol. According to HPLC/MS (High Performance Liquid Chromatography coupled to Mass Spectrometry), a redox process possibly occurs between caffeic acid and MC-LR. The optical microscopy and Scanning Electron Microscopy analyses revealed morphological changes that had been exposed to caffeic acid and vanillin, specifically in the cell division and presence of mucilage. Finally, assays in *Daphnia pulex* De Geer neonates indicated that caffeic acid had a non-toxic effect at concentrations as high as 100 mg/L at 48 h.

Keywords: blooms; cyanobacteria; control; toxins; phenyl-acyl compounds; caffeic acid; non-toxic; redox microcystin LR

---

1. Introduction

Currently, water bodies are threatened with eutrophication, which leads to an increased risk of blooms of potentially toxic cyanobacteria. The recognition of its impacts on ecosystems, and the implementation of sustainability programs in water management and bloom prevention, requires new alternatives to, and knowledge of, water treatment to avoid health risks [1]. In particular, the effects of cyanotoxins on human health are well-known. For example, 53 patients died due to intravenous exposure to several microcystins (MCs) during a hemodialysis treatment in Caruaru, Brazil [2].

To eliminate intact cyanobacteria and prevent the release of intracellular toxins, different approaches to water treatment have been used, such as flocculation, filtration, and coagulation. However, these methodologies sometimes do not remove all these compounds.

Some algicides, such as copper sulfate, potassium permanganate, and chlorine, are widely used. However, they are highly contaminating and lack specificity, as they cause the death of other organisms present in the same environment [3]. Copper sulfate is the most commonly used algicide because of its cost, ease of application, and effectiveness [4]. Nevertheless, its intensive application in lakes and reservoirs is a great problem due to copper accumulation and the consequent toxicity [5]. Therefore, new substances and organisms have been sought to control these blooms. For example, vitamin C has been applied to inhibit *Microcystis aeruginosa* Kützing [6]. Besides this, aquatic macrophytes affect photosynthesis and the antioxidant system of this cyanobacteria [7]. Other methods include adsorbent sediments [8], chitosan as a coagulant [9], and coagulation-flotation methods with pre-oxidation assistance [10].

It has been proposed that toxins, such as MCs, stimulate the formation of colonies of *M. aeruginosa* Kützing; thus, they may play a role in their persistence and dominance [11]. Therefore, the inhibition of the production of these toxins could be an alternative to control bloom formation as well avoid harmful effects in animals.

In this study, we assessed the effect of eight phenyl-acyl compounds, derived from natural sources (Figure 1), on the growth and morphology of *M. aeruginosa* Kützing blooms, as well as on the levels of the microcystin-LR (MC-LR) toxin. We also analyzed the toxicity of these compounds to *Daphnia pulex* De Geer.

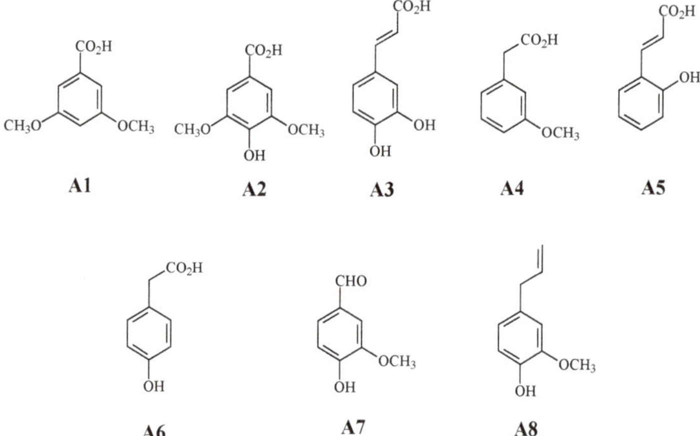

Figure 1. The phenyl-acyl compounds that were assayed in this work: 3,5-dimethoxybenzoic acid (A1, the code in this text), syringic acid (A2), caffeic acid (A3), 3-methoxyphenylacetic acid (A4), 2′-hydroxycinnamic acid (A5), 4′-hydroxyphenylacetic acid (A6), vanillin (A7), eugenol (A8), and, for copper sulfate, Cu(SO)$_4$ was assigned code A9.

2. Materials and Methods

2.1. Sampling and Culturing of M. aeruginosa

A cyanobacteria sample was collected from the Riogrande II reservoir (Antioquia, Colombia) in May 2016. *M. aeruginosa* Kützing was the most abundant cell in the blooms, accordingly microscopically analysis. For isolation, samples of 10 mL of bloom were incubated in 250 mL of Blue Green medium BG11 (SIGMA-ALDRICH, St. Louis, MO, USA) [12] in 1 L Erlenmeyer flasks with a photoperiod of 20 h/4 h light 930 lux/dark at 25 °C under continuous aeration using an air pump (Resum Air pump AC 9904, Cranbury, NJ, USA), for seven days. Then, 100 mL of culture were centrifuged at 6000 rpm for 5 min, and the supernatant was transferred to 300 mL of BG11 medium for another seven days to reduce contamination. Finally, 10 mL of the new culture were taken and incubated again in the same culture medium. For culture maintenance, the same initial conditions were followed and every 72 h the medium was replaced.

2.2. Substances

The following substances were obtained from SIGMA-ALDRICH (St. Louis, MO, USA): 3,5-dimethoxybenzoic acid (A1, the code in this text), syringic acid (A2), caffeic acid (A3), 3-methoxyphenylacetic acid (A4), 2′-hydroxycinnamic acid (A5), 4′-hydroxyphenylacetic acid (A6), vanillin (A7), eugenol (A8), and, for copper sulfate, $Cu(SO)_4$ was assigned code A9 (Figure 1). The MC-LR standard was obtained from MP Biomedicals, LLC (Illkirch, France).

2.3. The Growth Inhibition Test

The inoculants for the inhibition tests were prepared from eight-day cultures of *M. aeruginosa* Kützing in the exponential phase. For this purpose, 20 mL of the culture was taken, and then 200 mL of culture medium (BG11) containing each substance at different concentrations (100 mg/L (C3), 50 mg/L (C2), and 25 mg/L (C1)) were added; the cultures were maintained under the same initial conditions for 72 h. After that, samples were taken for growth measurements through cell counts and the Chl *a* level at 0 h and after 72 h of culture. The percentage of inhibition was calculated using the following formula: % cyanobacterial inhibition = $(1 -$ treatment/control$) \times 100$, where the treatment and control are the cell densities of *M. aeruginosa* Kützing exposed and not exposed to different substances, respectively. Finally, the samples were refrigerated for 12 h before lyophilization and then extracted with methanol for the detection of MCs by HPLC/DAD (Diode Array Detector) and HPLC/MS. Each treatment was performed in triplicate, and two growth controls were prepared, one of them with copper sulfate, and the other without substances.

2.3.1. Cell Counts

For the cell counts, 200 µL of *M. aeruginosa* Kützing cultures treated with substances were collected and diluted in distilled water to the desired concentration. The counts were performed using a Neubauer chamber under an optical microscope (Nikon, YS2-H, Tokyo, Japan). To calculate the number of cells per mL, the chamber dilution factor was considered. To determine the total number of cells after exposure to each substance, the number of final cells was subtracted from the initial cell number. The cell counts were evaluated by two independent observers and performed in triplicate and repeated four times.

2.3.2. The Chl *a* Measurement

Chl *a* fluorescence was used as a proxy for the estimation of *M. aeruginosa* Kützing biomass (µg Chl *a*/L) [13]. Chl *a* was measured using a Fluorometer (FluoroProbe III, bbe-Moldaenke, Schwentinental, Germany) with an integrated correction factor for colored dissolved organic matter (CDOM) to compensate for UV-LED radiation interference. This instrument measures the fluorescence of Chl *a* in the photosystem II after excitation with five light diodes (370, 470, 525, 570, 590, and 610 nm) with a resolution of 0.01 µg Chl *a*/L.

2.4. Morphological Observations under the Optical Microscope and the Scanning Electron Microscope (SEM)

Each culture that had been exposed to the different substances was observed under optical microscope to identify changes in cell morphology, colony appearance, and the presence of mucilage. For this purpose, acid staining using China ink was performed [14], and *M. aeruginosa* Kützing cells were placed on 1 cm^2 plates and fixed with 2.5% glutaraldehyde for 12 h for a SEM (JEOL, JEOL JSM 6490 LV, Tokyo, Japan) analysis. Then, cells were washed three times consecutively with Sorensen's phosphate buffer at neutral pH, followed by distilled water, and subsequently dehydrated by washing with 50%, 75%, 95%, and 100% ethanol. After that, the samples were fixed onto a graphite tape and a thin gold coating (Au) was applied (Denton Vacuum, Denton Vacuum Desk IV equipment, Moorestown, NJ, USA). Finally, they were analyzed by scanning electron microscopy to obtain high-resolution images. A secondary electron detector was used to evaluate the morphology and topography of the samples.

2.5. The Effect of Phenyl-Acyl Compounds on Toxin Concentration

The MCs were analyzed in a Gilson 305 and 306 HPLC with a diode array detector (DAD) using a Luna C-18 column (4.6 x 150 mm, 5 μm, Phenomenex, Torrance, CA, USA) with a mobile phase containing water acidified with 0.05% of TFA (trifluoroacetic acid) in pump A, and acetonitrile that had been acidified in the same way in pump B. The gradient was increased from 10% to 60% of B for 25 min, followed by 10 min at 100% of B, and, finally, returning to the initial conditions (10% of B) [15]. A 1 min flow was used, and each run lasted for 40 min. The absorbance was monitored at 238 nm, and the injection volumes were 20 μL. Each sample was injected in triplicate, and the analyses were performed using the Gilson UniPoint software v1.8 (UniPoint, Winnipeg, MB, Canada).

2.6. The Analysis of the Redox Effect of Caffeic Acid on MC-LR by LC/Q-TOF (Liquid Chromatography/Quadrupole-Time of Flight)

To establish the mechanism of action of caffeic acid, 50 mg of pure caffeic acid was added to 200 mL of Milli-Q water containing 50 mL of a MC-LR toxin sample that had been extracted from the culture of *M. aeruginosa* Kützing; the mixture was maintained for 72 h with constant agitation. Then, a 30 mL sample was taken at 0, 24, and 72 h. The samples were adsorbed in to solid phase extraction C18 cartridges, eluted with methanol, concentrated under reduced pressure, and injected into an HPLC/HR-MS using a UHR-QTOF Impact II-Bruker (Ultrahigh Resolution Quadrupole Time-of-Flight, Bruker Corporatio, Billerica, MA, USA). The extracts were separated using a Chromolith High Resolution C18 column, Germany (50 × 4.6 mm), which was maintained at 40 °C. The mobile phase comprised Milli-Q water plus 0.1% formic acid (A) and acetonitrile plus formic acid 0.1% (B). The separation was performed using a gradient that increased from 5% B to 10% B for 1 min, followed by 50% (B) up to 4 min and up to 95% (B) at 6 min, and that finally returned to the initial conditions at 8 min. The auto-sampler was maintained at 10 °C always. Data were acquired in positive ion electrospray scanning mode from 50 to 2000 m/z with a 2 s/scan and a 0.1 s inter-scan delay. The ion source capillary and sampling cone parameters were 2.9 and 25 V, respectively. The desolvation temperature was 220 °C, and the source temperature was 80 °C. The cone gas and desolvation gas flows were 8 L h^{-1}. A tuning mix solution was used as a calibrant. Instrument control and data acquisition (centroid) and processing were performed using the Bruker Compass DataAnalysis 4.3 software (Bruker Corporation, Billerica, MA, USA).

2.7. The Acute Toxicity Tests with D. pulex

Assays were carried out by exposing five neonates of *D. pulex*, aged <24 h old, to phenyl acyl compounds and copper sulfate. A total of 30 mL of growth medium, mixed with the substances, was added to each container containing *D. pulex* to a final concentration of 25, 50, and 100 mg/L. After 24, 48, and 72 h of exposure, the number of survivors in each test vessel was counted to identify the lethal concentration 50 (LC_{50}). Analyses were done in triplicate for a total of 15 neonates; in addition, water was used as a negative control.

2.8. Data Analysis

Initially, a goodness-of-fit test (the Kolmogorov–Smirnov test) was performed to establish statistically significant differences in the cell density measurements that were obtained from the tests using different substances. A parametric variance analysis with a significance level of 0.05 was applied. In cases wherein statistically significant differences were found, a comparison analysis of multiple ranges was performed to determine which samples were different from each other. To establish correlations between different cell density measurements, a Spearman correlation analysis with a significance level of 0.05 was applied using the statistical package Statgraphics Centurion XVI (Statgraphics.Net, Madrid, Spain).

The LC_{50} was determined from the acute toxicity results, and the IBM SPSS Statistics 21 software (IBM, Armonk, NY, USA) was used as an analytical tool.

3. Results

3.1. Growth Inhibition of M. aeruginosa Kützing

The effects of the substances on the cultures of *M. aeruginosa* Kützing were assessed in two ways (Figure 2): cell density measurements and Chl *a* quantities. Caffeic acid and eugenol were powerful inhibitors in both parameters at 50–100 mg/L; vanillin exhibited a medium inhibitory effect at both 50 mg/L and 100 mg/L.

However, a differential response was noticed when Chl *a* levels were considered. Syringic acid, 2′-hydroxycinnamic acid, and 4′-hydroxyphenylacetic acid were strong inhibitors of Chl *a* at the highest concentration. This seems to indicate a specific action in the levels of Chl *a*. Strangely, at a lower concentration, 3,5-dimethoxybenzoic acid, and especially syringic acid, were powerful promotors of *M. aeruginosa* Kützing growth. As might be expected, copper sulfate caused 100% inhibition; this compound has been conventionally used to control blooms in reservoirs for several years [16].

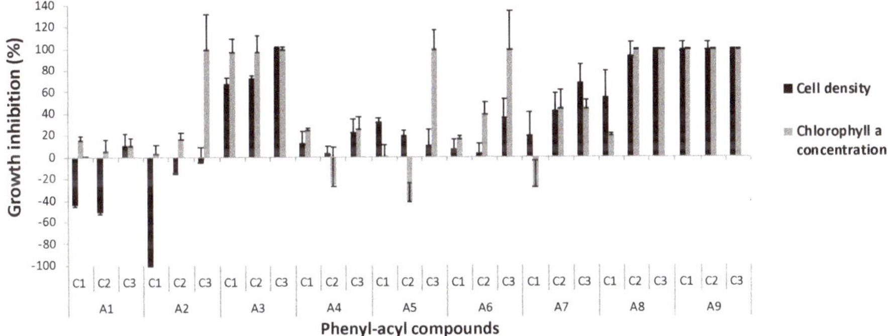

Figure 2. The effects of phenyl-acyl compounds on the growth of *Microcystis aeruginosa* Kützing at 72 h of growth according to cell density measurements and chlorophyll *a* concentration. 3,5-dimethoxybenzoic acid (A1), syringic acid (A2), caffeic acid (A3), 3-methoxyphenylacetic acid (A4), 2′-hydroxycinnamic acid (A5), 4-hydroxyphenylacetic acid (A6), vanillin (A7), eugenol (A8), and, for copper sulfate, a Cu(SO)4 control (A9). The concentrations assayed were: 100 mg/L (C3), 50 mg/L (C2), and 25 mg/L (C1). The letters and numbers refer to the structures shown in Figure 1.

3.2. The Analysis of Morphology of M. aeruginosa Kützing by Microscopy

The effect of the more active compounds (caffeic acid, eugenol, and vanillin) on *M. aeruginosa* Kützing was identified through a microscopy analysis. Several changes in the structure of the *M. aeruginosa* Kützing cells were observed under optical microscopy. A delayed cell division behavior was noticed 72 h after treatments with vanillin (Figure 3A) and eugenol (Figure 3B) at 100 mg/L. Moreover, optical microscopy evidenced changes in the cellular division of *M. aeruginosa* Kützing, since normal conditions produce symmetrical sister cells, but environmental perturbations increase the asymmetry between sister cells; when cells of *M. aeruginosa* Kützing were exposed to vanillin and eugenol, this effect was noticed as shown in Figure 3.

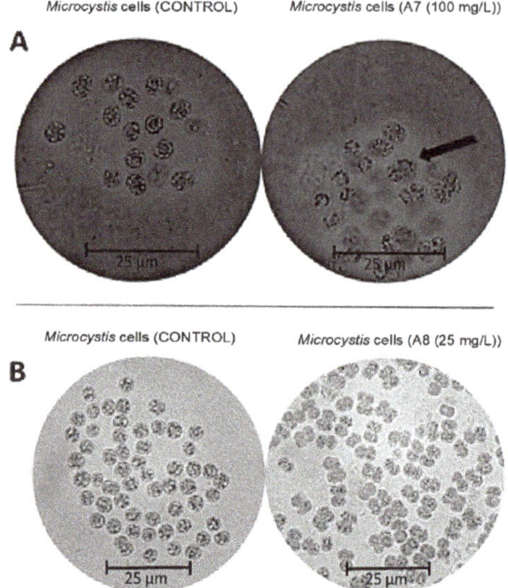

Figure 3. Aspects of *M. aeruginosa* Kützing cells under an optical microscope (40x) after treatment with vanillin (**A**) and eugenol (**B**) for 72 h. Scale bar: 25 µm.

In the SEM images, *M. aeruginosa* Kützing cultures treated with vanillin and caffeic acid showed significant changes in their morphology. In the first instance the control sample (Figure 4A) displayed well-formed and homogeneous cells embedded in a mucilage. In samples treated with vanillin (Figure 4B), the colony had partially lost its integrity and scattered cells can be seen in small groupings on a disintegrated mucilage. In cultures treated with caffeic acid (Figure 4C), the effects were more severe because mucilage was not appreciated, and the cells seem to have been lysed.

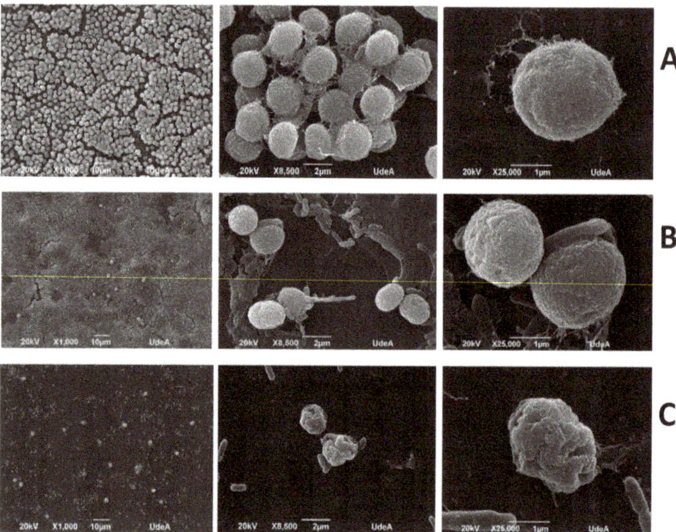

Figure 4. Scanning Electron Microscope (SEM) images showing *M. aeruginosa* Kützing colonies without treatment (**A**) and exposed to vanillin (**B**) and caffeic acid (**C**).

3.3. The Effect of Phenyl-Acyl Compounds on MC-LR Production

Through an HPLC-DAD analysis, the main toxin in *M. aeruginosa* Kützing was found to be MC-LR. To establish the effect of the compounds on the MC-LR level, we chose three compounds: syringic acid at non-growth inhibitory concentrations and caffeic acid and eugenol, which are powerful growth inhibitors and were found to reduce the levels of Chl *a*.

MC-LR concentrations were reduced after exposure to syringic acid by up to 77% at 50 mg/L and 84% at 100 mg/L, to caffeic acid by up to 89% at 50 mg/L and 100% at 100 mg/L, and after exposure to eugenol by up to 27% at 25 mg/L and 85% and 93% at 50 and 100 mg/L, respectively (Figure 5). A decrease in the toxin level could be explained based on the reduction of *M. aeruginosa* Kützing growth.

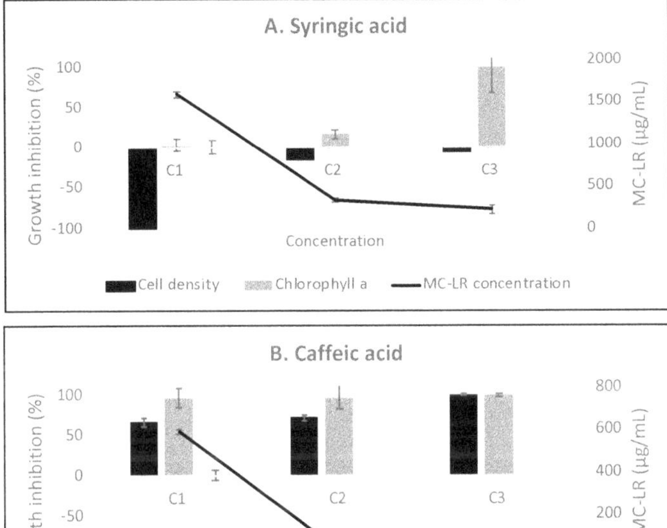

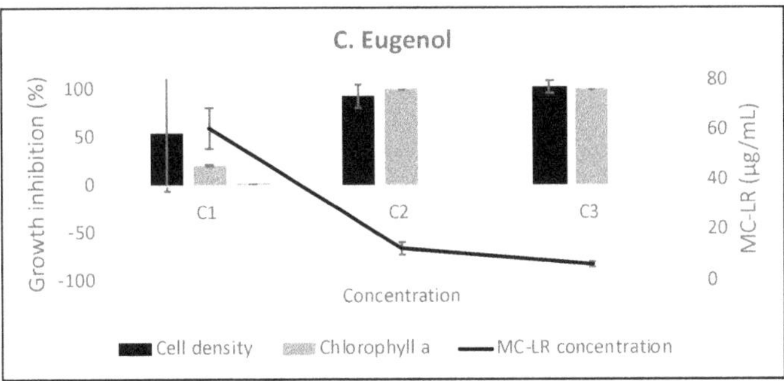

Figure 5. The effect of syringic acid (**A**), caffeic acid (**B**), and eugenol (**C**) on the concentration of microcystin-LR (MC-LR) in cultures of *M. aeruginosa* Kützing. The concentrations assayed were 100 mg/L (C3), 50 mg/L (C2), and 25 mg/L (C1).

3.4. The Analysis of the Redox Effect of Caffeic Acid on MC-LR by LC/Q-TOF

It is widely known that phenols possess a high redox capability, and because of that are tested like antioxidants. Several hours after mixing caffeic acid with MC-LR toxin, a grayish color had developed in the solution, which intensified over time, and changes in both compounds were observed through HPLC/MS. After 72 h, the presence of caffeic acid was detected by a peak at m/z 181.1; however, another substance with a peak at m/z 179.1 was also observed that corresponded to the quinone of caffeic acid by oxidation of the *orto*-dihydroxy system. Similarly, a peak at m/z 995.6 was assigned to the MC-LR toxin, and a peak at m/z 1001.5 to a new compound formed by reduction of the quinone of caffeic acid (Figure 6).

In addition, other molecules formed by degradation of the original toxin were identified, such as the peaks at m/z 88.05 and 135.1, which were assignable to an alanine and a part of the side chain ADDA, respectively. Besides this, the peaks at m/z 792.6 and 232.2 seem to indicate an oxidative breaking of ADDA.

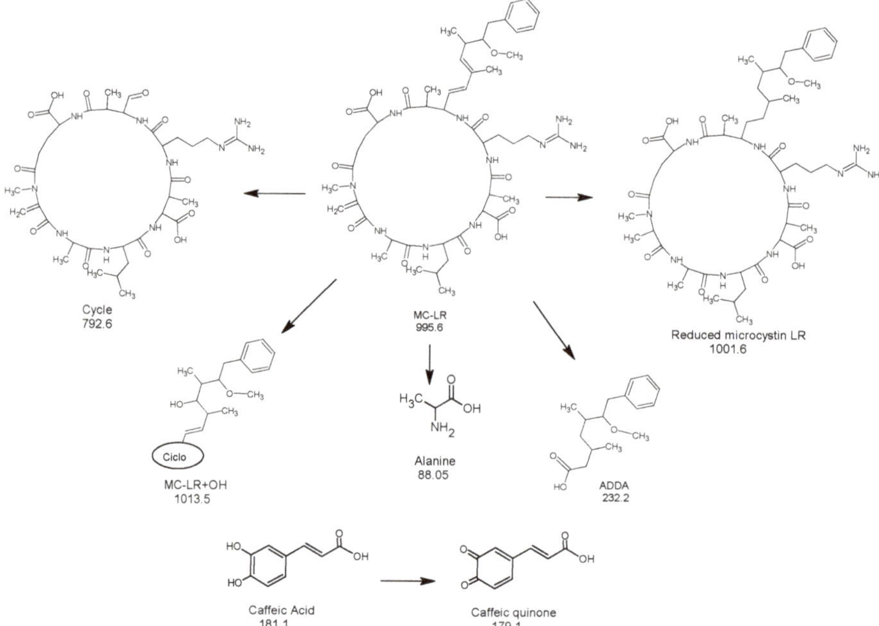

Figure 6. A proposed oxide-reductive pathway of the decomposition of MC-LR with caffeic acid.

3.5. Toxicological Effects

To establish the toxicity of the phenyl-acyl compounds, *D. pulex* neonates were used. 3,5-dimethoxybenzoic acid, syringic acid, caffeic acid, 3-methoxyphenylacetic acid, 2′-hydroxycinnamic acid, and 4′-hydroxyphenylacetic acid caused less than 10% mortality in the first 24 h. After 48 h, 4′-hydroxyphenylacetic acid caused close to 40% mortality, and the other compounds caused less than 20% mortality (Figure 7). However, eugenol and copper sulfate exhibited high neonate mortality: 80% to 100% at 24 h and 100% at 48 h at the concentrations of 25, 50, and 100 mg/L. For vanillin, mortality was 100% at 48 h at 100 mg/L. Table 1 shows the LC$_{50}$ results; some data were not estimated statistically because, under the treatment, all neonates either died (eugenol) or survived in the control (water).

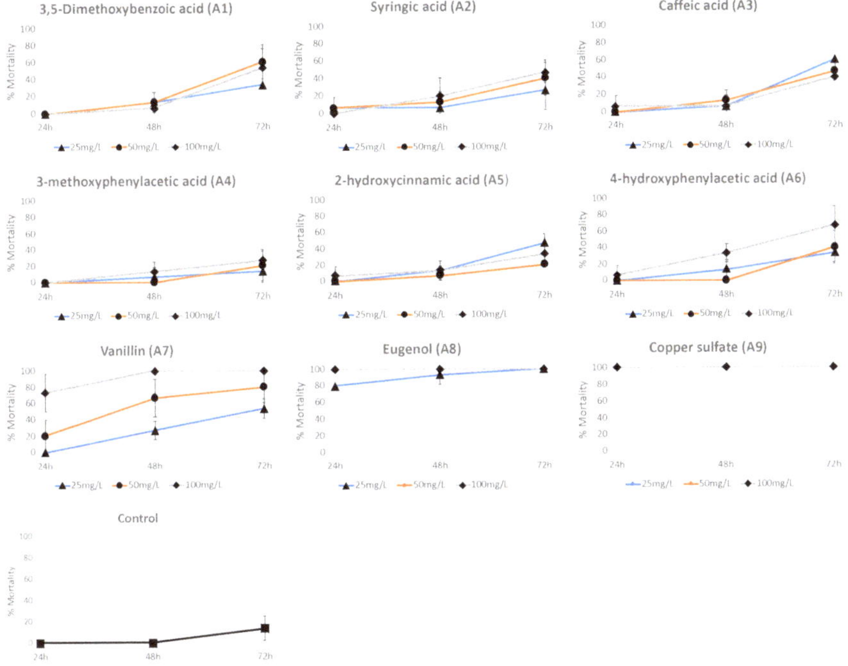

Figure 7. The toxicity of phenyl-acyl compounds and copper sulfate on *Daphnia pulex* De Geer neonates. Control = the culture medium without a compound.

Table 1. LC$_{50}$ values for *D. pulex* De Geer neonates exposed to different phenyl-acyl compounds at 24, 48, and 72 h. The confidence intervals (95%) are given in parenthesis.

Phenyl-Acyl	LC$_{50}$ 24 h	LC$_{50}$ 48 h	LC$_{50}$ 72 h
A1	NE	251.854	71.8 (52.98–108.68)
A2	760.41 (NE)	146.26 (101.15–436.72)	87.49 (66,86–132.07)
A3	131.62 (NE)	228.62 (NE)	84.30 (53.52–241.42)
A4	NE	179.88 (112.21–7471.96)	125.42 (88.37–261.20)
A5	131.62 (NE)	177.83 (112.21–7471.96)	125.63 (79.01–554.19)
A6	131.62 (NE)	126.25 (93.90–242.85)	69.37 (52.58–99.22)
A7	80.42 (67.26–98.41)	40.43 (32.73–51.42)	26.62 (22.,41–39.5)
A8	20.885 (NE)	13.23 (9.05–19.26)	7.1
A9	NE	NE	NE

NM, no mortality; NE, not estimated. 3,5-dimethoxybenzoic acid (A1), syringic acid (A2), caffeic acid (A3), 3-methoxyphenylacetic acid (A4), 2′-hydroxycinnamic acid (A5), 4′-hydroxyphenylacetic acid (A6), vanillin (A7), eugenol (A8), and, as a control, copper sulfate (A9).

4. Discussion

Currently, several strategies are used to control *Microcystis* blooms and toxins, ranging from the application of dangerous copper salts to ultrasound and adsorption on materials such as activated carbon [17]. In this research, eight phenyl-acyl compounds were assayed as inhibitors of *M. aeruginosa* Kützing growth and toxin production. Eugenol, caffeic acid, and vanillin showed inhibition activities of 100%, 75%, and 50%, respectively, at 50 mg/L. The effects on the growth of *M. aeruginosa* Kützing were assessed through cell count and Chl *a* analysis; however, these results were not comparable. Therefore, there was a lack of correlation between cell counts and Chl *a* levels, which was possibly due to the different mechanisms of action in *M. aeruginosa* Kützing.

The production of Chl *a* was completely inhibited by syringic acid, 2′-hydroxycinnamic acid, and 4′-hydroxyphenylacetic at 100 mg/L, while, according to the cell count, there was close to 45% inhibition. A probable mechanism of action of these compounds could involve oxidative damage to Chl *a* through the inhibition of antioxidant enzymes [18]. The other substances, 3,5-dimethoxybenzoic acid and 3-methoxyphenylacetic acid, were weak growth promotors; however, there are no data on the effect of these substances as inductors of cyanobacteria growth or their role in the formation of blooms. In the same way, some phenols, such as those that originate from the decomposition of organic matter, have also shown an effect on the development of cyanobacterial blooms [11,19].

According to these results, caffeic acid was the best *M. aeruginosa* Kützing growth inhibitor. It reduced the level of MC-LR and was practically a non-toxic substance, since overall survival was observed in *D. pulex* neonates, which survived when exposed to all concentrations; the effects that were noticed after 48 h may be caused by starvation. This harmless effect has been previously reported for this type of compound in *Danio rerio* [20]. However copper sulfate, which is recommended as an algicide, is toxic to many other species in aquatic environments [16,21].

Regarding the mechanism of action of these compounds, caffeic acid seems to involve a redox mechanism to degrade MC-LR, which was detected by means of direct observation of the color of the solution and HPLC/MS. Some authors have already proposed the use of phenolic acids to inhibit *M. aeruginosa* growth [20,22,23]. It has been suggested that the autoxidation of phenols induces inhibitory effects by the production of free radicals [24] with the generation of radical superoxide anions. In this way, they can induce a lipid peroxidation that affects cell membrane permeability and causes the death of *M. aeruginosa* Kützing [20]. Similarly, rice straw extracts that contain this same type of phenolic acid were shown to provoke changes in the antioxidant enzyme system in the cells of results cyanobacteria, such as *M. aeruginosa* Kützing, thereby suspending their growth and causing cell death [25].

In addition, several of the assayed molecules also seem to act at the stage of formation of the mucilage, which is mainly composed of exopolysaccharides, and whose main function is the agglutination and protection of the colonies [26,27]. These results are in accordance with the SEM images (Figure 4C) because damage to the membrane and changes in the mucilage's stability were also observed under caffeic acid treatment. In addition, vanillin and eugenol caused delays in cell division, since dividing cells were detected after 72 h of treatment (Figure 3A,B). These same substances also caused asymmetry between sister cells, an effect that has already been reported for *M. aeruginosa* Kützing under treatment of copper sulfate and the herbicide diuron, and could be an indicator of environmental disturbance [28].

About the effect of caffeic acid, the fragmentation patterns of MC-LR toxin agree with previous reports [29]. However, the exact mechanism of the MC-LR toxin's transformation needs to be elucidated, including the residues in the molecule that are reduced by caffeic acid and the kinetics of the process. These findings open new perspectives on the elimination of MC-LR toxins in potable reservoirs contaminated with cyanobacterial blooms, based on a chemical or physical redox process.

Cyanobacteria blooms develop quickly in large areas that contain millions of hectoliters of water; therefore, the application of substances for their control could be related to an early detection method. Otherwise, the amounts of required inhibitory substances would be too large and possibly affect the water's quality and increase costs. Therefore, knowledge about the effects and the mechanism of action of these types of substances can generate information about new control methods as well as provide additional data about the chemical, biological, and physical agents that trigger the blooms. Thus, exogenous substances and products from industrial activity or vegetal biomass decomposition in reservoirs could also elicit these blooms.

5. Conclusions

In this work, several effects were detected in *M. aeruginosa* Kützing after the application of several phenyl-acyl compounds. Some of them, such as caffeic acid, were inhibitory, while others were found to act as growth promotors. The activity of caffeic acid included changes in cell morphology and division and the formation of mucilage, while its mechanism of action on the MC-LR toxin seems to involve a degradative redox process. Molecules with promotor effects indicate the existence of other biochemical mechanisms in the blooms' formation that are different from those that have traditionally been considered, such as edaphic factors and nutrient availability.

Author Contributions: Methodology: N.H., F.E.; Investigation: N.H., J.P.V., M.T.F.; Interpretations: N.H., M.T.F., F.E.; Writing: N.H., F.E.; Supervision N.H., F.E.

Funding: COLCIENCIAS-COLOMBIA (grant FP44842-049-2016).

Conflicts of Interest: The authors declare no conflict of interest.

References

1. He, X.; Wert, E.C. Colonial cell disaggregation and intracellular microcystin release following chlorination of naturally occurring Microcystis. *Water Res.* **2016**, *101*, 10–16. [CrossRef] [PubMed]
2. Azevedo, S.M.F.O.; Carmichael, W.W.; Jochimsen, E.M.; Rinehart, K.L.; Lau, S.; Shaw, G.R.; Eaglesham, G.K. Human intoxication by microcystins during renal dialysis treatment in Caruaru — Brazil. *Toxicology* **2002**, *181–182*, 441–446. [CrossRef]
3. Su, J.; Yang, X.; Zhou, Y.; Zheng, T. Marine bacteria antagonistic to the harmful algal bloom species Alexandrium tamarense (Dinophyceae). *Biol. Control* **2011**, *56*, 132–138. [CrossRef]
4. Padovesi-Fonseca, C.; Philomeno, M.G. Effects of algicide (copper sulfate) application on short-term fluctuations of phytoplankton in Lake Paranoá, central Brazil. *Braz. J. Biol.* **2004**, *64*, 819–826. [CrossRef] [PubMed]
5. Carey, C. Resilience to Blooms Resilience to Blooms. *Science* **2015**, 10–13. [CrossRef]
6. Chen, Y.; Li, J.; Wei, J.; Kawan, A.; Wang, L.; Zhang, X. Vitamin C modulates *Microcystis aeruginosa* death and toxin release by induced Fenton reaction. *J. Hazard. Mater.* **2017**, *321*, 888–895. [CrossRef] [PubMed]
7. Mohamed, Z.A. Macrophytes-cyanobacteria allelopathic interactions and their implications for water resources management — A review. *Limnologica* **2017**, *63*, 122–132. [CrossRef]
8. Herrera, N.A.; Flórez, M.T.; Echeverri, L.F. Evaluación preliminar de la reducción de microcistina-LR en muestras de florecimientos a través de sistemas sedimentarios. *Rev. Int. Contam. Ambient.* **2015**, *31*, 405–414.
9. Mucci, M.; Noyma, N.P.; de Magalhães, L.; Miranda, M.; van Oosterhout, F.; Guedes, I.A.; Huszar, V.L.M.; Marinho, M.M.; Lürling, M. Chitosan as coagulant on cyanobacteria in lake restoration management may cause rapid cell lysis. *Water Res.* **2017**, *118*, 121–130. [CrossRef]
10. Lin, J.L.; Hua, L.C.; Hung, S.K.; Huang, C. Algal removal from cyanobacteria-rich waters by preoxidation-assisted coagulation–flotation: Effect of algogenic organic matter release on algal removal and trihalomethane formation. *J. Environ. Sci. (China)* **2018**, *63*, 147–155. [CrossRef]
11. Lin, Y.; Chen, A.; Wu, G.; Peng, L.; Xu, Z.; Shao, J. Growth, microcystins synthesis, and cell viability of *Microcystis aeruginosa* FACHB905 to dissolved organic matter originated from cattle manure. *Int. Biodeterior. Biodegrad.* **2017**, *118*, 126–133. [CrossRef]
12. Stanier, R.Y.; Kunisawa, R.; Mandel, M.; Cohen-Bazire, G. Purification and properties of unicellular blue-green algae (Order Chroococcales). *Bacteriol. Rev.* **1971**, *35*, 171–205. [CrossRef] [PubMed]
13. Gregor, J.; Geriš, R.; Maršálek, B.; Heteša, J.; Marvan, P. In situ quantification of phytoplankton in reservoirs using a submersible spectrofluorometer. *Hydrobiologia* **2005**, *548*, 141–151. [CrossRef]
14. Sant'Anna, C.L.; Azevedo, M.T.D.P.; Senna, P.A.C.; Komárek, J.; Komárková, J. Planktic Cyanobacteria from São Paulo State, Brazil: Chroococcales. *Rev. Bras. Botânica* **2004**, *27*, 213–227. [CrossRef]
15. Aguete, E.C.; Gago-Martínez, A.; Leão, J.M.; Rodríguez-Vázquez, J.A.; Menàrd, C.; Lawrence, J.F. HPLC and HPCE analysis of microcystins RR, LR and YR present in cyanobacteria and water by using immunoaffinity extraction. *Talanta* **2003**, *59*, 697–705. [CrossRef]
16. Kinley, C.M.; Iwinski, K.J.; Hendrikse, M.; Geer, T.D.; Rodgers, J.H. Cell density dependence of *Microcystis aeruginosa* responses to copper algaecide concentrations: Implications for microcystin-LR release. *Ecotoxicol. Environ. Saf.* **2017**, *145*, 591–596. [CrossRef]

17. Şengül, A.B.; Ersan, G.; Tüfekçi, N. Removal of intra- and extracellular microcystin by submerged ultrafiltration (UF) membrane combined with coagulation/flocculation and powdered activated carbon (PAC) adsorption. *J. Hazard. Mater.* **2018**, *343*, 29–35. [CrossRef]

18. Ni, L.; Rong, S.; Gu, G.; Hu, L.; Wang, P.; Li, D.; Yue, F.; Wang, N.; Wu, H.; Li, S. Inhibitory effect and mechanism of linoleic acid sustained-release microspheres on *Microcystis aeruginosa* at different growth phases. *Chemosphere* **2018**, *212*, 654–661. [CrossRef]

19. Wang, W.; Zhang, Y.; Shen, H.; Xie, P.; Yu, J. Changes in the bacterial community and extracellular compounds associated with the disaggregation of Microcystis colonies. *Biochem. Syst. Ecol.* **2015**, *61*, 62–66. [CrossRef]

20. Jin, P.; Wang, H.; Huang, W.; Liu, W.; Fan, Y.; Miao, W. The allelopathic effect and safety evaluation of 3,4-Dihydroxybenzalacetone on *Microcystis aeruginosa. Pestic. Biochem. Physiol.* **2017**. [CrossRef]

21. Zhou, C.; Huang, J.C.; Liu, F.; He, S.; Zhou, W. Effects of selenite on Microcystis aeruginosa: Growth, microcystin production and its relationship to toxicity under hypersalinity and copper sulfate stresses. *Environ. Pollut.* **2017**, *223*, 535–544. [CrossRef] [PubMed]

22. Mecina, G.F.; Dokkedal, A.L.; Saldanha, L.L.; Chia, M.A.; Cordeiro-Araújo, M.K.; do Carmo Bittencourt-Oliveira, M.; da Silva, R.M.G. Response of *Microcystis aeruginosa* BCCUSP 232 to barley (Hordeum vulgare L.) straw degradation extract and fractions. *Sci. Total Environ.* **2017**, *599–600*, 1837–1847. [CrossRef] [PubMed]

23. Zhang, C.; Ling, F.; Yi, Y.L.; Zhang, H.Y.; Wang, G.X. Algicidal activity and potential mechanisms of ginkgolic acids isolated from Ginkgo biloba exocarp on Microcystis aeruginosa. *J. Appl. Phycol.* **2014**, *26*, 323–332. [CrossRef]

24. Nakai, S.; Inoue, Y.; Hosomi, M. Algal growth inhibition effects and inducement modes by plant-producing phenols. *Water Res.* **2001**, *35*, 1855–1859. [CrossRef]

25. Hua, Q.; Liu, Y.-G.; Yan, Z.-L.; Zeng, G.-M.; Liu, S.-B.; Wang, W.-J.; Tan, X.-F.; Deng, J.-Q.; Tang, X.; Wang, Q.-P. Allelopathic effect of the rice straw aqueous extract on the growth of *Microcystis aeruginosa. Ecotoxicol. Environ. Saf.* **2018**, *148*, 953–959. [CrossRef]

26. Kehr, J.-C.; Dittmann, E. Biosynthesis and Function of Extracellular Glycans in Cyanobacteria. *Life* **2015**, *5*, 164–180. [CrossRef] [PubMed]

27. Pereira, S.; Zille, A.; Micheletti, E.; Moradas-Ferreira, P.; De Philippis, R.; Tamagnini, P. Complexity of cyanobacterial exopolysaccharides: Composition, structures, inducing factors and putative genes involved in their biosynthesis and assembly. *FEMS Microbiol. Rev.* **2009**, *33*, 917–941. [CrossRef]

28. Costas, E.; Lopez-Rodas, V. Copper sulphate and DCMU-herbicide treatments increase asymmetry between sister cells in the toxic cyanobacteria *Microcystis aeruginosa*: Implications for detecting environmental stress. *Water Res.* **2006**, *40*, 2447–2451. [CrossRef]

29. Duan, X.; Sanan, T.; De La Cruz, A.; He, X.; Kong, M.; Dionysiou, D.D. Susceptibility of the Algal Toxin Microcystin-LR to UV/Chlorine Process: Comparison with Chlorination. *Environ. Sci. Technol.* **2018**, *52*, 8252–8262. [CrossRef]

Article

Seasonal Succession of Phytoplankton Functional Groups and Driving Factors of Cyanobacterial Blooms in a Subtropical Reservoir in South China

Lingai Yao [1,2], Xuemin Zhao [2,*], Guang-Jie Zhou [3], Rongchang Liang [2], Ting Gou [2], Beicheng Xia [1,*], Siyang Li [2] and Chang Liu [2]

[1] School of Environmental Sciences and Engineering, Sun Yat-sen University, Guangzhou 510275, China; yaolingai102@163.com
[2] South China Institute of Environmental Science, Ministry of Ecology and Environment, Guangzhou 510345, China; liangrongchang@scies.org (R.L.); gouting@scies.org (T.G.); lisiyang@scies.org (S.L.); liuchang@scies.org (C.L.)
[3] The Swire Institute of Marine Science and School of Biological Sciences, The University of Hong Kong, Pokfulam, Hong Kong, China; zhougj01@gmail.com
* Correspondence: xiabch@mail.sysu.edu.cn (B.X.); zhaoxuemin@scies.org (X.Z.); Tel.: +86-20-8411-4591 (B.X.); +86-20-2911-9680 (X.Z.)

Received: 12 March 2020; Accepted: 16 April 2020; Published: 19 April 2020

Abstract: Freshwater phytoplankton communities can be classified into a variety of functional groups that are based on physiological, morphological, and ecological characteristics. This classification method was used to study the temporal and spatial changes in the phytoplankton communities of Gaozhou Reservoir, which is a large municipal water source in South China. Between January 2015 and December 2017, a total of 155 taxa of phytoplankton that belong to seven phyla were identified. The phytoplankton communities were classified into 28 functional groups, nine of which were considered to be representative functional groups (relative biomass > 10%). Phytoplankton species richness was greater in the summer and autumn than in the winter and spring; cyanobacterial blooms occurred in the spring. The seasonal succession of phytoplankton functional groups was characterized by the occurrence of functional groups P (*Staurastrum* sp. and *Closterium acerosum*) and Y (*Cryptomonas ovata* and *Cryptomonas erosa*) in the winter and spring, and functional groups NA (*Cosmarium* sp. and *Staurodesmus* sp.) and P (*Staurastrum* sp. and *Closterium acerosum*) in the summer and autumn. The temperature, nitrogen, and phosphorus levels were the main factors driving seasonal changes in the phytoplankton communities of Gaozhou Reservoir. The functional group M (*Microcystis aeruginosa*) dominated the community during the cyanobacterial blooms in spring 2016, with the maximum algal cell density of 3.12×10^8 cells L^{-1}. Relatively low temperature (20.8 °C), high concentrations of phosphorus (0.080–0.110 mg L^{-1}), suitable hydrological and hydrodynamic conditions (e.g., relatively long retention time), and relatively closed geographic location in the reservoir were the key factors that stimulated the cyanobacterial blooms during the early stages.

Keywords: subtropical reservoir; functional groups; phytoplankton; seasonal succession; environmental factors; cyanobacterial bloom

---

1. Introduction

Phytoplankton, as primary producers in aquatic ecosystems, are important for the maintenance of a stable aquatic environment [1,2], and they strongly affect the total productivity of aquatic ecosystems [3]. Phytoplankton are particularly sensitive to physical, chemical, and biological changes in the aquatic environment [4], and increases in nutrient availability often lead to severe phytoplankton proliferation [5]. Once blooms form, aquatic ecosystem services, such as drinking water quality, fisheries, and landscape, become damaged [6]. The process of bloom formation has a destructive effect on phytoplankton community structure, due to the dominance of certain phytoplankton species [7]. Changes in the composition and structure of the phytoplankton community (e.g., species composition, diversity index, community structure, and quantitative distribution) can be used to evaluate the eutrophication status of reservoirs [8]. Thus, these metrics represent important indicators of the environmental quality of aquatic ecosystems [9].

Reynolds et al. [10] and Padisák et al. 2009 [11] proposed functional groups to describe both phytoplankton community structure and the changes to that structure. The species of phytoplankton from specific habitats with similar sensitivities are classified into the same functional group; in total 31 functional groups were devised for freshwater phytoplankton [12]. Phytoplankton functional groups have been used to study the effects of changes in aquatic ecology on the physiological, morphological, and ecological characteristics of phytoplankton in various rivers, lakes, and reservoirs worldwide [13–16]. Studies have shown that phytoplankton functional groups H, X1, LO, and S1 are the most representative during the rainy and dry seasons in semi-arid reservoirs in Brazil [17], while phytoplankton functional groups LM, P, T, and Y exhibit strong seasonal variation in Erhai Lake, Yunnan, China [18]. The application of phytoplankton functional group classification methods in aquatic ecosystems provides key data that reflect the dynamics of phytoplankton communities [19,20].

There are many methods for assessing the nutritional status of aquatic environments. The assessment of eutrophication in aquatic ecosystems is essentially a multivariate comprehensive decision-making process for aquatic ecosystems [21–23]. Based on the EU Water Framework Directive [19], Padisák et al. developed the Q index method, based on the taxonomy of the phytoplankton functional groups, to evaluate the ecological status of different types of water bodies [24]. Phytoplankton functional groups and Q index evaluation methods have been widely used in phytoplankton related research studies [17,25]. In addition, the comprehensive trophic level index (TLI (Σ)) has been widely used for evaluating eutrophication in reservoirs [26,27]. TLI (Σ) index, which is based on water quality parameters, provides a continuous value that represents the trophic state of a reservoir. Continuous numerical changes provide a basis that can be used to investigate the mechanisms underlying eutrophication [26]. In this study, two evaluation methods, Q index and TLI (Σ) index, were used to study phytoplankton community succession and the factors driving succession in a reservoir in southern China (i.e., Gaozhou Reservoir).

Gaozhou Reservoir is a large artificial reservoir in western Guangdong, China, with an annual water supply of 50 billion cubic meters. This reservoir is located in the monsoon region of China, between the northern tropics and southern subtropics. The climate in this region is warm, with abundant sunshine and rainfall. However, rainfall is unevenly distributed and it changes substantially with season. The rainfall in the spring and summer is usually twice that in autumn and winter [28,29]. In recent years, pollution originating from domestic sewage and agricultural non-point sources has increased in the catchment areas of Gaozhou Reservoir, resulting in serious eutrophication and cyanobacterial blooms [30,31]. Cyanobacterial blooms have been recorded four times since 2009 (in 2010, 2011, 2013, and 2016) [32,33], threatening the water quality security of the reservoir. Although many studies have investigated the seasonal succession of phytoplankton functional groups in rivers and lakes, the seasonal succession of phytoplankton functional groups in drinking water reservoirs, as well as the associated cyanobacterial blooms, remain largely unstudied.

The objectives of this study were (1) to identify the phytoplankton composition in the reservoir; (2) to determine the phytoplankton functional groups and their seasonal succession; (3) to analyze the driving factors for the occurrence of cyanobacterial blooms; and, (4) to understand and assess the status of the aquatic environment in the reservoir, while using the phytoplankton functional group classification method, as well as the Q index and the TLI (Σ) index. The study provides basic and useful data regarding seasonal succession of phytoplankton communities and environmental determinants of algal species and abundance, which can be effectively used to develop strategies for the management of cyanobacterial blooms in the reservoir.

2. Materials and Methods

2.1. Study Period and Site Description

The Gaozhou Reservoir is located in Maoming City, Guangdong Province, China. This region has a subtropical climate. The reservoir has a canyon shape, with a storage capacity of 1.15×10^9 cubic meters and a catchment area of 1022 km^2 [34]. The water retention time (WRT) in the Gaozhou reservoir was about 451 days [34]. Three rivers (the Shenzhen River, the Pengqing River, and the Guding River) feed into the rain collection area of the two sub-areas of the reservoir: the Liangde and Shigu areas (Figure 1).

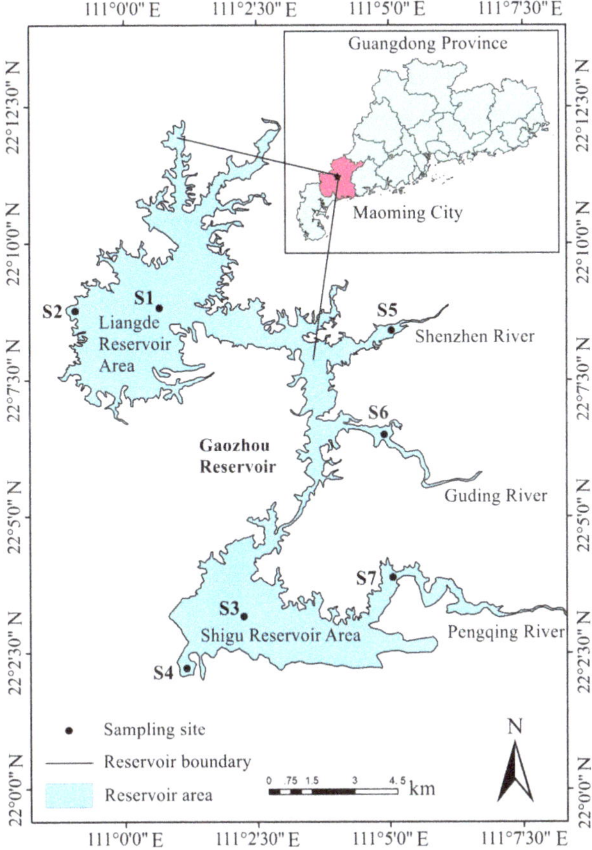

Figure 1. of the Gaozhou Reservoir and of the sampling sites in the reservoir.

The water samples were collected at seven sites in the Gaozhou Reservoir (Figure 1) in January, April, July, and November of 2015, 2016 and 2017 (12 sampling events with 84 samples in total). Sites S1 and S2 were located in the Liangde reservoir district; sites S3 and S4 were located in the Shigu reservoir district; and, sites S5, S6, and S7 were located at the entrances of the Shenzhen River, the Guding River, and the Pengqing River, respectively.

2.2. Sample Analysis

A portable Professional Plus Multiparameter Instrument (Pro Plus, YSI, USA) was used to measure the water temperature (WT), pH, dissolved oxygen (DO), and electrical conductivity (EC). Samples (5 L each) were taken at 0.5 m water depth at each site. The measurements of water quality parameters were performed according to the Chinese National Standard Method [35] (similar to those of the American Public Health Association [36]). The total nitrogen (TN), total phosphorus (TP), permanganate index (COD_{Mn}), and suspended solids (SS) were measured with unfiltered water, while nitrate (NO_3-N) and ammonium (NH_4-N) were measured with filtered water. The chlorophyll-a concentrations were determined after extraction with acetone [35,37].

At each site, a phytoplankton net (No.25, Beijing Purity Instrument) was dragged at 0.5 m water depth for five minutes. Subsequently, a 25-mL filtered water sample was collected, fixed by adding 1 mL of formalin solution, and then inspected using the eyepiece visual field counting method under a microscope. At each site, an additional 1 L water sample was collected at 0.5 m water depth and 10 mL of Lugol's solution was added on site in order to fix the sample. After sedimentation for 24 h, the water sample was concentrated to 30 mL, and then examined under a microscope to quantify cell densities [38]. Phytoplankton biomass was calculated according to the method that was described by Hillebrand et al. [39], while assuming that 1 mm^3 equals 1 mg [40].

2.3. Data Analysis

The European Water Framework Directive (European Parliament and Council, 2000) developed a Q index evaluation method that is based on the interactions between functional groups and environmental characteristics [41]. The Q index primarily uses the composition of functional groups and phytoplankton biomass to characterize the eutrophic state of the water body, and it reflects aquatic ecological health status [11]. The Q index is calculated, as follows:

$$Q = \sum_{i=1}^{n} \left(\frac{n_i}{N} \times F_i \right),$$

(1)

where N is the total biomass; n is the number of functional groups; n_i is the biomass of the i-th functional group; and, F_i is the value of the i-th functional group [24].

We assigned an F value that ranges from 0 to 5, with higher values for the more-pristine assemblages in the reservoir, and lower values for assemblages that are typical of less-pristine conditions. Following Reynolds et al. [10], species representing > 10% of the total biomass were considered to be representative functional groups (i.e., the functional groups that made a substantial contribution to total biomass).

The Q index was divided into five levels: the values 0–1, 1–2, 2–3, 3–4, and 4–5 correspond to bad, tolerable, medium, good, and excellent, respectively [24]. Smaller Q values indicated a higher degree of eutrophication.

We used chlorophyll a (Chla), TP, TN, transparency (SD), and the COD_{Mn} to calculate the Comprehensive Trophic Level Index (TLI (Σ)), which was used to evaluate the eutrophication status of Gaozhou Reservoir [27,42]:

$$TLI\ (\Sigma) = \sum_{j=1}^{m} W_j \times TLI\ (j),$$

(2)

where TLI (Σ) is the comprehensive trophic level index; TLI (j) is the nutrition status index representing the *j*th parameter; and, W*j* is the relevant weight of the nutrition status index of the jth parameter.

In order to calculate the nutritional status index of each individual item, we used the following formulas:

$$TLI\ (Chl\ a) = 10 \times (2.5 + 1.086 \ln\ (Chl\text{-}a)),$$

(3)

$$TLI\ (TP) = 10 \times (9.436 + 1.624 \ln\ (TP)),$$

(4)

$$TLI\ (TN) = 10 \times (5.453 + 1.694 \ln\ (TN)),$$

(5)

$$\text{TLI (SD)} = 10 \times (5.118 - 1.940 \ln (\text{SD})), \tag{6}$$

$$\text{TLI (COD}_{Mn)} = 10 \times (0.109 + 2.661 \ln (\text{COD}_{Mn})). \tag{7}$$

Here, Chla is expressed in mg m^{-3}; SD is expressed in m; and, all other variables are expressed in mg L^{-1}.

The values of the TLI (Σ) index were divided into six levels: oligotrophic (0–30), mesotrophic (30–50), light-eutrophic (50–60), mid-eutrophic (60–70), high-eutrophic (70–80), and hyper-eutrophic (>80) [23,43].

One-way analysis of variance (one-way ANOVA) was used to compare the difference of environmental factors among seasons. We took three samplings of each sampling site as three replicates for each season. Redundancy analysis (RDA) was used to reveal the contribution of 13 environmental factors to variations of nine representative functional groups of phytoplankton for each season. Statistical analysis was performed while using the SPSS 19.0 statistical package software (IBM, Armonk, NY, USA), and RDA analysis was analyzed using the Canoco 5.04 software.

3. Results

3.1. Phytoplankton Dynamics

In total, 155 phytoplankton taxa were identified across all of the samples collected from the Gaozhou Reservoir between January 2015 and December 2017. These taxa included cyanobacteria, diatoms, chlorophytes, dinoflagellates, cryptophytes, euglenophytes, and chrysophytes. There were 103, 109, 126, and 120 taxa in samples collected in the winter, spring, summer, and autumn, respectively; species richness in the summer and autumn was greater than that in the winter and spring (Table 1).

Table 1. Species of phytoplankton and their respective proportion (%) in different seasons in the Gaozhou Reservoir.

	Winter		Spring		Summer		Autumn		All seasons	
	Taxa	Proportion	Taxa	Proportion	Taxa	Proportion	Taxa	Proportion	Taxa	Proportion
Cyanobacteria	10	9.71%	15	13.76%	19	15.08%	15	12.50%	20	12.90%
Cryptophytes	4	3.88%	4	3.67%	4	3.17%	4	3.33%	4	2.58%
Dinoflagellates	4	3.88%	5	4.59%	5	3.97%	5	4.17%	5	3.23%
Chrysophytes	1	0.97%	1	0.92%	1	0.79%	1	0.83%	1	0.65%
Diatoms	26	25.24%	26	23.85%	29	23.02%	28	23.33%	30	19.35%
Euglenophytes	2	1.94%	3	2.75%	3	2.38%	3	2.50%	3	1.94%
Chlorophytes	56	54.37%	55	50.46%	65	51.59%	64	53.33%	92	59.35%
Total taxa	103	100%	109	100%	126	100%	120	100%	155	100%

The phytoplankton taxa in the reservoir fell into 28 of the 31 groups previously described: M, S1, SN, H1, H2, X1, X2, X3, Y, LM, LO, K, TC, E, D, C, B, A, MP, W1, W2, WO, G, J, F, T, NA, and P (Table 2). Here, nine groups were considered to be representative functional groups (marked with "a" in Table 2).

During the study period, the average densities of algal cells in the winter, spring, summer, and autumn were 5.42 $\times$ 10^6 cells L^{-1}, 29.7 $\times$ 10^6 cells L^{-1}, 13.2 $\times$ 10^6 cells L^{-1}, and 11.3 $\times$ 10^6 cells L^{-1}, respectively. The density of algal cells was highest in the spring, followed by the summer, autumn, and winter (Figure 2). In the spring of 2016, cyanobacterial blooms occurred at S3 and S4, where S3 is near the water supply outlet of the reservoir. The algal cell densities at S3 and S4 reached a maximum of 3.12 $\times$ 10^8 cells L^{-1} and 1.51 $\times$ 10^8 cells L^{-1}, respectively.

Table 2. Main phytoplankton taxa, functional groups, and respective F factors in the samples collected from the Gaozhou Reservoir from 2015 to 2017.

Functional Groups	Phytoplankton Species	Taxonomic Group	F Factor
M	*Microcystis aeruginosa* [a]	Cyanobacteria	0
S1	*Pseudanabaena* sp.	Cyanobacteria	0
SN	*Cylindrospermopsis raciborskii*	Cyanobacteria	0
H1	*Aphanizomenon flos-aquae*	Cyanobacteria	0
H2	*Dolichospermum circinale*	Cyanobacteria	2.0
X1	*Ankistrodesmus falcatus, Monoraphidium* sp.	Chlorophytes	3.5
X2	*Chroomonas acuta* [a]	Cryptophytes	5.0
X3	*Schroederia* sp.	Chlorophytes	5.0
Y	*Cryptomonas ovata* [a], *Cryptomonas erosa* [a], *Gymnodinium aeruginosum* [a]	Cryptophytes Dinoflagellates	3.0
LM	*Ceratium hirundinella*	Dinoflagellates	4.0
LO	*Peridiniopsis borgei* [a] *Chroococcus* sp., *Merismopedia glauca*	Dinoflagellates Cyanobacteria	4.0
K	*Aphanocapsa* sp.	Cyanobacteria	0
TC	*Gloeocapsa punctata*	Cyanobacteria	4.0
E	*Dinobryon divergens*	Chrysophytes	5.0
D	*Synedra acus, Nitzschia sublinearis*	Diatoms	2.0
C	*Cyclotella meneghiniana, Cymbella perpusilla, Navicula* sp., *Diploneis* sp.	Diatoms	3.0
B	*Cyclotella bodanica*	Diatoms	4.0
A	*Rhizosolenia* sp., *Attheya zachariasi*	Diatoms	4.0
MP	*Achnanthes exigua, Cocconeis placentula*	Diatoms	4.0
W1	*Euglena* sp., *Phacus* sp.	Euglenophytes	0
W2	*Trachelomonas* sp.	Euglenophytes	1.0
WO	*Chlamydomonas globosa*	Chlorophytes	0
G	*Eudorina elegans, Pandorina morum*	Chlorophytes	2.0
J	*Tetraëdron trigonum* [a], *Pediastrum duplex* var. *gracillimum, Scenedesmus* sp., *Chodatella* sp., *Crucigenia* sp., *Coelastrum* sp.	Chlorophytes	2.0
F	*Haematococcus pluvialis, Planktosphaeria gelotinosa, Quadrigula chodatii, Elakatothrix gelatinosa* [a], *Selenastrum dibraianum, Kirchneriella lunaris, Oocystis lacustis*	Chlorophytes	5.0
T	*Mougeotia gracillima* [a]	Chlorophytes	5.0
NA	*Cosmarium* sp. [a], *Staurodesmus* sp. [a], *Euastrum* sp., *Staurastrum* sp. [a], *Closterium acerosum* [a]	Chlorophytes	3.0
P	*Melosira varians* [a], *Fragilaria* sp. [a]	Chlorophytes Diatoms	2.0

[a] Descriptor species (>10% of the total biomass).

In total, 18 representative functional groups were identified in the winter and spring, and 20 were identified in the summer and autumn (Figure 2). The functional groups P (*Staurastrum* sp. and *Closterium acerosum*) and Y (*Cryptomonas ovata* and *Cryptomonas erosa*) were the dominant phytoplankton taxa in the winter and spring. Functional group X2 (*Chroomonas acuta*) was more abundant in the winter, while functional group M (*Microcystis aeruginosa*) grew faster in the spring. Functional groups NA (*Cosmarium* sp. and *Staurodesmus* sp.) and P (*Staurastrum* sp. and *Closterium acerosum*) dominated the communities in the summer and autumn. In spring 2016, the community at site S3 was dominated by functional group M (*Microcystis aeruginosa*), comprising 98.5% of the total biomass, while the community at site S4 was dominated by both functional group M (*Microcystis aeruginosa*) and functional group P (*Melosira varians*), comprising, respectively, 86.4% and 12.1% of the total biomass.

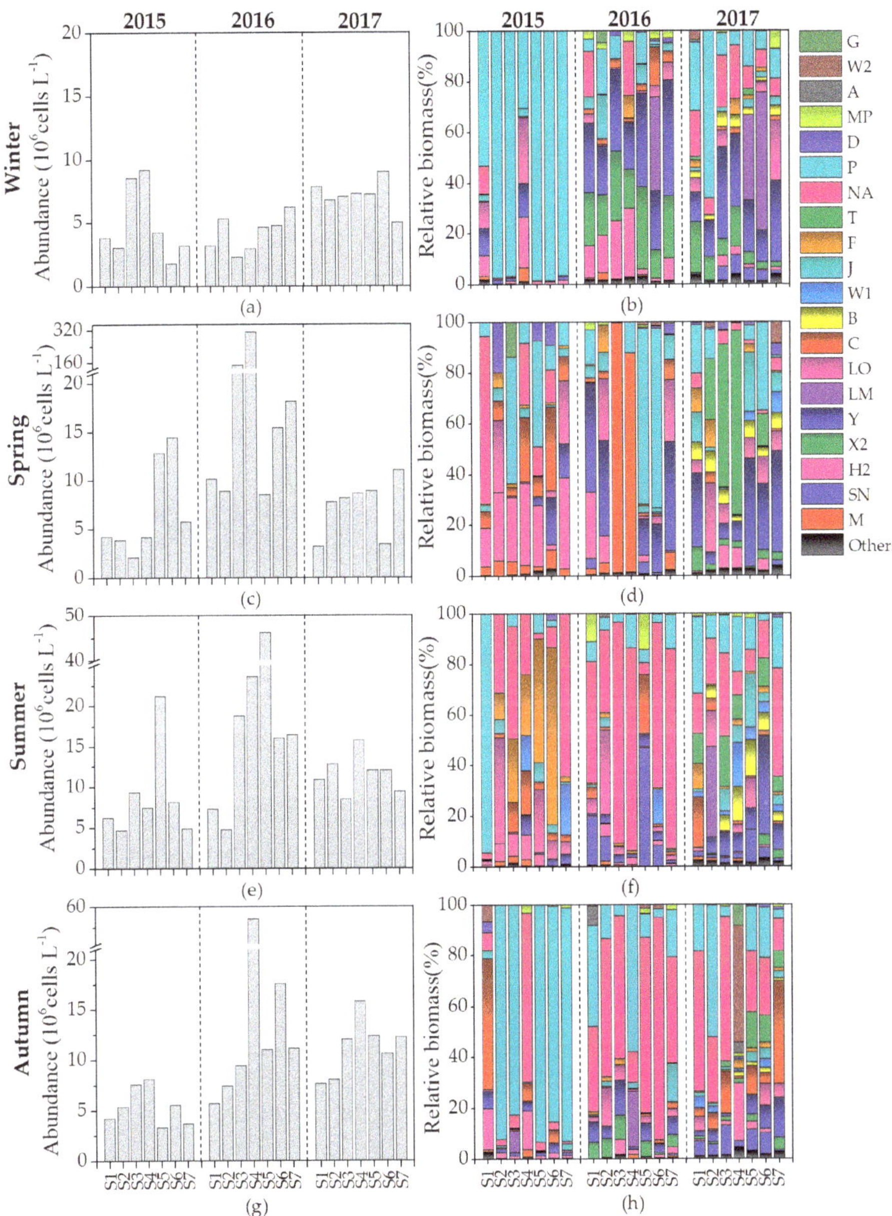

Figure 2. Composition and abundance of the representative functional groups of phytoplankton in Gaozhou Reservoir from 2015 to 2017. (**a**) abundance in winter; (**b**) relative biomass in winter; (**c**) abundance in spring; (**d**) relative biomass in spring; (**e**) abundance in summer; (**f**) relative biomass in summer; (**g**) abundance in autumn; and (**h**) relative biomass in autumn.

3.2. Variations in Environmental Factors

The average WTs in the winter, spring, summer, and autumn were 20.8 °C, 28.0 °C, 31.2 °C, and 26.5 °C, respectively. The pH was 8.63 in the summer, which was significantly higher than that in all other seasons ($p < 0.05$). However, there were no significant differences in pH among the samples that were collected during the other seasons. DO and EC were highest in the spring, measuring 10.03 mg L^{-1} and 66.1 μs cm^{-1}, respectively. These values were significantly higher than those that were measured during the other three seasons ($p < 0.05$). Across all samples, the TN concentrations ranged from 0.20 mg L^{-1} to 2.60 mg L^{-1}, with an average of 0.66 mg L^{-1}. The TN concentrations of the spring samples (average: 0.94 mg L^{-1}; maximum: 1.15 mg L^{-1}) were significantly higher than those of the other three seasons ($p < 0.05$). Across all of the samples, the TP concentrations ranged from 0.008 mg L^{-1} to 0.110 mg L^{-1}, with an average of 0.025 mg L^{-1}. The average TP concentration was the highest in the spring of 2016 (0.056 mg L^{-1}), but there was no significant difference in TP concentration between spring and each of the other three seasons. Throughout the survey period, the ratio of nitrogen concentration to phosphorus concentration at each sampling site was greater than 16, and this value did not differ significantly among seasons. Seasonal rainfall was 50–1116 mm, with the most rain falling in the summer, followed by the spring, autumn, and winter (Table 3).

Table 3. Mean and range (Min–Max) of environmental factors in the Gaozhou Reservoir for each season from 2015 to 2017.

	Winter		Sping		Summer		Autumn	
	Mean	Range	Mean	Range	Mean	Range	Mean	Range
WT (°C)	20.8	17.8–23.9	28.0	22.9–32.2	31.2	28.8–33.3	26.5	24.8–27.8
pH	7.33	6.73–8.36	7.32	6.29–8.33	8.63	8.10–9.28	7.61	6.84–8.93
DO (mg L^{-1})	8.11	6.12–10.71	10.03	8.14–14.38	8.78	7.74–9.90	7.40	5.54–9.25
EC (μs cm^{-1})	58.7	54.4–67.0	66.1	59.8–88.3	58.8	53.1–64.7	56.9	53.8–60.9
SS (mg L^{-1})	2.1	0.3–5.2	9.0	2.0–59.0	3.5	1.4–8.0	2.5	1.4–5.0
TN (mg L^{-1})	0.58	0.37–0.95	0.94	0.36–2.60	0.58	0.20–1.06	0.56	0.38–0.85
NH$_4$-N (mg L^{-1})	0.064	0.028–0.154	0.069	0.013–0.135	0.060	0.027–0.178	0.055	0.029–0.097
NO$_3$-N (mg L^{-1})	0.35	0.19–0.68	0.38	0.08–0.85	0.22	0.08–0.62	0.25	0.12–0.49
TP (mg L^{-1})	0.023	0.008–0.071	0.035	0.010–0.110	0.022	0.009–0.040	0.022	0.009–0.048
COD$_{Mn}$ (mg L^{-1})	1.9	0.8–4.2	3.3	0.6–12.2	2.2	1.4–3.5	1.9	1.2–2.8
Chla (mg m^{-3})	8.9	3.1–36.3	25.8	5.2–193.3	16.8	4.4–53.8	14.8	6.8–32.4
WL (m)	60.6	60.3–61.3	61.0	60.4–61.8	60.7	60.4–61.2	60.7	60.2–61.2
Rainfall (mm)	239	50–428	504	428–684	888	633–1116	337	210–517

WT: water temperature; DO: dissolved oxygen; EC: electrical conductivity; SS: suspended solid; TN: total nitrogen; NH$_4$-N: ammonium nitrogen; NO$_3$-N: nitrate nitrogen; TP: total phosphorus; COD$_{Mn}$: permanganate index; Chla: chlorophyll a concentration; WL: water level.

Among all of the environmental factors, measured during the survey period, TP was most strongly affected by season. The average TP in spring of 2016 was 0.056 mg L^{-1}. TP values at S3 and S4 were higher (0.080 mg L^{-1} and 0.110 mg L^{-1}) than those at other sites, and these values were 4–5 times greater than those in spring 2015 and 2017 (Figure 3).

3.3. Changes in Q and TLI (Σ)

The average Q index across all of samples from 2015 to 2017 was 1.87–3.48 (Figure 4), indicating that the water quality of the reservoir was primarily good and medium. However, water quality in spring of 2016 was obviously bad. The average TLI (Σ) index for all samples from 2015 to 2017 was 34.4–52.6 (Figure 4), implying that water quality was lightly eutrophic. The average TLI (Σ) index was substantially lower in the samples that were collected in the spring of 2016, as compared to those collected in other seasons.

The Q index and the TLI (Σ) index were especially abnormal in the spring of 2016, coinciding with local cyanobacterial blooms. The TLI (Σ) indexes of the samples from sites S3 and S4 were 63.7 and 71.6, respectively, while the Q indexes were 0.21 and 0.24, respectively. The degree of eutrophication in the water of the reservoir was abnormally high.

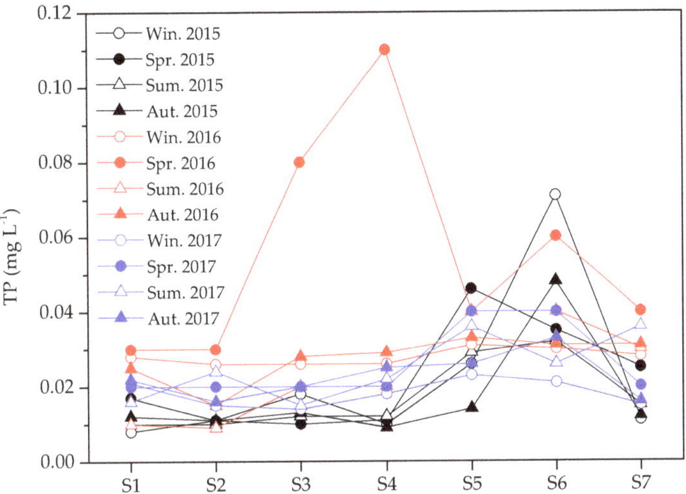

Figure 3. Seasonal changes in total phosphorus (TP) at different sampling sites in Gaozhou Reservoir from 2015 to 2017.

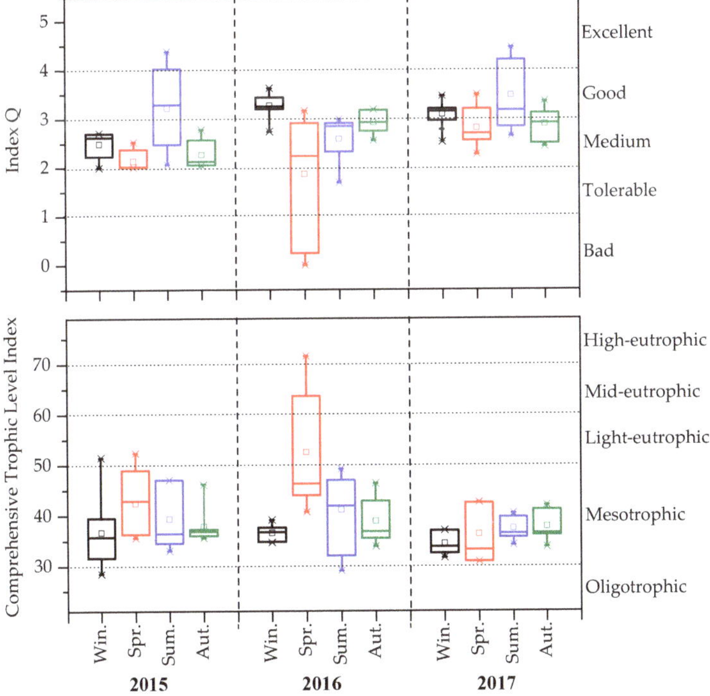

Figure 4. Evaluation of the ecological status of Gaozhou Reservoir from 2015 to 2017 based on the Q and TLI(Σ) indexes.

3.4. Redundancy Analysis

In this study, the functional groups with greater than 10% relative biomass were defined as the dominant functional groups. Nine groups were identified in Gaozhou Reservoir: M, X2, Y, LO, J, F, T, NA, and P (Table 2). Figure 5 shows the results of the RDA, based on dominant functional groups and major environmental factors.

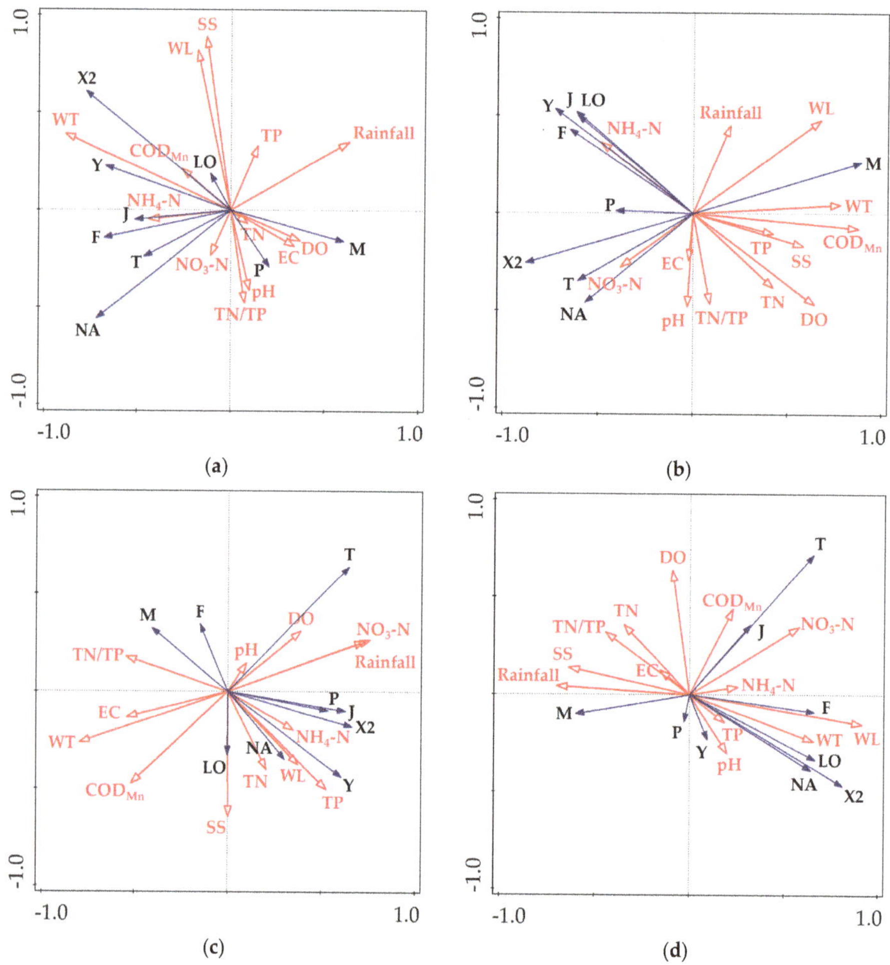

Figure 5. Redundancy analysis of the representative functional groups of phytoplankton and the environmental factors in Gaozhou Reservoir. (**a**) winter; (**b**) spring; (**c**) summer; and, (**d**) autumn.

WT, nitrogen, and phosphorus were the most important environmental factors affecting the phytoplankton functional groups (Figure 5). However, the influencing factors were different in different seasons. In the winter, WT and COD_{Mn} were positively associated with groups X2 and Y, but they were negatively associated with group M (Figure 5a). In the spring, WT was the main environmental factor affecting group M (Figure 5b). In the summer, TP, TN, and NH_4-N were the main environmental factors affecting groups Y and NA (Figure 5c). In the autumn, WT, pH, and TP were the main environmental factors affecting groups X2 and NA (Figure 5d). The redundant environmental factors differed among seasons. The redundant environmental factors were TP and rainfall in the winter; COD_{Mn}, DO, and TN in the spring; WT, COD_{Mn}, and EC in the summer; and, TN, rainfall, and EC in the autumn.

4. Discussion

4.1. Phytoplankton Seasonal Dynamics

We found that phytoplankton species richness in Gaozhou Reservoir was greater in the summer and autumn than in the winter and spring. Generally, the mechanisms regulating the changes in phytoplankton communities are complex [44,45]. For example, the nutrient resources of aquatic ecosystems strongly affect ecosystem diversity [46]. Species are season-specific, which is a basic manifestation of phytoplankton adaptations to natural environments [10,47].

Functional groups P (*Staurastrum* sp. and *Closterium acerosum*) and Y (*Cryptomonas ovata* and *Cryptomonas erosa*) prefer weaker light intensity [10,11]. Light intensity is greater in the summer and autumn, but weaker in the winter and spring, as the Gaozhou Reservoir is located in the subtropics. Therefore, functional groups P and Y dominated the communities during the winter and spring [16,48,49]. Similarly, functional group X2 (*Chroomonas acuta*), which tolerates low temperatures [50], grew faster in the winter and spring.

The functional groups NA and P were dominant in the summer and autumn, when the nitrogen concentrations in Gaozhou Reservoir were lower than in the winter and spring. *Cosmarium* sp. and *Staurodesmus* sp. were the dominant taxa in the aquatic environment when the nutrient levels were low to moderate [11].

Functional group M had clear advantages at sites S3 and S4 in spring 2016, when cyanobacterial blooms were occurring at these two sites. Functional group P had strong advantages at most of the sampling sites in the winter and autumn of 2015, and functional group H2 dominated the community in the summer and autumn of 2016.

4.2. Environmental Driving Factors

Our results were consistent with previous studies that reported that temperature and nutrient availability strongly influenced phytoplankton species and biomass in aquatic ecosystems [51,52]. For example, the RDA results showed that WT was highly correlated with different functional groups in the winter as compared to the spring. WT was correlated with functional groups Y and X2 in the winter; with functional group M in the spring; and, with functional groups F, LO, X2, and NA in the autumn. In the winter and spring, NO_3-N was highly correlated with the functional group F, and NH_4-N was highly correlated with the functional groups T and NA. In the summer, NH_4-N and TP affected additional functional groups, including Y, NA, X2, P, and J. TP affected the functional groups NA and X2 in the summer and autumn.

In winter, environmental parameters, including TP, rainfall, and SS, were located in the first quadrant (Figure 5a), but no functional groups were located in this quadrant, which indicated that these parameters had no direct impact on the functional groups. In the spring, quadrant II had no functional groups (Figure 5b), but several parameters, including TP, TN, and DO, were observed in this quadrant. There were no functional groups in quadrant III in the summer, and the environmental factors in this quadrant were WT, COD_{Mn}, and EC (Figure 5c). There were no functional groups in quadrant IV in the autumn, and the environmental factors in this quadrant were TN, DO, rainfall, and EC (Figure 5d). This seasonal difference in environmental factors among quadrants indicated that the redundancy of environmental factors differed among seasons, and the relationship between environmental factors and functional groups also fluctuated seasonally. For example, the environmental factors that were positively affected in the winter were redundant variables, which included TP. This suggested that the constraints associated with other factors (such as WT) would limit the occurrence and development of functional groups, even if TP were suitable. In the summer, the negatively affected environmental factors were also redundant variables, including WT. Obviously, water temperature is the highest in the summer and, thus, WT no longer acts as a restrictive environmental condition to limit the development of functional groups. Under these conditions, the environmental factors affecting the phytoplankton community structure are nitrogen and phosphorus nutrient levels.

4.3. Analysis of Cyanobacterial Blooms

Phytoplankton blooms are a comprehensive manifestation of the long-term aquatic ecological risks that are associated with a regional water environment [5,53]. The cyanobacterium *Microcystis aeruginosa* is induced at low temperatures and is dormant at high temperatures [54,55]. Guo et al. (2016) found that *M. aeruginosa* was induced at 15 °C, grew well at 19 °C, and entered dormancy at 31 °C in Dianchi Lake, China [55]. In Gaozhou Reservoir, the *M. aeruginosa* bloom occurred after the WT increased from 20.8 °C in winter 2016 to 28.0 °C in spring 2016; *M. aeruginosa* entered dormancy in the summer of 2016 at 31.2 °C. In addition, the reservoir also experienced blooms in the spring seasons of 2009, 2010, and 2013. The environmental conditions before the blooms had several similarities, such as low temperatures in the winter and increasing temperatures in the spring [32]. The Gaozhou Reservoir is rainy in the spring and, thus, nutrient concentrations in the reservoir increase due to the input of rainwater runoff. Previous studies recorded high concentrations of phosphorus at sampling sites S3 and S4 in the spring of 2016; these high levels of phosphorus were associated with the rainfall and the hydrodynamic conditions at these two sites. Sampling sites S3 and S4 are located in the bay of the Reservoir. The Gaozhou reservoir has a relatively long hydraulic retention time (approximately 451 days). In this study, S3 and S4 were two sites with serious cyanobacterial bloom. S4 site was around by the shoreside of the reservoir and closer to the dam, while S3 site might be affected by S4 site, due to the short distance between the two sites. Hence, the areas with cyanobacterial bloom were relatively closed, with lower flow velocity and less water exchange when compared to other areas of the reservoir [34]. These conditions may result in higher concentrations of phosphorus at these sites, which may induce cyanobacterial blooms. In addition, the prevailing winds blew many phytoplankton cells into the reservoir bay, resulting in a higher density of phytoplankton cells in this area when compared to surrounding locations. Therefore, environmental conditions and changes increased the aggregation of cyanobacterial cells at sites S3 and S4.

4.4. Environmental Implications

Here, data and analysis suggested that the low temperatures in Gaozhou Reservoir in the winter and spring, coupled with the dramatic difference in temperature between the winter and spring, resulted in the rapid growth of *Microcystis aeruginosa*. In addition, the high concentrations of phosphorus in the reservoir in the spring might have provided sufficient nutrients in order support the explosive proliferation of cyanobacteria. Finally, the specific hydrodynamic conditions of the water body might have inhibited dispersion, allowing for the growing cyanobacterial cells to further aggregate.

5. Conclusions

This study demonstrated that there was seasonal succession of phytoplankton species and functional groups in the Gaozhou Reservoir. The phytoplankton diversity was higher in the summer and autumn than in the winter and spring. Functional groups P (*Staurastrum* sp. and *Closterium acerosum*) and Y (*Cryptomonas ovata* and *Cryptomonas erosa*) were dominant in the winter and spring, while functional groups NA (*Cosmarium* sp. and *Staurodesmus* sp.) and P (*Staurastrum* sp. and *Closterium acerosum*) were dominant in summer and autumn. The temperature, nitrogen, and phosphorus levels were the important factors that were associated with the seasonal changes in phytoplankton community structure in the reservoir. The key factors leading to the blooms of *Microcystis aeruginosa* were temperature and phosphorus levels. The low temperatures in the winter of 2016 and the high concentrations of phosphorus in the spring of 2016, together with the favorable rainfall and hydrodynamic conditions in the reservoir, were the major factors triggering the phytoplankton blooms.

Author Contributions: Conceptualization, L.Y., X.Z. and B.X.; data curation, L.Y., R.L. and T.G.; investigation, L.Y., R.L. and T.G.; methodology, L.Y., X.Z. and T.G.; validation, B.X., G.-J.Z., S.L. and C.L.; writing—Original draft, L.Y., X.Z., G.-J.Z., B.X. and S.L.; writing—Review and editing, L.Y., B.X., G.-J.Z. and X.Z. All authors have read and agreed to the published version of the manuscript.

Funding: This work was funded by the Key-Area Research and Development Program of Guangdong Province (2019B110205004), National Natural Science Foundation of China (41977353, 41401115), the International Science & Technology Cooperation Program of Guangzhou (201704030110) and the Special Fund of Chinese Central Government for Basic Scientific Research Operations in Commonweal Research Institutes (PM-zx703-201602-048, PM-zx097-201904-131).

Acknowledgments: We thank Kunying Li and Yanhui Feng for water sample collection and analysis.

Conflicts of Interest: The authors declare no conflict of interest.

References

1. Wilken, S.; Soares, M.; Urrutia-Cordero, P.; Ratcovich, J.; Ekvall, M.K.; Donk, E.V.; Hansson, L.A. Primary producers or consumers? Increasing phytoplankton bacterivory along a gradient of lake warming and browning: Increasing phytoplankton bacterivory. *Limnol. Oceanogr.* **2018**, *63*, S142–S155. [CrossRef]

2. Zhao, X.; Drakare, S.; Johnson, R.K. Use of taxon-specific models of phytoplankton assemblage composition and biomass for detecting impact. *Ecol. Indic.* **2019**, *97*, 447–456. [CrossRef]

3. Santana, R.M.D.C.; Dolbeth, M.; de Lucena Barbosa, J.E.; Patrício, J. Narrowing the gap: Phytoplankton functional diversity in two disturbed tropical estuaries. *Ecol. Indic.* **2018**, *86*, 81–93. [CrossRef]

4. Dong, X.; Li, B.; He, F.; Gu, Y.; Sun, M.; Zhang, H.; Tan, L.; Xiao, W.; Liu, S.; Cai, Q. Flow directionality, mountain barriers and functional traits determine diatom metacommunity structuring of high mountain streams. *Sci. Rep.* **2016**, *6*, 24711. [CrossRef] [PubMed]

5. Chirico, N.; António, D.C.; Pozzoli, L.; Marinov, D.; Malagó, A.; Sanseverino, I.; Beghi, A.; Genoni, P.; Dobricic, S.; Lettieri, T. Cyanobacterial Blooms in Lake Varese: Analysis and Characterization over Ten Years of Observations. *Water* **2020**, *12*, 675. [CrossRef]

6. Michalak, A.M. Study role of climate change in extreme threats to water quality. *Nature* **2016**, *535*, 349–350. [CrossRef] [PubMed]

7. Kozak, A.; Budzyńska, A.; Dondajewska-Pielka, R.; Kowalczewska-Madura, K.; Gołdyn, R. Functional Groups of Phytoplankton and Their Relationship with Environmental Factors in the Restored Uzarzewskie Lake. *Water* **2020**, *12*, 313. [CrossRef]

8. Liu, X.; Qian, K.; Chen, Y.; Gao, J. A comparison of factors influencing the summer phytoplankton biomass in China's three largest freshwater lakes: Poyang, Dongting, and Taihu. *Hydrobiologia* **2017**, *792*, 283–302. [CrossRef]

9. Biggs, J.; von Fumetti, S.; Kelly-Quinn, M. The importance of small waterbodies for biodiversity and ecosystem services: Implications for policy makers. *Hydrobiologia* **2017**, *793*, 3–39. [CrossRef]

10. Reynolds, C.S.; Huszar, V.; Kruk, C.; Naselliflores, L.; Melo, S. Towards a functional classification of the freshwater phytoplankton. *J. Plankton Res.* **2002**, *24*, 417–428. [CrossRef]

11. Padisák, J.; Crossetti, L.O.; Naselli-Flores, L. Use and misuse in the application of the phytoplankton functional classification: A critical review with updates. *Hydrobiologia* **2009**, *621*, 1–19. [CrossRef]

12. Costa, L.S.; Huszar, V.L.M.; Ovalle, A.R. Phytoplankton Functional Groups in a Tropical Estuary: Hydrological Control and Nutrient Limitation. *Estuar. Coast.* **2009**, *32*, 508–521. [CrossRef]

13. Kruk, C.; Devercelli, M.; Huszar, V.L.M.; Hernández, E.; Beamud, G.; Diaz, M.; Silva, L.H.S.; Segura, A.M. Classification of Reynolds phytoplankton functional groups using individual traits and machine learning techniques. *Freshw. Biol.* **2017**, *62*, 1681–1692. [CrossRef]

14. Rodrigues, L.C.; Pivato, B.M.; Vieira, L.C.G.; Bovo-Scomparin, V.M.; Bortolini, J.C.; Pineda, A.; Train, S. Use of phytoplankton functional groups as a model of spatial and temporal patterns in reservoirs: A case study in a reservoir of central Brazil. *Hydrobiologia* **2018**, *805*, 147–161. [CrossRef]

15. Becker, V.; Huszar, V.L.M.; Crossetti, L.O. Responses of phytoplankton functional groups to the mixing regime in a deep subtropical reservoir. *Hydrobiologia* **2009**, *628*, 137–151. [CrossRef]

16. Becker, V.; Caputo, L.; Ordóñez, J.; Marcé, R.; Armengol, J.; Crossetti, L.O.; Huszar, V.L.M. Driving factors of the phytoplankton functional groups in a deep Mediterranean reservoir. *Water Res.* **2010**, *44*, 3345–3354. [CrossRef]

17. Souza, M.D.C.D.; Crossetti, L.O.; Becker, V. Effects of temperature increase and nutrient enrichment on phytoplankton functional groups in a Brazilian semi-arid reservoir. *Acta Limnol. Bras.* **2018**, *30*, e215. [CrossRef]

18. Cao, J.; Hou, Z.; Li, Z.; Chu, Z.; Yang, P.; Zheng, B. Succession of phytoplankton functional groups and their driving factors in a subtropical plateau lake. *Sci. Total Environ.* **2018**, *631*, 1127–1137. [CrossRef]

19. Santana, L.M.; Crossetti, L.O.; Ferragut, C. Ecological status assessment of tropical reservoirs through the assemblage index of phytoplankton functional groups. *Braz. J. Bot.* **2017**, *40*, 695–704. [CrossRef]

20. Cupertino, A.; Gücker, B.; Von Rückert, G.; Figueredo, C.C. Phytoplankton assemblage composition as an environmental indicator in routine lentic monitoring: Taxonomic versus functional groups. *Ecol. Indic.* **2019**, *101*, 522–532. [CrossRef]

21. Cai, Q.; Liu, J.; King, L. A comprehensive model for assessing lake eutrophication. *Chin. J. Appl. Ecol.* **2002**, *13*, 1674–1678.

22. Trolle, D.; Spigel, B.; Hamilton, D.P.; Norton, N.; Sutherland, D.; Plew, D.; Allan, M.G. Application of a Three-Dimensional Water Quality Model as a Decision Support Tool for the Management of Land-Use Changes in the Catchment of an Oligotrophic Lake. *Environ. Manag.* **2014**, *54*, 479–493. [CrossRef] [PubMed]

23. Huo, S.; Ma, C.; Xi, B.; Su, J.; Zan, F.; Ji, D.; He, Z. Establishing eutrophication assessment standards for four lake regions, China. *J. Environ. Sci. China* **2013**, *25*, 2014–2022. [CrossRef]

24. Padisák, J.; Borics, G.; Grigorszky, I.; Soróczki-Pintér, É. Use of Phytoplankton Assemblages for Monitoring Ecological Status of Lakes within the Water Framework Directive: The Assemblage Index. *Hydrobiologia* **2006**, *553*, 1–14. [CrossRef]

25. Yang, C.; Nan, J.; Li, J. Driving Factors and Dynamics of Phytoplankton Community and Functional Groups in an Estuary Reservoir in the Yangtze River, China. *Water* **2019**, *11*, 1184. [CrossRef]

26. Lu, X.; Song, S.; Lu, Y.; Wang, T.; Liu, Z.; Li, Q.; Zhang, M.; Suriyanarayanan, S.; Jenkins, A. Response of the phytoplankton community to water quality in a local alpine glacial lake of Xinjiang Tianchi, China: Potential drivers and management implications. *Environ. Sci. Process. Impacts* **2017**, *19*, 1300–1311. [CrossRef]

27. Zhang, J.; Ni, W.; Luo, Y.; Jan Stevenson, R.; Qi, J. Response of freshwater algae to water quality in Qinshan Lake within Taihu Watershed, China. *Phys. Chem. Earth Parts A/B/C* **2011**, *36*, 360–365. [CrossRef]

28. Yao, L.; Zhao, X.; Ma, Q.; Liang, R.; Xia, B.; Zhuo, Q. Nitrogen and Phosphorous Adsorption Characteristics of Suspended Solids Input into a Drinking Water Reservoir via Typhoon Heavy Rainfall. *Environ. Sci.* **2018**, *39*, 3622–3630.

29. Yao, L.; Zhao, X.; Ma, Q.; Liang, R.; Xia, B.; Gou, T. Simulation of the impact of typhoon-induced suspended solids precipitation on water quality in a reservoir for drinking water. *Ecol. Environ. Sci.* **2018**, *27*, 1900–1907.

30. Xu, Y.; Gu, J.; Yang, Y.; Xiao, L. Seasonal dynamics of phytoplankton morphological characters and driving factors in tropical reservoirs: A case study from Gaozhou Reservoir. *J. Lake Sci.* **2019**, *31*, 825–836.

31. Li, S.; Han, Z.; Xu, Z.; Zhao, X.; Yao, L.; Wei, D.; Zhang, J.; Hu, F. Study on the Structure of Phytoplankton Community and Its Relationships with Environmental Factors in Gaozhou reservoir. *J. Hydroecology* **2013**, *34*, 16–24.

32. Yao, L.; Zhao, X.; Zhou, G.; Wanyan, H.; Cai, L.; Hu, G.; Xu, Z. Preliminary regulating factors of spring cyanobacterial bloom in Gaozhou Reservoir, Guangdong Province. *J. Lake Sci.* **2011**, *23*, 534–540.

33. Ma, T.; Huang, Y. Characteristics change of nitrogen and phosphorus nutrients and water strategies in Gaozhou Reservoir. *Ecol. Sci.* **2015**, *34*, 31–37.

34. Xiao, L.; Hu, R.; Peng, L.; Lei, L.; Feng, Y.; Han, B. Dissimilarity of phytoplankton assemblages in two connected tropical reservoirs: Effects of water transportation and environmental filtering. *Hydrobiologia* **2016**, *764*, 127–138. [CrossRef]

35. The State Environmental Protection Administration. *Water and Wastewater Monitoring and Analysis Method*, 4th ed.; China Environmental Science Press: Beijing, China, 2002.

36. Rice, E.W.; Baird, R.B.; Eaton, A.D.; Clesceri, L.S. *Standard Methods for the Examination of Water and Wastewater*; American Public Health Association: Washington, DC, USA, 2012; Volume 10.

37. Zeng, Q.; Qin, L.; Bao, L.; Li, Y.; Li, X. Critical nutrient thresholds needed to control eutrophication and synergistic interactions between phosphorus and different nitrogen sources. *Environ. Sci. Pollut. Res.* **2016**, *23*, 21008–21019. [CrossRef]

38. Lund, J.W.G.; Kipling, C.; Cren, E.D.L. The inverted microscope method of estimating algal numbers and the statistical basis of estimations by counting. *Hydrobiol* **1958**, *11*, 143–170. [CrossRef]

39. Hillebrand, H.; Du Rselen, C.; Kirschtel, D.; Zohary, U.P.A.T. Biovolume calculation for pelagic and benthic microalgae. *J. Phycol.* **1999**, *35*, 403–424. [CrossRef]

40. Wetzel, R.G.; Likens, G.E. *Limnological Analyses*, 3rd ed.; Springer: Berlin/Heidelberg, Germany, 2000.

41. Nixdorf, B.; Rektins, A.; Mischke, U. Standards and Thresholds of the EU Water Framework Directive (WFD)—Phytoplankton and Lakes. In *Standards and Thresholds for Impact Assessment*; Springer: Berlin/Heidelberg, Germany, 2008; pp. 301–314.

42. Tu, Q.; Jin, X. *The Standard Methods in Lake Eutrophication Investigation*, 2nd ed.; China Environmental Science Press: Beijing, China, 1990.

43. Zheng, B.; Xu, Q.; Zhu, Y. Primary study on enacting the lake nutrient control standard. *Environ.Sci.* **2009**, *30*, 2497–2501. [CrossRef]

44. Lenard, T.; Ejankowski, W.; Poniewozik, M. Responses of Phytoplankton Communities in Selected Eutrophic Lakes to Variable Weather Conditions. *Water* **2019**, *11*, 1207. [CrossRef]

45. Dai, J.; Wu, S.; Wu, X.; Xue, W.; Yang, Q.; Zhu, S.; Wang, F.; Chen, D. Effects of Water Diversion from Yangtze River to Lake Taihu on the Phytoplankton Habitat of the Wangyu River Channel. *Water* **2018**, *10*, 759. [CrossRef]

46. Hardikar, R.; Haridevi, C.K.; Chowdhury, M.; Shinde, N.; Ram, A.; Rokade, M.A.; Rakesh, P.S. Seasonal distribution of phytoplankton and its association with physico-chemical parameters in coastal waters of Malvan, west coast of India. *Environ. Monit. Assess.* **2017**, *189*, 151. [CrossRef] [PubMed]

47. Huang, G.; Wang, X.; Chen, Y.; Xu, L.; Xu, D. Seasonal succession of phytoplankton functional groups in a reservoir in central China. *Fundam. Appl. Limnol. /Arch. Für Hydrobiol.* **2018**, *192*, 1–14. [CrossRef]

48. Yu, H.; Wu, J.; Ma, C.; Qin, X. Seasonal dynamics of phytoplankton functional groups and its relationship with the environment in river: A case study in northeast China. *J. Freshw. Ecol.* **2012**, *27*, 429–441. [CrossRef]

49. Lv, J.; Wu, H.; Chen, M. Effects of nitrogen and phosphorus on phytoplankton composition and biomass in 15 subtropical, urban shallow lakes in Wuhan, China. *Limnologica* **2011**, *41*, 48–56. [CrossRef]

50. Tian, C.; Hao, D.; Pei, H.; Doblin, M.A.; Ren, Y.; Wei, J.; Feng, Y. Phytoplankton Functional Groups Variation and Influencing Factors in a Shallow Temperate Lake. *Water Environ. Res.* **2018**, *90*, 510–519. [CrossRef] [PubMed]

51. Xiao, L.; Wang, T.; Hu, R.; Han, B.; Wang, S.; Qian, X.; Padisák, J. Succession of phytoplankton functional groups regulated by monsoonal hydrology in a large canyon-shaped reservoir. *Water Res.* **2011**, *45*, 5099–5109. [CrossRef]

52. Yang, J.; Wang, F.; Lv, J.; Liu, Q.; Nan, F.; Liu, X.; Xu, L.; Xie, S.; Feng, J. Interactive effects of temperature and nutrients on the phytoplankton community in an urban river in China. *Environ. Monit. Assess.* **2019**, *191*, 688. [CrossRef]

53. Jiang, Z.; Du, P.; Liao, Y.; Liu, Q.; Chen, Q.; Shou, L.; Zeng, J.; Chen, J. Oyster farming control on phytoplankton bloom promoted by thermal discharge from a power plant in a eutrophic, semi-enclosed bay. *Water Res.* **2019**, *159*, 1–9. [CrossRef]

54. Shang, L.; Feng, M.; Liu, F.; Xu, X.; Ke, F.; Chen, X.; Li, W. The establishment of preliminary safety threshold values for cyanobacteria based on periodic variations in different microcystin congeners in Lake Chaohu, China. *Environ. Sci. Proc. Imp.* **2015**, *17*, 728–739. [CrossRef]

55. Guo, W.; Liu, Q.; Peng, X.; Liu, C. Principle of Vernalization in Microcystis aeruginosa in Dianchi Lake and Improvement of Gene Model on Controlling the Vernalization. *Ecol. Environ. Sci.* **2016**, *25*, 2028–2034.

Article

Ecological Connectivity in Two Ancient Lakes: Impact Upon Planktonic Cyanobacteria and Water Quality

Matina Katsiapi [1,2,*], Savvas Genitsaris [3], Natassa Stefanidou [1], Anastasia Tsavdaridou [4], Irakleia Giannopoulou [5], Georgia Stamou [5], Evangelia Michaloudi [5], Antonios D. Mazaris [4] and Maria Moustaka-Gouni [1,*]

[1]   Department of Botany, School of Biology, Aristotle University of Thessaloniki, 54124 Thessaloniki, Greece; natasa.stefanidou@gmail.com

[2]   EYATH SA, Water Supply Division/Drinking Water Treatment Facility, 57008 Nea Ionia, Greece

[3]   School of Science and Technology, International Hellenic University, 57001 Thermi, Greece; s.genitsaris@ihu.edu.gr

[4]   Department of Ecology, School of Biology, Aristotle University of Thessaloniki, 54124 Thessaloniki, Greece; atsavda@bio.auth.gr (A.T.); amazaris@bio.auth.gr (A.D.M.)

[5]   Department of Zoology, School of Biology, Aristotle University of Thessaloniki, 54124 Thessaloniki, Greece; giannopoulou.ira@gmail.com (I.G.); gstamouc@bio.auth.gr (G.S.); tholi@bio.auth.gr (E.M.)

*   Correspondence: matinakatsiapi@gmail.com (M.K.); mmustaka@bio.auth.gr (M.M.-G.)

Received: 29 November 2019; Accepted: 16 December 2019; Published: 19 December 2019

Abstract: The ancient lakes Mikri Prespa and Megali Prespa are located in SE Europe at the transnational triangle and are globally recognized for their ecological significance. They host hundreds of flora and fauna species, and numerous types of habitat of conservational interest. They also provide a variety of ecosystem services. Over the last few decades, the two lakes have been interconnected through a surface water channel. In an attempt to explore whether such a management practice might alter the ecological properties of the two lakes, we investigated a series of community metrics for phytoplankton by emphasizing cyanobacteria. Our results demonstrate that the cyanobacterial metacommunity structure was affected by directional hydrological connectivity and high dispersal rates, and to a lesser extent, by cyanobacterial species sorting. Cyanobacterial alpha diversity was twofold in the shallow upstream Lake Mikri Prespa (Simpson index average value: 0.70) in comparison to downstream Lake Megali Prespa (Simpson index average value: 0.37). The cyanobacterial assemblage of the latter was only a strict subset of that in Mikri Prespa. Similarly, beta diversity components clearly showed a homogenization of cyanobacteria, supporting the hypothesis that water flow enhances fluvial translocation of potentially toxic and bloom-forming cyanobacteria. Degrading of the water quality in the Lake Megali Prespa in anticipation of improving that of the Lake Mikri Prespa is an issue of great concern for the Prespa lakes' protection and conservation.

Keywords: man-made surface water channel; transboundary; nestedness; Balkan; *Dolichospermum lemmermannii*; *Microcystis aeruginosa*

1. Introduction

One can view lakes as islands in a terrestrial world. In lakes connected to each other by direct water flow, the physical transport through water is unidirectional, resulting in a dominance of colonization from the upstream lake [1]. This way, the drift of phytoplankton, and particularly of cyanobacteria (both the resting stages and the individuals of active populations), are very effective means of dispersal and are not accidental episodes [2]. For cyanobacteria, high dispersal levels and global warming are emergent drivers of their community assembly in lakes [3]. In a lake that is already populated by a given phytoplankton species, competition and predation by zooplankton makes it difficult for a new invader to establish. However, this "priority effect" sensu De Meester et al. [4] does not hold in cases where local communities are not saturated [5]. Recent invasions and proliferation of toxic cyanobacteria in diverse aquatic habitats, a well-known worldwide phenomenon [6], shows that new invaders can establish in a populated lake. Studying the species' dispersal patterns, recognising species' replacements (i.e., turnover) and recognising losses/additions of species in an ecosystem (nestedness) [7], could serve as critical steps towards implementing management practices [8].

The increase and dominance of cyanobacteria in a water body is indicative of water quality degradation, since cyanobacteria are implicated in food-web disturbances, oxygen depletion and animal mortality; and they have adverse effects on human health and on ecosystem services in general [9]. Furthermore, cyanobacterial species with allelopathic characteristics can alter phytoplankton composition and biodiversity [10]. It is, therefore, globally acknowledged that the management of lakes should aim at maintaining environmental heterogeneity while preventing further eutrophication and expansion of toxic and allelopathic cyanobacteria. This management practice could favour the maintenance of high phytoplankton beta and gamma diversity [11].

The ancient neighboring lakes Megali Prespa and Mikri Prespa, SE Europe, are connected by a man-made surface water channel with temporal flow. Over the period of 1984–2011, the two lakes had an almost identical range in water isotope composition, reflecting their hydrological connection [12]. The inflow from Lake Mikri Prespa is about 9% of the water inflows into the Lake Megali Prespa [13]. Thus, the community properties and water quality of both lakes could be driven by the plankton species of the Lake Mikri Prespa [14]. Still, no study exists on the impacts of the inflows from the Lake Mikri Prespa on the cyanobacteria and phytoplankton assemblages and the water quality of the Lake Megali Prespa. Notwithstanding, the Lake Mikri Prespa has a history of cyanobacterial blooms formed by potentially toxic species [15,16]. In 2014, microcystins were measured in lake water at a concentration that posed a high risk of adverse human health effects [17].

In the present work, we examined past hydrological connectivity's cumulative effect on the Prespa lakes' cyanobacteria assemblages and the lake's water quality. Towards those ends, we studied the two lakes' phytoplankton communities over a year without water transfer, focusing on cyanobacteria. We examined temporal variation in composition, biomass, alpha diversity, total beta diversity and the species' turnover and nestedness as the means to delineate the potential role of hydrological connectivity upon the phytoplankton assemblages. Species network interactions within each lake were also explored as the means to identify co-occurrence patterns and negative interactions, while the multimetric index PhyCoI$_{GP}$ [18] was applied to compare water quality of the two lakes.

2. Materials and Methods

2.1. Study Site

The lakes Mikri Prespa and Megali Prespa are of two the oldest lakes on earth, both with an estimated age of 2–5 million years [19]. These two lakes are European endangered monuments hosting high richness of endemic and internationally important endangered species. The two lakes are also included in the Natura 2000 network that constitutes the cornerstone of the conservation policy of the European Union. Since 1975, Lake Mikri Prespa was also designated as a Ramsar Site. Both lakes are situated at the northwestern part of Greece, SE Europe (Megali Prespa: 40°45′ N, 21°01′ E, Mikri Prespa: 40°46′ N, 21°05′ E) at an altitude of ≈850 m above sea level (Figure 1). The Lake Mikri Prespa is also a transboundary lake between Greece and Albania. The Lake Megali Prespa is located at the southernmost tip of the Alpine biogeographical region of Europe [20], and is shared by three countries: Greece, Albania and North Macedonia.

Figure 1. Study area and sampling sites, the lakes Mikri and Megali Prespa, indicated with red circles.

Lake Mikri Prespa is a shallow (maximum depth: 9 m), moderately large (surface: 46.7 km^2) polymictic, eutrophic lake experiencing toxic cyanobacterial blooms [17,21]. Lake Megali Prespa is a large (surface 256 km^2), deep (maximum depth 55 m), warm momomicitc lake, less impacted by humans [22,23]. The two lakes (Figure 1) are connected by a man-made channel which used to be narrow and quite shallow until the 1950s, but was re-constructed in 1969. Since 2005, the Prespa Park Management Body controls the water level by regulating the outflow discharge to the Lake Megali Prespa in order to sustain the Lake Mikri Prespa's good ecological status and the local economy [24]. During the twelve years of controlled discharge (2005–2016), a total of 212.6 × 10^6 m^3 of surface lake water of the Lake Mikri Prespa was discharged to the Lake Megali Prespa based on the water level adjustments [25].

2.2. Field Work—Microscopy Analysis

Phytoplankton samplings were conducted in the lakes Mikri and Megali Prespa, from September 2015 to August 2016, on a monthly basis. Water samples were collected at the deeper parts of the lakes in the vicinity of the connection points of the lakes to the water channel (Mikri Prespa outlet, Megali Prespa inlet; Figure 1) using a Niskin sampler. The samples were taken at discrete depths every one meter from the surface of the lakes to the bottom of the euphotic zone (defined as 2.5× Secchi Depth). The set of these samples in each sampling site were mixed in a plastic container, and an integrated sample from the euphotic zone was thus obtained. Subsamples of 500 mL for phytoplankton analysis were preserved immediately after sampling with Lugol's solution and were kept in the dark till microscopy analysis. Also, fresh (non-preserved) sub-samples were kept in the dark in a portable refrigerator. The fresh samples were transported immediately at the laboratory and checked (within 4–6 h from sampling) for the microscopic identification of species with limited diagnostic features.

The phytoplankton data were based on the microscopic analysis of the preserved samples. For this, sub-samples were placed in sedimentation chambers of Hydrobios of different volumes (3, 5, 10 and 25 mL) based on phytoplankton density according to the inverted microscope method. These were examined using an inverted microscope (Nikon Eclipse TE 2000-S, Melville, NY, USA). Species identification was carried using appropriate taxonomic keys [26]. Phytoplankton counting was performed using Utermöhl's sedimentation method. At least 400 individuals were counted in each sample. For biomass estimation, the dimensions of 30 individuals of dominant species were measured using the relevant tools of a digital microscope camera (Nikon DS-L1, Melville, NY, USA). Mean cell or filament volume estimates were calculated using appropriate geometric formulae [27].

2.3. Data Analysis

Alpha diversity estimators (i.e., species richness and the Simpson index) for the whole phytoplankton community and for cyanobacteria were calculated with the PAST3 software [28]. To compute beta diversity of the whole phytoplankton community and of the cyanobacteria assemblages in each sampling, we used the "betapart" R package, version 1.5.1 [29]. Beta diversity was portioned into its spatial turnover and nestedness components, following Baselga's approach [30]. This approach [7,30] suggests that Sorensen pair-wise dissimilarity (bSOR) should be partitioned into two components: spatial turnover in species composition, measured as Simpson pair-wise dissimilarity index (bSIM); and variation in species composition due to nestedness (bNES) measured as the nestedness-fraction of Sorensen pair-wise dissimilarity. The above analyses were run in the R 3.5.3 environment (R core team 2018).

We performed network analysis for each lake to explore significant relationships (positive, which indicate co-occurrence patterns; or negative, which provide evidence of exclusion) among cyanobacterial and other phytoplankton species. The relationships were characterized through Maximal Information-based Nonparametric Exploration (MINE) statistics by computing the Maximal Information Coefficient (MIC), based on the species biomass per sample, in species pairwise comparisons (see Supplementary Table S1 for the species included) [31]. The matrix of MIC values corresponding to a *p*-value < 0.01, based on pre-computed *p*-values of various MIC scores at different sample sizes, was used (MIC > 0.68 in this case) in order to visualize networks of species' associations with Cytoscape 3.5.1 [32]. We identified the negative or positive type of relationship between each pair of species included in the network according to Hernández-Ruiz et al. [33].

2.4. Water Quality Assessment

The phytoplankton modified $PhyCoI_{GP}$ index [18] was used to assess ecological water quality of each lake. For the index calculation, we combined six metrics/sub-indices; i.e., the total phytoplankton biomass, the cyanobacterial biomass (according to World Health Organization Guidelines), the modified

Nygaard sub-index based on the biomass of indicator taxonomic groups, the modified Nygaard sub-index based on species richness of indicator taxonomic groups, the quality group species Index and the grazing potential of zooplankton. The ecological water quality assessment was based on the data of the sampling period June–September. This period is used for the lake ecological water quality assessments based on phytoplankton in the Greece/Mediterranean region [18].

3. Results

In total, 119 species were identified in the phytoplankton communities of the Prespa lakes; 111 species were found in Lake Mikri Prespa and 75 phytoplankton species were identified in Lake Megali Prespa. Sixty-seven (67) phytoplankton species were found in both lakes (Supplementary Figure S1). Chlorophytes were the richest phytoplankton taxonomic group (47 species). A total of 30 species of Cyanobacteria were detected in both lakes (20 species in Lake Megali Prespa and 30 in Lake Mikri Prespa) (Supplementary Table S1). The cyanobacterial assemblage in Lake Megali Prespa was a strict subset of the cyanobacterial assemblage of Lake Mikri Prespa. All nostocalean species (*Dolichospermum* cf. *flos-aquae*, *D. lemmermannii*, *D. viguieri*, *Aphanizomenon gracile*) were found in both lakes. Most of the cyanobacteria that were not observed in the Lake Megali Prespa (eight out of 10) were chroococcalean; two of them were *Microcystis* species (*M. flos-aquae* and *M. wesenbergii*). The maximum cyanobacteria richness was detected in September and the minimum in February (Table 1).

Table 1. Total phytoplankton biomass (TB), cyanobacterial assemblage biomass (CB), species richness of phytoplankton community (SR-*ph*), species richness of cyanobacterial assemblage (SR-*cy*), Simpson index of phytoplankton community (SI-*ph*), Simpson index of cyanobacterial assemblage (SI-*cy*) and the modified phytoplankton water quality index (PhyCoI$_{GP}$) in the lakes Mikri and Megali Prespa during the period from September 2015 to August 2016. Values for Lake Megali Prespa are in bold.

Date	TB (mg L^{-1})		CB (mg L^{-1})		SR-*ph*		SR-*cy*		SI-*ph*		SI-*cy*		PhyCoI$_{GP}$	
September 2015	3.07	**1.18**	2.47	**0.29**	64	**36**	26	**12**	0.80	**0.69**	0.70	**0.73**	2.7	**3.1**
October 2015	2.37	**1.72**	1.93	**0.07**	57	**28**	19	**9**	0.84	**0.65**	0.77	**0.55**	2.7	**2.7**
November 2015	1.18	**1.46**	0.86	**0.03**	56	**28**	20	**9**	0.81	**0.54**	0.67	**0.26**	3.0	**3.1**
December 2015	0.92	**0.89**	0.67	**0.003**	43	**19**	15	**2**	0.78	**0.18**	-	-	3.3	**3.6**
January 2016	0.23	**0.10**	0.08	**0.001**	36	**17**	11	**3**	0.90	**0.69**	-	-	3.5	**4.2**
February 2016	0.72	**0.12**	0.17	**0.001**	33	**19**	9	**2**	0.57	**0.45**	-	-	3.5	**3.8**
March 2016	0.70	**0.58**	0.11	**0.002**	51	**18**	17	**3**	0.78	**0.20**	-	-	3.6	**4.3**
April 2016	1.54	**0.41**	0.38	**0.05**	46	**18**	17	**4**	0.63	**0.75**	0.72	**0.19**	3.3	**3.9**
June 2016	0.87	**4.10**	0.54	**2.63**	66	**27**	22	**5**	0.86	**0.53**	0.71	**0.004**	3.7	**2.6**
July 2016	1.07	**0.72**	0.43	**0.10**	54	**15**	19	**6**	0.85	**0.72**	0.77	**0.43**	3.5	**2.9**
August 2016	1.93	**0.76**	0.91	**0.23**	54	**20**	19	**8**	0.81	**0.75**	0.57	**0.43**	3.4	**3.2**

During the study period, the phytoplankton biomass ranged from 0.10 to 4.1 mg L^{-1} in the two lakes (Figure 2). The highest value was recorded in Megali Prespa in June 2016, whereas in Lake Mikri Prespa the maximum biomass reached 3.07 mg L^{-1} in September 2015. The cyanobacterial biomass reached 2.47 mg L^{-1} in Lake Mikri Prespa and 2.63 mg L^{-1} in Lake Megali Prespa. The highest annual contribution of cyanobacteria to phytoplankton biomass was observed in Lake Mikri Prespa (58.6%). We found that the cyanobacterium *Dolichospermum lemmermannii* was forming water blooms in both lakes, but in different periods, and contributed up to 23.1% of the annual cyanobacterial biomass in Lake Megali Prespa (Figure 2). The cyanobacterium *Microcystis aeruginosa*, was rarely detected in Lake Megali Prespa, but dominated the cyanobacterial biomass in Lake Mikri Prespa (34.6% annual contribution, which was followed by *D. lemmermannii* (14.0%) and *Microcystis panniformis* (5.4%), another rarely recorded species in Lake Megali Prespa where the second species in terms of biomass was *Aphanizomenon gracile* (2.4% annual contribution).

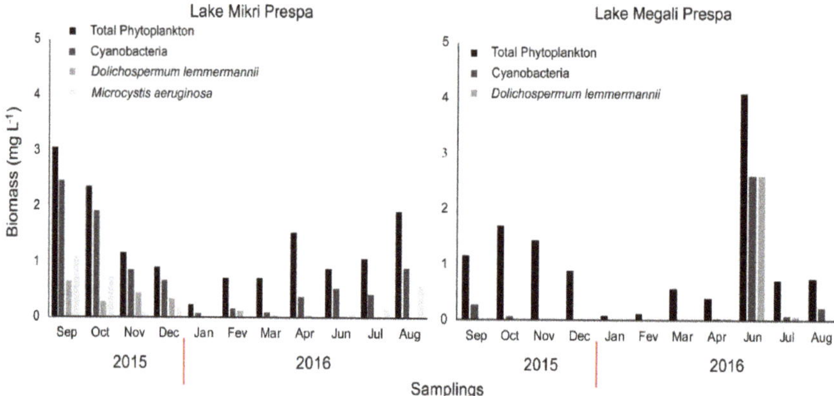

Figure 2. Total phytoplankton biomass, cyanobacterial assemblage biomass and *Dolichospermum lemmermannii* and *Microcystis aeruginosa* biomasses in the lakes Mikri and Megali Prespa during the period from September 2015 to August 2016. *M. aeruginosa* biomass was counted only in the samples of Lake Mikri Prespa.

Phytoplankton alpha diversity (Simpson index) varied between 0.18 and 0.90 in both lakes, and the cyanobacterial assemblage alpha diversity during their growth period (September–November 2015 and April–August 2016) varied between 0.004 and 0.73 in Lake Megali Prespa (average 0.37) and 0.57 and 0.77 (average 0.7) in Lake Mikri Prespa. Phytoplankton beta diversity (bSOR) ranged from 0.43 to 0.63 (Figure 3) and cyanobacterial beta diversity ranged from 0.38 to 0.7. Phytoplankton community compositional variation was attributed equally to the species turnover (50.1%) and nestedness (49.9%) while cyanobacterial variation during their growth period was mostly attributed to nestedness (77.4%) rather than species turnover (22.6%). For total phytoplankton, both components had substantial monthly fluctuations, while for cyanobacteria, nestedness was always higher (Figure 3).

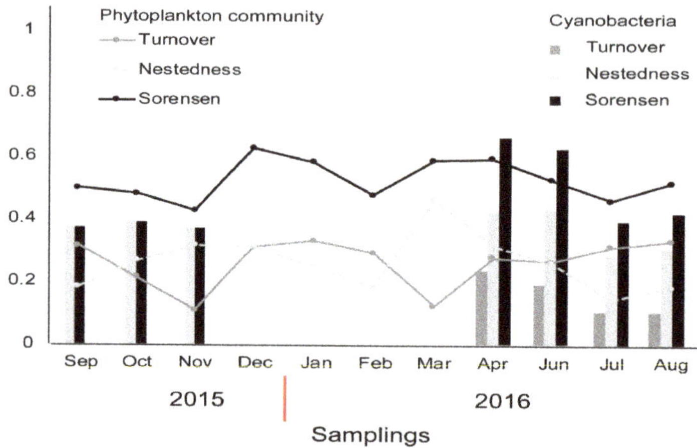

Figure 3. Phytoplankton and cyanobacterial assemblage beta diversity (bSOR), and their turnover and (bSIM) and nestedness (bNES) components values in the lakes Mikri and Megali Prespa.

The strong connections between cyanobacterial/phytoplankton species according to MIC correlation coefficients (MIC values corresponding to *p*-values < 0.01) are visualized in networks (Figure 4). Several negative and positive connections between cyanobacteria and other phytoplankton taxa were observed. The observed links were attributed to Cyanobacteria and Chlorophyta. In particular, the dominant cyanobacteria in Lake Mikri Prespa, *M. aeruginosa* and *D. lemmermannii* were negatively connected. In Lake Megali Prespa, *D. lemmermannii* had negative connections both with the other nostocalean species *Dolichospermum viguieri* and *Aphanizomenon gracile* and the oscillatorealean *Planktolyngbya limnetica*.

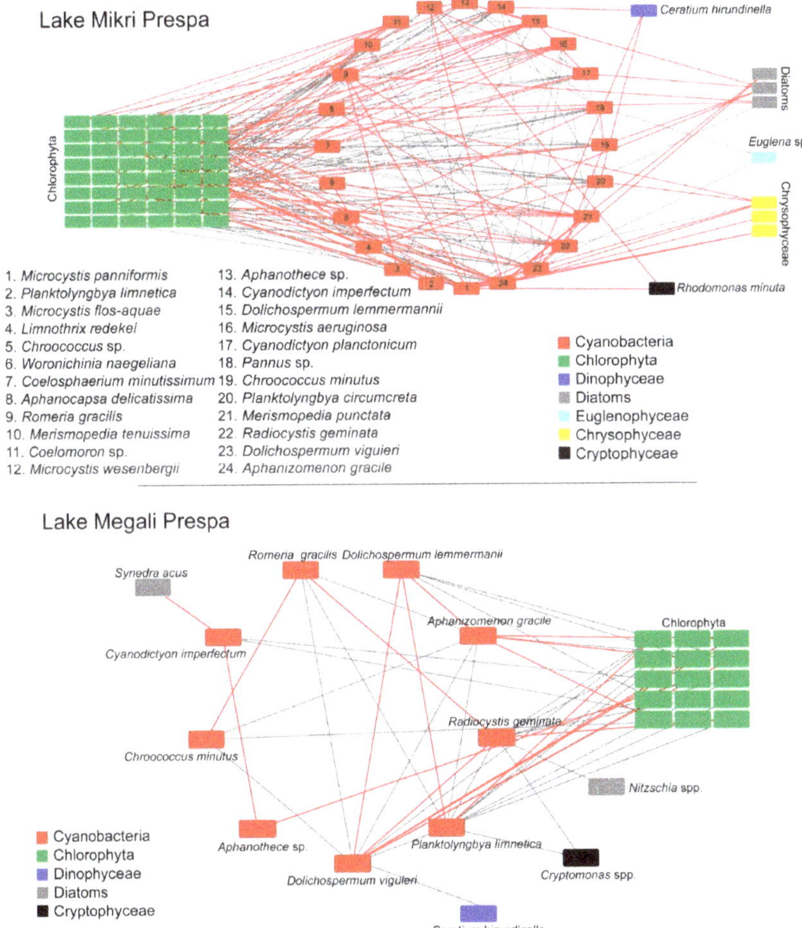

Figure 4. Network diagrams of the phytoplankton species (nodes) showing correlations (edges) with highly significant connections (*p*-values < 0.01) based on the maximal information coefficient (MIC) scores in the lakes Mikri and Megali Prespa. A positive relationship between each pair of phytoplankton species is depicted with grey edges while negative with red edges. Different node colors represent different taxonomic groups according to the color key.

The values of the water quality index PhyCoI$_{GP}$ ranged from 2.6 to 4.3 in Lake Megali Prespa and from 2.7 to 3.7 in Lake Mikri Prespa (Table 1). Ecological water quality assessment indicated a similar level for both lakes; i.e., around the good–moderate boundary. In particular, Lake Megali Prespa was classified in moderate water quality (average index value 2.9) and Lake Mikri Prespa was classified in the good class (average index value 3.3).

4. Discussion

Our focus on cyanobacterial species spatial distribution and community properties in two hydrologically connected ancient European lakes was motivated by the scarce information on hydrological connectivity as an emergent driver for cyanobacterial species homogenization in global expansion studies of toxic and non-toxic species [3,34]. The species pool of phytoplankton communities was found different in the two lakes with the average value of Simpson index almost double in the shallow, eutrophic Lake Mikri Prespa [21]. This result was rather expected because of the different typology (i.e., depth, surface area, stratification type) of the two lakes as well as their different trophic states, all significant determinants of phytoplankton alpha diversity [22,35]. However, we detected a high similarity in cyanobacterial diversity between these two different lakes: the cyanobacterial species of Lake Megali Prespa were a strict subset of the species pool inhabiting the shallow Lake Mikri Prespa. The increased similarity in cyanobacterial composition could be explained, to some extent, by the high occurrence of nostocalean species, bearing akinetes, cyanobacteria. These dormant cells survive long and extreme dispersion routes, assuring perennial germination and proliferation of toxic and non-toxic blooms [36].

The spread of nostocalean species from Lake Mikri Prespa to Lake Megali Prespa can be supported by the fluvial immense seeds of their cells (dormant and vegetative). Planktic nostocalean cyanobacteria have the ability to produce most types of cyanotoxins [37] and are frequently referred to as successful invaders [6]. *Dolichospermum lemmermannii*, the dominant nostocalean species in both lakes in terms of biomass, is known for its ability to produce various toxins (e.g., microcystins, anatoxins, saxitoxin) [38]. Also, *D. lemmermannii* has widened its geographic distribution across temperate lakes. Moreover, its ability to fix nitrogen underlies its competitive advantages in a lake with low inorganic nitrogen [39]. It has been reported to flourish in Lake Mikri Prespa since 1990 [15]. In this lake, in 2014, when *D. lemmermannii* and *Microcystis aeruginosa* formed a bloom, microcystins were measured at a concentration that posed a high risk of adverse human health effects [17]. Successful dispersal of nostocaleans could be favoured by their large population sizes reaching bloom densities [40].

However, all cyanobacteria species were not nostocalean nor identical in both lakes. The differences in cyanobacterial composition could be explained to some extent by the lakes' environmental filtering and species traits. For example, the other bloom-forming toxic cyanobacterium in Lake Mikri Prespa, *M. aeruginosa*, was found in low numbers in Lake Megali Prespa, whereas other *Microcystis* species (*M. flos-aquae* and *M. wesenbergii*) were not detected. *M. aeruginosa* is differentiated from the co-existing nostocalean species by its preference for higher phosphorus concentrations and for higher mixing depth/euphotic depth and nitrogen/light conditions ratios in the neighboring Lake Kastoria [41]. It seems that the lower trophic state and different lake typology characteristics are the environmental barriers for the *Microcystis* population increase in Lake Megali Prespa. The negative connection between *M. aeruginosa* and *D. lemmermannii*, shown in the network of Lake Mikri Prespa, may suggest that these species have temporally separated niches, indicating different species traits and preferences [1]. Lake environmental filtering and cyanobacterial species traits (dispersal abilities and competitive advantages), as described above, may play an important role in determining the Prespa lakes' metacommunity structure [42]. Geographic expansion of chroococcalean cyanobacteria has been frequently reported, as intensive blooms of toxic *Microcystis* have been attributed to global warming and regional eutrophication [6]. In addition, the influence of hydrological connectivity on toxic *Microcystis* spreading between two drowned river mouth lakes resulted in *Microcystis* dominance in both lakes [43]. Nevertheless, local conditions in each lake were important in determining which cyanobacterial species could maintain a viable population. The other chroococcalean species did not co-occur in the Prespa lakes. This might indicate different environmental preferences and habitat checkboards [44].

Phytoplankton, and particularly, cyanobacteria, are subjected to directional dispersal (water flow), a regional factor leading different habitat communities to be highly connected [45]. Thus, habitat connectivity and dispersal interact to structure metacommunity. The cyanobacterial dispersal because of the directional water flow could potentially override habitat control leading to species occupying unfavourable habitats [46]. Our analyses demonstrated a moderate to low dissimilarity in phytoplankton and cyanobacteria in the Prespa lakes. This is in contrast with results from other lakes and reservoirs of the world with strong environmental heterogeneity [11,47]. High beta diversity in these freshwaters was mainly explained by species turnover. In the Prespa lakes, beta diversity was explained equally by the species turnover and nestedness, while cyanobacterial assemblage beta diversity was mainly attributed to nestedness. In an analysis of freshwater plankton, bacteria were found to have a higher degree of nestedness than phytoplankton did, whereas the degree of nestedness was related mainly to habitat quality [44]. A meta-analysis of nestedness and turnover components of beta diversity across organisms and ecosystems by Soininen et al. [48] showed that passively dispersed organisms (e.g., phytoplankton) had lower turnover and total beta diversity than flying organisms but still much higher values than in the Prespa lakes.

In the Prespa lakes, different typologies and trophic states would initially lead to an assumption of low cyanobacterial nestedness. However, long-term hydrological connectivity of the two lakes appears to override habitat control through source-sink dynamics, leading to cyanobacterial species also occupying unfavourable habitats [46]. Furthermore, the higher cyanobacteria nestedness in relation to other phytoplankton taxa found in our study might also show the cyanobacteria's ability to adapt to unfavourable conditions. Kraus et al. [49] studying hydrological variations and ecological phytoplankton patterns in Amazonian floodplain lakes, explained the absence of phytoplankton dissimilarities to the ability of cyanobacteria to adapt to contrasted ecological conditions. No dispersal limitation and high levels of stochastic cyanobacterial establishment with a weak influence of phosphorus has been found in peri-Alpine lakes [3]. Cyanobacterial biodiversity patterns over one century of eutrophication and climate change showed clear signs of beta diversity loss with homogenization across sites in recent decades. This suggests that potentially toxic and bloom-forming cyanobacteria can widen their geographic distribution in the European temperate region [3]. Monchamp et al. [3] also found that *Dolichospermum* species were able to colonize a wide range of lakes. *Dolichospermum* species' distribution in the Prespa lakes agrees with these findings. The expansion of the bloom-forming, potentially toxic *D. lemmermannii* in several deep southern subalpine lakes has also been recently reported [40]. The species is considered as one of the most problematic algae in the subalpine Lake District, raising serious concerns because of the impacts on the tourist economy and the potential toxigenic effects. There is also an interesting expansion of this species between the 40th parallel and the Arctic Circle, which is in contrast to the prevailing south to north dispersion paths typical of other Nostocales [40]. In this case, our study provides important information about *D. lemmermannii*'s southernmost distribution in a deep and large lake (i.e., Lake Megali Prespa).

Based on the results of the application of the PhyCoI$_{GP}$ index for Mediterranean lakes, the water quality of Lake Megali Prespa cannot be considered good. The water quality of Lake Mikri Prespa showed considerable improvement characterized by much lower total phytoplankton and cyanobacterial biomas than during the period of 1990–2010 [15,22]. Moreover, based on phytoplankton biomass data of the period 2012–2014 and using the Alpine classification system for lake types L-AL3 and L-AL4, Lake Megali Prespa was also classified at lower than good ecological class [22]. The use of Alpine classification system was based on the location of Lake Megali Prespa at the southernmost tip of the Alpine biogeographical region of Europe [20].

We conclude that the Prespa lakes' cyanobacterial metacommunity structure was mainly affected by directional hydrological connectivity and high dispersal rates (Mikri Prespa source—Megali Prespa sink), a mass effect paradigm. Furthermore, it was affected to a less extent by species sorting (mostly for chroococcalean species) reflected by low beta diversity and high nestedness.

The hydrological connectivity is an emergent driver for cyanobacterial species homogenization, water quality homogenization and potentially toxic bloom expansion. The degrading of the water quality in Lake Megali Prespa in anticipation of improving that of Lake Mikri Prespa is an issue of great concern for lakes management. Our study supports the fact that the management of the Prespa lakes should aim at maintaining environmental heterogeneity while preventing the species of Lake Mikri Prespa, which include potentially toxic cyanobacteria, flooding into Lake Megali Prespa.

Supplementary Materials: The following are available online at http://www.mdpi.com/2073-4441/12/1/18/s1. Figure S1: Venn diagram showing the number of unique and common phytoplankton and cyanobacterial species between the lakes Mikri and Megali Prespa. Table S1: Phytoplankton species list that was recorded from the lakes Mikri and Megali Prespa during the period September 2015 to August 2016. X indicates presence; $ indicates participation in the networks.

Author Contributions: Conceptualization, M.K., A.D.M. and M.M.-G.; data curation, M.K., S.G. and N.S.; resources, E.M. and M.M.-G.; writing—original draft, M.K., S.G., N.S., A.D.M. and M.M.-G.; writing—review and editing, M.K., S.G., N.S., A.T., E.M., A.D.M. and M.M.-G.; visualization, M.K., S.G., N.S., A.T., I.G. and G.S.; supervision, A.D.M., E.M. and M.M.-G.; project administration, A.D.M. and M.M.-G.; funding acquisition, A.D.M., E.M. and M.M.-G. All authors have read and agreed to the published version of the manuscript.

Funding: This research was funded by ProLife, a project co-funded by the European Union and by national funds of the participating countries under the IPA Cross-Border Programme "Greece—The Republic of North Macedonia 2007–2013." The views expressed in this publication do not necessarily reflect the views of the European Union, the participating countries and the Managing Authority.

Conflicts of Interest: The authors declare no conflict of interest.

References

1. Lampert, W.; Sommer, U. *The Ecology of Lakes and Streams*, 2nd ed.; Oxford University Press: Oxford, UK, 2007; p. 32.

2. Chrisostomou, A.; Moustaka-Gouni, M.; Sgardelis, S.; Lanaras, T. Air-dispersed phytoplankton in a Mediterranean river-reservoir system (Aliakmon-Polyphytos, Greece). *J. Plankton Res.* **2009**, *31*, 877–884. [CrossRef]

3. Monchamp, M.-E.; Spaak, P.; Pomati, F. High dispersal levels and lake warming are emergent drivers of cyanobacterial community assembly in peri-Alpine lakes. *Sci. Rep.* **2019**, *9*, 7366. [CrossRef] [PubMed]

4. De Meester, L.; Gómez, A.; Okamura, B.; Schwenk, K. The Monopolization Hypothesis and the dispersal-gene flow paradox in aquatic organisms. *Acta Oecol.* **2002**, *23*, 121–135. [CrossRef]

5. Whittaker, R.J.; Willis, K.J.; Field, R. Scale and species richness: Towards a general, hierarchical theory of species diversity. *J. Biogeogr.* **2001**, *28*, 453–470. [CrossRef]

6. Sukenik, A.; Quesada, A.; Salmaso, N. Global expansion of toxic and non-toxic cyanobacteria: Effect on ecosystem functioning. *Biodivers. Conserv.* **2015**, *24*, 889–908. [CrossRef]

7. Baselga, A.; Orme, C.D.L. Betapart: An R package for the study of beta diversity. *Methods Ecol. Evol.* **2012**, *3*, 808–812. [CrossRef]

8. Baselga, A. Partitioning the turnover and nestedness components of beta diversity. *Glob. Ecol. Biogeogr.* **2010**, *19*, 134–143. [CrossRef]

9. Moustaka-Gouni, M.; Sommer, U.; Katsiapi, M.; Vardaka, E. Monitoring of Cyanobacteria for Water Quality: Doing the Necessary Right or Wrong? Available online: http://www.publish.csiro.au/mf/MF18381 (accessed on 18 December 2019).

10. Muhl, R.M.W.; Roelke, D.L.; Zohary, T.; Moustaka-Gouni, M.; Sommer, U.; Borics, G.; Gaedke, U.; Withrow, F.G.; Bhattacharyya, J. Resisting annihilation: Relationships between functional trait dissimilarity, assemblage competitive power and allelopathy. *Ecol. Lett.* **2018**, *21*, 1390–1400. [CrossRef]

11. Maloufi, S.; Catherine, A.; Mouillot, D.; Louvard, C.; Couté, A.; Bernard, C.; Troussellier, M. Environmental heterogeneity among lakes promotes hyper β-diversity across phytoplankton communities. *Freshw. Biol.* **2016**, *61*, 633–645. [CrossRef]

12. Leng, M.J.; Wagner, B.; Boehm, A.; Panagiotopoulos, K.; Vane, C.H.; Snelling, A.; Haidon, C.; Woodley, E.; Vogel, H.; Zanchetta, G.; et al. Understanding past climatic and hydrological variability in the Mediterranean from Lake Prespa sediment isotope and geochemical record over the last glacial cycle. *Quat. Sci. Rev.* **2013**, *66*, 123–136. [CrossRef]

13. Hadjisolomou, E.; Stefanidis, K.; Papatheodorou, G.; Papastergiadou, E. Assessment of the eutrophication-related environmental parameters in two Mediterranean lakes by integrating statistical techniques and self-organizing maps. *Int. J. Environ. Res. Public Health* **2018**, *15*, 547. [CrossRef] [PubMed]

14. Hollis, G.E.; Stevenson, A.C. The physical basis of the Lake Mikri Prespa systems: Geology, climate, hydrology and water quality. *Hydrobiologia* **1997**, *351*, 1–19. [CrossRef]

15. Tryfon, E.; Moustaka-Gouni, M.; Nikolaidis, G.; Tsekos, I. Phytoplankton and physical-chemical features of the shallow Lake Mikri Prespa, Macedonia, Greece. *Arch. Hydrobiol.* **1994**, *131*, 477–494.

16. Vardaka, E.; Moustaka-Gouni, M.; Cook, C.M.; Lanaras, T. Cyanobacterial blooms and water quality in Greek waterbodies. *J. Appl. Phycol.* **2005**, *17*, 391–401. [CrossRef]

17. Christophoridis, C.; Zervou, S.-K.; Manolidi, K.; Katsiapi, M.; Moustaka-Gouni, M.; Kaloudis, T.; Triantis, T.M.; Hiskia, A. Occurrence and diversity of cyanotoxins in Greek lakes. *Sci. Rep.* **2018**, *8*, 17877. [CrossRef] [PubMed]

18. Stamou, G.; Katsiapi, M.; Moustaka-Gouni, M.; Michaloudi, E. Grazing potential—A functional plankton food web metric for ecological water quality assessment in Mediterranean lakes. *Water* **2019**, *11*, 1274. [CrossRef]

19. Hampton, S.E.; McGowan, S.; Ozersky, T.; Virdis, S.G.P.; Vu, T.T.; Spanbauer, T.L.; Kraemer, B.M.; Swann, G.; Mackay, A.W.; Powers, S.M.; et al. Recent ecological change in ancient lakes. *Limnol. Oceanogr.* **2018**, *63*, 2277–2304. [CrossRef]

20. European Environment Agency. Biogeographical Regions. Available online: http://www.eea.europa.eu/data-and-maps/figures/biogeographical-regions-in-europe-1 (accessed on 5 November 2019).

21. Tryfon, E.; Moustaka-Gouni, M. Species composition and seasonal cycles of phytoplankton with special reference to the nanoplankton of Lake Mikri Prespa. *Hydrobiologia* **1997**, *351*, 61–75. [CrossRef]

22. Moustaka-Gouni, M.; Sommer, U.; Economou-Amilli, A.; Arhonditsis, G.B.; Katsiapi, M.; Papastergiadou, E.; Kormas, K.A.; Vardaka, E.; Karayanni, H.; Papadimitriou, T. Implementation of the Water Framework Directive: Lessons learned and future perspectives for an ecologically meaningful classification based on phytoplankton of the status of Greek lakes, Mediterranean region. *Environ. Manag.* **2019**, *64*, 675–688. [CrossRef]

23. Katsiapi, M.; Michaloudi, E.; Moustaka-Gouni, M.; Pahissa López, J. First ecological evaluation of the ancient Balkan Lake Megali Prespa based on plankton. *J. Biol. Res.* **2012**, *17*, 51–56.

24. Parisopoulos, G.; Sapountzakis, G.; Georgiou, P.; Giamouri, M.; Malakou, M. Development of rule curves for sluice gate operation and water level management in lake Micro Prespa. In Proceedings of the 4th Conference on Water Observation and Information System for Decision Support (BALWOIS), Ohrid, Republic of Macedonia, 25–29 May 2010.

25. Parisopoulos, G. *Review of the Sluice Gate Operation and Water Level Management in Lake Micro Prespa*; Technical Report; Society for the Protection of Prespa: Agios Germanos, Greece, 2016.

26. Komárek, J.; Anagnostidis, K. Cyanoprokaryota 1. Chroococcales. In *Süsswasserflora von Mitteleuropa*; Ettl, H., Gärtner, G., Heynig, H., Mollenhauer, D., Eds.; Spektrum Akademischer Verlag: Heidelberg, Germany, 1998; pp. 1–548.

27. Hillebrand, H.; Dürselen, C.-D.; Kirschtel, D.; Pollingher, U.; Zohary, T. Biovolume calculation for pelagic and benthic microalgae. *J. Phycol.* **1999**, *35*, 403–424. [CrossRef]

28. Hammer, Ø.; Harper, D.A.T.; Ryan, P.D. Past: Paleontological statistics software package for education and data analysis. *Palaeontol. Electron.* **2001**, *4*, 9.

29. Baselga, A. Partitioning abundance-based multiple-site dissimilarity into components: Balanced variation in abundance and abundance gradients. *Methods Ecol. Evol.* **2017**, *8*, 799–808. [CrossRef]

30. Baselga, A. Multiplicative partition of true diversity yields independent alpha and beta components; additive partition does not. *Ecology* **2010**, *91*, 1974–1981. [CrossRef] [PubMed]

31. Reshef, D.N.; Reshef, Y.A.; Finucane, H.K.; Grossman, S.R.; McVean, G.; Turnbaugh, P.J.; Lander, E.S.; Mitzenmacher, M.; Sabeti, P.C. Detecting novel associations in large data sets. *Science* **2011**, *334*, 1518–1524. [CrossRef]

32. Smoot, M.E.; Ono, K.; Ruscheinski, J.; Wang, P.-L.; Ideker, T. Cytoscape 2.8: New features for data integration and network. *Bioinformatics* **2011**, *27*, 431–432. [CrossRef]

33. Hernández-Ruiz, M.; Barber-Lluch, E.; Prieto, A.; Alvarez-Salgado, X.A.; Logares, R.; Teira, E. Seasonal succession of small planktonic eukaryotes inhabiting surface waters of a coastal upwelling system. *Environ. Microbiol.* **2018**, *20*, 2955–2973. [CrossRef]

34. Monchamp, M.-E.; Spaak, P.; Domaizon, I.; Dubois, N.; Bouffard, D.; Pomati, F. Homogenization of lake cyanobacterial communities over a century of climate change and eutrophication. *Nat. Ecol. Evol.* **2018**, *2*, 317–324. [CrossRef]

35. Stomp, M.; Huisman, J.; Mittelbach, G.G.; Litchman, E.; Klausmeier, C.A. Large-scale biodiversity patterns in freshwater phytoplankton. *Ecology* **2011**, *92*, 2096–2107. [CrossRef]

36. Moustaka-Gouni, M.; Hiskia, A.; Genitsaris, S.; Katsiapi, M.; Manolidi, K.; Zervou, S.-K.; Christophoridis, C.; Triantis, T.M.; Kaloudis, T.; Orfanidis, S. First report of *Aphanizomenon favaloroi* occurrence in Europe associated with saxitoxins and a massive fish kill in Lake Vistonis, Greece. *Mar. Freshw. Res.* **2017**, *68*, 793–800. [CrossRef]

37. Cirés, S.; Wörmer, L.; Ballot, A.; Agha, R.; Wiedner, C.; Velásquez, D.; Casero, M.C. Phylogeography of cylindrospermopsin and paralytic shellfish toxin-producing Nostocales cyanobacteria from Mediterranean Europe (Spain). *Appl. Environ. Microb.* **2014**, *80*, 1359–1370. [CrossRef] [PubMed]

38. Lepistö, L.; Kauppila, P.; Rapala, J.; Pekkarinen, M.; Sammalkorpi, I.; Villa, L. Estimation of reference conditions for phytoplankton in a naturally eutrophic shallow lake. *Hydrobiologia* **2006**, *568*, 55–66. [CrossRef]

39. Salmaso, N.; Capelli, C.; Shams, S.; Cerasino, L. Expansion of bloom-forming *Dolichospermum lemmermannii* (Nostocales, Cyanobacteria) to the deep lakes south of the Alps: Colonization patterns, driving forces and implications for water use. *Harmful Algae* **2015**, *50*, 76–87. [CrossRef]

40. Finlay, B.J. Global dispersal of free-living microbial eukaryote species. *Science* **2002**, *296*, 1061–1063. [CrossRef]

41. Moustaka-Gouni, M.; Vardaka, E.; Tryfon, E. Phytoplankton species succession in a shallow Mediterranean lake (L. Kastoria, Greece): Steady-state dominance of *Limnothrix redekei*, *Microcystis aeruginosa* and *Cylindrospermopsis raciborskii*. *Hydrobiologia* **2007**, *575*, 129–140. [CrossRef]

42. Pujoni, D.G.F.; Barros, C.F.D.A.; Dos Santos, J.B.O.; Maia-Barbosa, P.M.; Barbosa, F.A.R. Dispersal ability and niche breadth act synergistically to determine zooplankton but nor phytoplankton metacommunity structure. *J. Plankton Res.* **2019**, *41*, 479–490. [CrossRef]

43. Xie, L.; Hagar, J.; Rediske, R.R.; O'Keefe, J.; Dyble, J.; Hong, Y.; Steinman, A.D. The influence of environmental conditions and hydrologic connectivity on cyanobacteria assemblages in two drowned river mouth lakes. *J. Gt. Lakes Res.* **2011**, *37*, 470–479. [CrossRef]

44. Soininen, J.; Köngäs, P. Analysis of nestedness in freshwater assemblages-patterns across species and trophic levels. *Freshw. Sci.* **2012**, *31*, 1145–1155. [CrossRef]

45. Moritz, C.; Meynard, C.N.; Devictor, V.; Guizien, K.; Labrune, C.; Guarini, J.-M. Disentangling the role of connectivity, environmental filtering, and spatial structure on metacommunity dynamics. *Oikos* **2013**, *122*, 1401–1410. [CrossRef]

46. Tonkin, J.D.; Stoll, S.; Jähnig, S.C.; Haase, P. Contrasting metacommunity structure and beta diversity in an aquatic-floodplain system. *Oikos* **2016**, *125*, 686–697. [CrossRef]

47. Wojciechowski, J.; Heino, J.; Bini, L.M.; Padial, A.A. Temporal variation in phytoplankton beta diversity patterns and metacommunity structures across subtropical reservoirs. *Freshwater Biol.* **2017**, *62*, 751–766. [CrossRef]

48. Soininen, J.; Heino, J.; Wang, J. A meta-analysis of nestedness and turnover components of beta diversity across organisms and ecosystems. *Glob. Ecol. Biogeogr.* **2018**, *27*, 96–109. [CrossRef]

49. Kraus, C.N.; Bonnet, M.-P.; Miranda, C.A.; de Souza Nogueira, I.; Garnier, J.; Vieira, L.C.G. Interannual hydrological variations and ecological phytoplankton patterns in Amazonian floodplain lakes. *Hydrobiologia* **2019**, *830*, 135–149. [CrossRef]

Article

Cyanobacterial Blooms in Lake Varese: Analysis and Characterization over Ten Years of Observations

Nicola Chirico [1], Diana C. António [1], Luca Pozzoli [1], Dimitar Marinov [1], Anna Malagó [1], Isabella Sanseverino [1], Andrea Beghi [2], Pietro Genoni [2], Srdan Dobricic [1] and Teresa Lettieri [1,*]

[1]   European Commission, Joint Research Centre (JRC), 21027 Ispra, Italy; nicola.chirico@uninsubria.it (N.C.); diana_conduto@hotmail.com (D.C.A.); luca.pozzoli@ec.europa.eu (L.P.); dimitar.marinov@ext.ec.europa.eu (D.M.); anna.malago@ec.europa.eu (A.M.); isabella.sanseverino@ec.europa.eu (I.S.); srdan.dobricic@ec.europa.eu (S.D.)

[2]   Agenzia Regionale per la Protezione dell'Ambiente della Lombardia, Regione Lombardia, 21100 Varese, Italy; a.beghi@arpalombardia.it (A.B.); p.genoni@arpalombardia.it (P.G.)

*   Correspondence: teresa.lettieri@ec.europa.eu

Received: 30 November 2019; Accepted: 19 February 2020; Published: 1 March 2020

Abstract: Cyanobacteria blooms are a worldwide concern for water bodies and may be promoted by eutrophication and climate change. The prediction of cyanobacterial blooms and identification of the main triggering factors are of paramount importance for water management. In this study, we analyzed a comprehensive dataset including ten-years measurements collected at Lake Varese, an eutrophic lake in Northern Italy. Microscopic analysis of the water samples was performed to characterize the community distribution and dynamics along the years. We observed that cyanobacteria represented a significant fraction of the phytoplankton community, up to 60% as biovolume, and a shift in the phytoplankton community distribution towards cyanobacteria dominance onwards 2010 was detected. The relationships between cyanobacteria biovolume, nutrients, and environmental parameters were investigated through simple and multiple linear regressions. We found that 14-days average air temperature together with total phosphorus may only partly explain the cyanobacteria biovolume variance at Lake Varese. However, weather forecasts can be used to predict an algal outbreak two weeks in advance and, eventually, to adopt management actions. The prediction of cyanobacteria algal blooms remains challenging and more frequent samplings, combined with the microscopy analysis and the metagenomics technique, would allow a more conclusive analysis.

Keywords: freshwater; cyanobacteria; bloom; air temperature; nutrients; model

---

1. Introduction

Cyanobacteria are photosynthetic bacteria that occur naturally in fresh, brackish, marine waters, and terrestrial environments [1]. Cyanobacteria are a worldwide problem [2] for their ability to form massive blooms that can produce a wide range of harmful toxins [3]. Bloom events are likely to be promoted by eutrophication and climate change, and the number and intensity of these blooms increased globally over the last decades [4]. Cyanobacteria can cause a whole range of problems for human health and the environment. Fish killed by anoxia caused by the decay of the cyanobacteria biomass is a notorious effect of cyanobacteria blooms [5]. Moreover, in some conditions, cyanobacteria can produce cyanotoxins (that includes hepatotoxins, neurotoxins, cytotoxins, and dermotoxins) negatively impacting the survival of aquatic organisms [6]. In addition to wild animals, intoxication can also occur for domestic animals and humans, either by direct ingestion of cyanobacteria cells and/or through consumption of drinking water containing cyanotoxins [7], leading to public health concerns [3]. Cyanobacteria blooms also generate bad-smelling mucilaginous scum that both impact the recreational use of water bodies and prevent their use for drinking water. Considering the negative impacts of cyanobacterial blooms on ecological, economical, and human health, their monitoring and forecast is of paramount importance for lake management [8]. Several monitoring approaches and predictive models were developed to provide accurate and timely information regarding the development of cyanobacterial bloom in the waterbodies.

Different modeling techniques are known and used for algal bloom prediction, the most common being multiple linear regression (MLR) and artificial neural networks (ANNs). MLR is the simplest technique used to develop linear models and used, for example, to predict cyanobacteria [9,10] or chlorophyll-a [11–13] abundance, while ANNs are complex machine learning techniques, which mimic the neural adaptation behavior in order to "learn" how to solve a problem.

Eutrophication has been cited as a major cause of increasing harmful cyanobacterial/algal blooms, in particular in the Mediterranean area [14], and factors including light, temperature, quiescent water and nutrients, mainly total nitrogen (TN) and total phosphorus (TP), are considered among the main drivers of cyanobacterial blooms [15,16]. Albeit, it is well known that important predictors for cyanobacteria dominance and biomass are TP and TN [17,18], there are increasing evidences that water temperature (WT) is an important factor [9,19–24]. Warming waters intensify the vertical stratification and lengthen the period of seasonal stratification, which is one of the main physical variables determining the occurrence of algal bloom outbreaks [25]. The most evident effect of stratification is the changing availability of nutrients, which may be accumulated in surface layers or mixed in the entire water column. The increasing global air temperature may increase the strength and depth of stratification, with possible influences on the seasonal timing and changes in the phytoplankton phenology and community succession. Stratification leads cyanobacteria to outcompete other phytoplankton groups, both because of their buoyancy regulation ability [26,27] and because cyanobacteria are positively affected by the increase of water temperature [22]. In addition, it has been shown that the meteorological variables as air temperature, wind speed, and relative humidity, could be drivers of hypolimnetic anoxia, which is an indirect consequence of thermal stratification [28].

Previous studies found conflicting results on the response of cyanobacteria to climate and nutrients [23,29]. Here, we aim to identify the most important and efficient predictor for cyanobacteria blooms in Lake Varese, an eutrophic lake in northern Italy and one of the first and most glaring examples of eutrophication in Europe [30]. Initially classified as hypertrophic lake, following remedial actions aimed at reduction of the P loading, the lake is now in eutrophic status with bloom events occurring every year during the summer and early autumn. Lake Varese is "naturally productive" due to its morphometric characteristics and the geology of its drainage basin [31]. However, the further increase of human activities in the area accelerated the degradation of its water quality. The analysis of carotenoid stratigraphy of the sediments showed also that some phytoplanktonic groups are particularly well adapted in environments with high organic content.

The first signals of summer anoxia in the Lake were documented in 1957 and the fishery activities ended in 1975 [32]. Albeit, many studies testified the deterioration of the water quality, phytoplankton studies in Lake Varese were carried out only occasionally [33]. The first detailed analysis dates back to 1979 [34] and ten years later a further comprehensive study [35] showed that the eutrophication process was not showing any sign of reversion, and that the P release from sediments is a major factor constraining the recovery of lake ecosystems [36].

In this study, we firstly analyzed the cyanobacteria community with a detailed ten-year picture of the dynamic composition. We further explored the possible role of chemical and physical parameters triggering cyanobacteria blooms, introducing a novel approach to find possible relationships between meteorological data, lake stratification, and cyanobacteria abundance. We developed a simple approach that can be applied to other lakes using relatively few data and weather forecast data to put in place an early warning system for cyanobacteria blooms.

2. Materials and Methods

2.1. Sites Description

The selected study area is Lake Varese (45° 49′ N 8° 44′ E), which is located in northern Italy at the feet of the Alps mountain range at a mean altitude of 236 m above sea level (Figure 1). Lake Varese is a monomictic and eutrophic shallow lake, with a mean depth of 11 m, a maximum depth of 26 m, a surface area of 14.8 km^2, a volume of 153×106 m^3, and a theoretical renewal time of 1.7–1.9 years [37,38]. Its catchment, with a surface area of 115.5 km^2, hosts an average population density of 700 inhabitant/km^2 and is associated with many industrial and commercial activities. The lake has two tributaries: The Brabbia channel and the Tinella stream, with annual average discharges of 23×10^6 and 10×10^6 m^3 yr^{-1}, respectively, and one effluent, the Bardello stream, with annual average discharge of 80.4×10^6 m^3 yr^{-1} [30]. Information related to watershed characteristics of Lake Varese can be retrieved at the following link: https://www.regione.lombardia.it/wps/portal/istituzionale/HP/ aqst-lago-di-varese/documenti-e-atti-istitutivi.

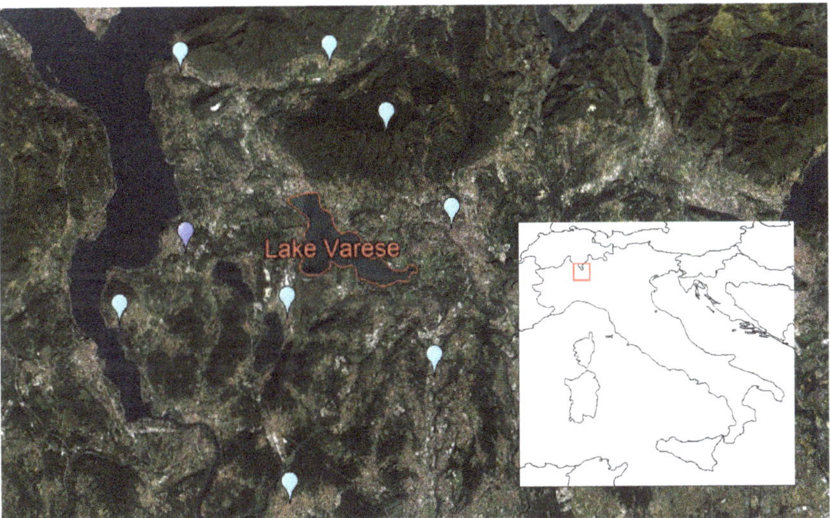

Figure 1. Location of Lake Varese (red). The positions of the available meteorological stations in the region are indicated by the light blue dots (Regional Environmental Protection Agency Lombardia, ARPA), and the dark blue dot (the Joint Research Center). The red polygon in the lower right corner map indicates the location of the model grid-cell from the ERA-Interim atmospheric reanalysis, where Lake Varese is located.

2.2. Sampling and Analysis

Physical, chemical, and phytoplankton measurements of Lake Varese were provided by the department of the Regional Environmental Protection Agency of Lombardia (ARPA). The samplings were performed at least six times a year, as foreseen in the national reference methods [39], by using the multiprobe (Ocean Seven 316 plus, IDRONAUT) to measure pH, conductivity, redox potential, photosynthetically active radiation (PAR), oxygen, depth, and temperature. Water transparency was determined using a Secchi disk. The euphotic region was determined as 2.5 times the Secchi disk depth or the region where PAR was larger than 1% of the radiation determined immediately below the water surface, and applied for the collection of integrated samples used for phytoplankton analysis.

The complete dataset is composed of 90 sampling campaigns distributed over ten years, from 2004 to 2014, including measures of chemical and physical parameters of lake water sampled at different depths in the euphotic zone. Measurements of nutrients and other chemical parameters were reported at only three depths (surface, 13 m and bottom) until 2011, while from 2012 they were also reported at 4/5 of the epilimnium. This vertical resolution is rather coarse to analyze the vertical stratification of these parameters and for this reason we only provide an overview of these data by reporting mean values and their ranges (minimum and maximum). In general, the concentrations of nutrients (phosphorus and nitrogen) and oxygen are related to the presence of cyanobacteria blooms (i.e., nitrogen compounds and phosphorus are consumed/depleted from the surface while oxygen is enriched due to photosynthesis). Table 1 summarizes the physical–chemical parameters with a sufficient number of valid measurements used for this study. All parameters with missing values or below the limit of quantification (LOQ) for more than 20% of the samples were excluded. For the remaining variables, values below LOQ were arbitrarily halved, resulting in a preliminary list of nine potential variables for modeling cyanobacteria abundance. The variables may be all related, directly or indirectly, to the abundance of cyanobacteria and include both nutrient concentrations (such as: Total phosphorus, TP; ammonium nitrogen, AN; and reactive silicates, RS) and physical parameters (such as: Water temperature, WT; pH; conductivity, CD; dissolved oxygen, DO; oxygen saturation, OS; and Secchi disk depth, SD). Total nitrogen (TN) was not calculated, therefore this parameter is not available in the dataset.

Table 1. List of water surface physical–chemical variables and total cyanobacteria biovolume/density measured at Lake Varese during the period 2004–2014. The mean, median, and range (minimum–maximum) values are reported for each variable. The number of samples above and below the level of quantification (LOQ), and missing values are reported. The number of cyanobacteria (present/absent) over the total sampling campaigns is shown.

Parameter and Unit of Measures		Mean	Median	Range	Above LOQ	Below LOQ	Missing
Total phosphorous (µg/L)	(TP)	42.3	28	2.5–110	80	9	1
Ammonium nitrogen (mg/L)	(AN)	0.12	0.048	0.0075–0.69	75	14	1
Reactive silicates (SiO2) (mg/L)	(RS)	1.16	0.92	0.05–3.9	82	6	2
pH	(pH)	8.18	8.2	7.5–9.6	90	0	0
Conductivity (µS/cm 20 °C)	(CD)	257	258	175–310	90	0	0
Dissolved oxygen (mg/L)	(DO)	9.64	9.8	3.7–14.8	87	0	3
Oxygen saturation (%)	(OS)	101	100	50–173	88	0	2
Water temperature (°C)	(WT)	16.3	18	3.6–30	89	0	1
Secchi disk depth (m)	(SD)	4.1	3.6	1.1–9.6	90	0	0
		Mean	**Median**	**Range**	**Present**	**Absent**	
Cyanobacteria biovolume (mm^3/m^3)	(CyanoBV)	1.62×10^3	3.05×10^2	0–3.94×10^4	83	7	
Cyanobacteria density (cells/L)	(CyanoD)	9.69×10^6	3.62×10^5	0–1.26×10^8	84	6	

Phytoplakton Analysis

Collected samples were fixed with a Lugol Acid solution and stored in the dark until analysis was performed. Phytoplankton community was evaluated according to Utermöhl's method [40]. Shortly, a fixed sample was placed on a sedimentation tower during 24 h before microscopy analysis (Olympus model IX71 with 200X and 400X magnification). For each sample a minimum of 200 cells were identified to the genus or species level and enumerated to determine cell density. Biovolume was calculated using geometric similar models for each identified cell [41].

Cyanobacteria were measured at the level of genera/species as cell densities (cells/L) and biovolumes (mm^3/m^3), measured as integrated samples from the surface to 2.5 times the SD (considered as the limit of the euphotic zone). The total cyanobacteria cell density (CyanoD) and biovolume (CyanoBV) was calculated as the sum of all reported genera/species for each sampling date. Due to the large range of measured total biovolumes, spanning over several order of magnitudes, we applied the log based 10 transformation to predict the cyanobacteria biovolume (LOG CyanoBV). Variability associated with multiple cell counters could be present, however, a Phytoplankton Proficiency Test organized by EQAT Phytoplankton (External Quality Assessment Trials) was successfully completed in 2013 and 2016.

2.3. Meteorological Data

The meteorological data were obtained from the European Centre Medium-Range Weather Forecast (ECMWF) ERA-Interim atmospheric reanalysis product [42]. The reanalysis is a global model simulation of the atmosphere with a data assimilation system which includes a 4-dimensional variational analysis (4D-var). A large number of atmospheric observations, ground-based and from satellites, are included in the model variational analysis every 12 h, constraining the atmospheric weather simulation towards real observational data. ERA-Interim data are available starting from year 1989 and continuously updated once a month, with a delay of two months to allow for quality assurance. A large variety of atmospheric variables are available at three hourly time intervals, and with horizontal spatial resolution of about 80×80 km. Data for Lake Varese were extracted from the model grid cell containing its latitude/longitude coordinates (red polygon on the map in the lower right corner of Figure 1). We initially considered atmospheric variables which may have an impact on the physical, hydrological, and biogeochemical properties of the lake according to literature studies [22,25,43–45], such as surface air temperature, wind speed, photosynthetically active radiation (PAR), and total precipitation. Statistics of ERA-Interim variables from 2004 to 2014 at the Lake Varese grid cell are reported in Table 2.

Table 2. List of meteorological variables extracted from the grid cell of the ERA-Interim reanalysis where Lake Varese is located. The mean, median, and range (minimum and maximum value) calculated over the period 2004–2014 are reported for each variable.

Variable and Unit of Measures	Mean	Median	Range
Surface air temperature (°C)	9	8.97	−11.6–24.1
Photosynthetically active radiation (W/m^2)	126	122	2.17–273
Wind speed (m/s)	1.96	1.76	0.46–6.37
Total precipitation (mm/day)	6.07	0.84	0–163

The ERA-Interim reanalysis provides a series of advantages, including continuous data for a long period and representativeness of the region of interest around Lake Varese compared to scattered measurements not always directly located at the lake, which may reflect specific local conditions not representative of the lake conditions. Eight meteorological stations are available in a distance range of maximum 15 km from lake Varese from ARPA (light blue dots in Figure 1), and one station is located at the Joint Research Center, about 8 km far from Lake Varese (dark blue dot in Figure 1). Generally, the simulated atmospheric variables are in good agreement with the observations in the region, as shown for example in Figure A1 in Appendix A, where the ERA-Interim data are compared to available meteorological stations for the period April 2014–October 2014. A further advantage of using ERA-Interim is given by the continuous update of the simulations, which may be useful to further evaluate the model in the future and to use weather forecast data to define an early warning system for algal blooms.

Using this dataset, we calculated the stratification strength (i.e., the maximum slope of the thermocline °C/m) from the vertical temperature profiles measured at Lake Varese by probes from 2004 to 2014 (see Figure A2 in Appendix A). The averages of the air temperature and wind speed were calculated from 1 to 28 days (named from T1 to T28 for simplicity) preceding every water sampling campaign date included in the dataset.

2.4. Statistical Analysis

The relationships between cyanobacteria biovolume and the selected environmental parameters (physical and chemical water properties, Table 1, and meteorological data, Table 2) were analyzed through an exploratory stepwise regression (SLR) approach followed by a multiple linear regression (MLR) method. In particular, the SLR models were applied for investigating the correlation with environmental parameters. We looked for correlations with air temperature and wind speed near the surface from the ERA-Interim reanalysis data. We also studied the standard deviation in water temperature vertical profiles as a proxy of water stratification with similar results [46]. The Pearson correlation coefficient was calculated between the thermocline slopes and the average surface temperatures and separately between the thermocline slopes and the average wind speeds.

After that, the MLR model was used as a forecasting approach for predicting the cyanobacteria growth. In particular, we used physical–chemical parameters of lake surface water as predictors with potential influence on cyanobacteria growth and we chose cyanobacteria biovolume as a measure of cyanobacteria growth and as response variable, expressed in ten-based logarithm value (LOG CyanoBV). MLR was selected for its simplicity and because the parameters were not enough for properly training the ANNs method without incurring in overfitting problems.

MLR calculations and statistical analysis were performed using the software environment for statistical computing and graphics R v.3.3.2 (R Core Team 2019). The performance of the MLR models was assessed using the coefficient of determination (R^2), adjusted coefficient of determination (R^2 adj), and root mean square error (RMSE). The best relationship was finally validated using an independent dataset from the European Commission Joint Research Centre (JRC) (see Section 3.2.).

3. Results

3.1. Occurrence, Magnitude, and Timing of Cyanobacteria

Microscopic analysis of the samples collected from 2004 to 2014 were analyzed to characterize the community distribution and dynamics along the years. Lake Varese is stratified for most of the year and it happens to be vertically mixed only in the period between November and February when the surface water temperature decreases below 10 °C.

Cyanobacteria were detected at Lake Varese in 84 of the 90 sampling campaign (93%), with a median density of 362,000 cells/L and median biovolume of 305 mm^3/m^3, with maximum of 126 million of cells/L and 39,000 mm^3/m^3, respectively (Table 1). As expected in eutrophic environments, cyanobacteria were over represented in the lake phytoplanktonic community, accounting for more than 50% of the total cell abundance, in average. Over those 10 years, Oscillatoriales and Chroococcales accounted for 50% of the total CyanoBV, while among Nostocales (43%), Aphanizomenon accounted for 38% (data not shown).

The cell number and biovolume were analyzed as relative annual abundance per phylum, as shown in Figure 2. Considering the community abundance based on biovolume, cyanobacteria abundance may not appear so representative, although the ultraplankton could contribute significantly to the cyanobacteria community (see Figure 2b).

The occurrence of Oscillatoriales had visibly increased since 2010 while Chroococcales showed an inverted trend (Figure 3). The peaks of cyanobacteria abundance were mainly seen during Summer-Spring seasons (Figure 3a, see ss2007, ss2008 and from 2011 to 2013), although a Winter-Autumn peak was reported in 2011 (Figure 3a, see wa2011). This episode occurred during the last four years of the considered period, where Oscillatoriales were dominant.

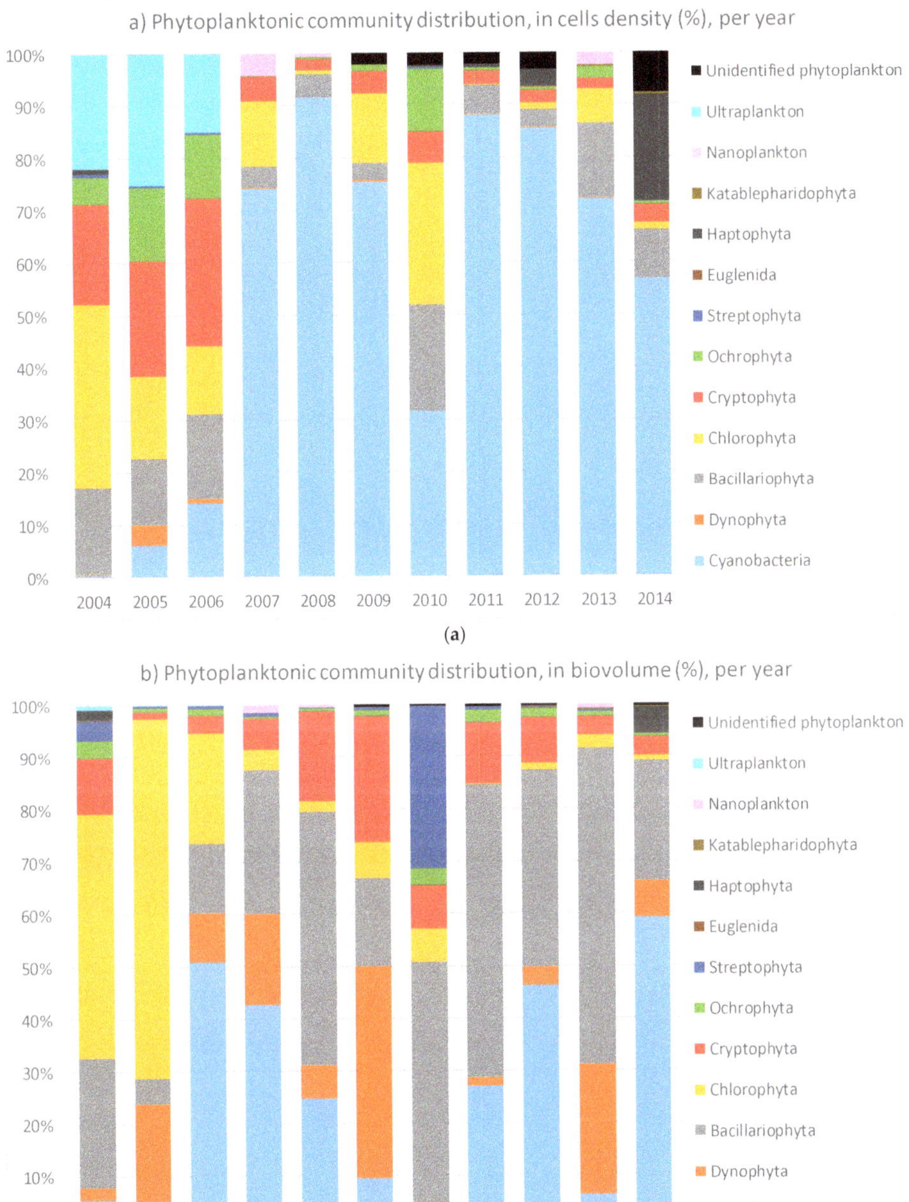

Figure 2. Distribution of the total phytoplankton community expressed in percentage (Y-axis) of the average of the annual campaign from 2004 to 2014. (**a**) On the top plot, the distribution is reported as percentage of cell density (cells/L); (**b**) on the bottom the distribution is shown as percentage of biovolume (mm^3/m^3). The total cell density and biovolume values for each year are reported in Table A1 in Appendix A.

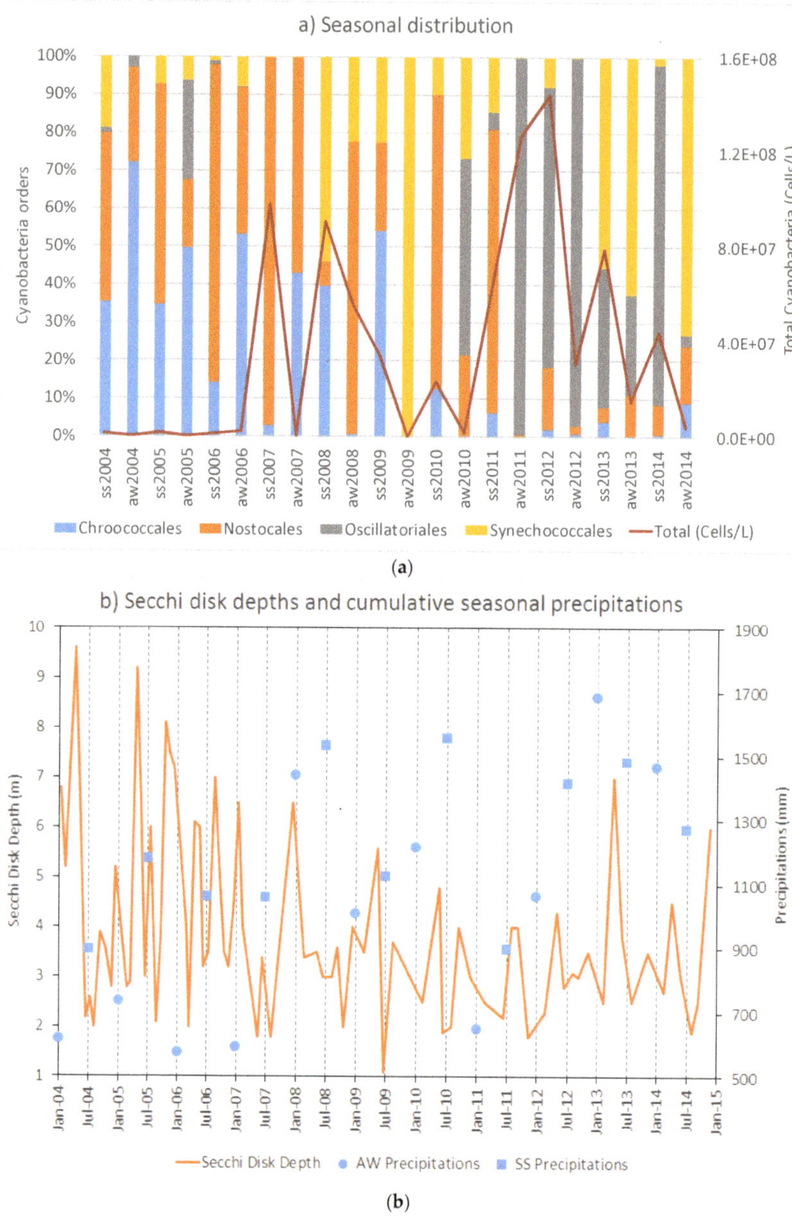

Figure 3. (**a**) The top graph shows a representation of total cyanobacteria community distribution as percentage, based on biovolume (mm^3/m^3). Community distribution is presented as the average of the annual campaigns data, grouped as Summer-Spring (ss) and Autumn-Winter (aw) seasons, and by order. The red line reports the total cyanobacteria number per data set; (**b**) bottom graph represents the Secchi disk depth (continuous line in orange) at each sampling campaign and the six-month cumulative precipitations, Autumn-Winter (aw, blue circles) and Spring-Summer (ss, blue squares).

Considering that cyanobacteria abundance is generally favored by the thermocline stability and the increasing of the temperature, the dynamic of the cyanobacteria community was evaluated compared to the available physical and meteorological parameters, such as air and water temperature, photic region maximum depth or precipitation. Air and water temperature revealed a repetitive seasonal pattern (data not shown), although the euphotic region showed an overall decrease since 2008 (Secchi disk depths in Figure 3b). The cumulative seasonal precipitations on the other hand showed a slight increase over time. In the present study, SD was found to be directly correlated with cyanobacteria abundance. Water turbidity can vary according to the abundance of organisms or suspended organic matter. Precipitation may promote bleaching of sediments from the surrounding area decreasing the turbidity, and consequently the SD, but this phenomenon did not occur in Lake Varese, as can be seen in Figure 3b. However, precipitation may play a role on cyanobacteria dynamics in this lake, considering that the community shifted from Chroococcales and Nostocales to Oscillatoriales and Synechoccocales dominance along time, as the total precipitation increased. Nostocalles have competitive advantages such as capacity to fix atmospheric nitrogen and to produce dormant cells, resistant to adverse conditions [47]. However, Oscillatoriales have been reported to often dominate shallow polymictic eutrophic lakes showing cyclic successions between Microcystis (Chroococcales) and Planktothrix (Oscillatoriales) [48].

Other parameters which may influence the cyanobacteria growth and dominance are the nutrients' bioavailability. The nutrients known to strongly impact the cyanobacteria dynamics and abundance are nitrogen (in the form of ammonia, nitrite, and nitrate), phosphorus (or orthophosphate) and iron [45]. In this study, all these nutrients except for the iron, were measured and made then available for further evaluation. Indeed, together with temperature, nitrogen, and phosphorus showed higher potential for the predictive model development (see Section 3.2.), based on biovolume abundance.

3.2. Relationships between Cyanobacteria and Environmental Parameters

Applying a first preliminary SLR analysis between the cyanobacteria biovolume (LOG CyanoBV) and the measured water temperature at lake surface, we found the highest Pearson correlation coefficient (0.5), which was also expected from previous studies [9,19–24]. For the average wind speeds, we observed negligible correlations with the thermocline slopes (data not shown), while strong correlations were found for the average surface temperatures and the thermocline slopes, as shown on Figure A3 in Appendix A. The highest value of the correlation coefficient was found for the average temperature of the 14 preceding days (T14) and no evident outliers or bias were detected as shown in Figure 4. Thus, the strong relationship between the stratification at Lake Varese and the T14 surface air average temperature may be used as predictor of lake stratification and algal/cyanobacterial bloom outbreaks, with the advantage of using an air average temperature from reanalysis data, without the need of continuous water temperature profile measurements. The 14 days' time lag between air surface temperature and water stratification may be also very useful as a predictive variable using weather forecasted data.

A further analysis including the new T14 variable (average air surface temperature of the last 14 days) showed the highest correlation with LOG CyanoBV. Thus, keeping T14 and alternating the variables listed in Table 1, initially eight MLR models with two variables were tested (see in Appendix A). Thereafter, since from the lake management standpoint is known that the nutrients (TP and TN) and WT are good predictors of cyanobacteria biomass [22], several of the initial models were dismissed (see Table A2 in Appendix A). For instance, the models with DO and OS were excluded because these variables are more an effect than a triggering parameter of a cyanobacterial outbreak. Similarly, RS was not expected as a relevant variable for cyanobacterial biomass prediction. Thus, the final MLR analysis was performed considering the remaining four variables (TP, AN, CD, and SD) with a dataset slightly increased up to 78 samples (Table A3 in Appendix A). The resulting MLR models are shown in Table 3 in descending order according to their coefficient of determination. Although Model No.1 has the highest R^2, because of the relationship of CD with cyanobacteria, the outbreak may be more difficult to interpret from the biological and physical–chemical point of view. Model No.2 was therefore selected among the two-variable models as the most suitable for LOG CyanoBV prediction in the Lake of Varese among the two-variable models.

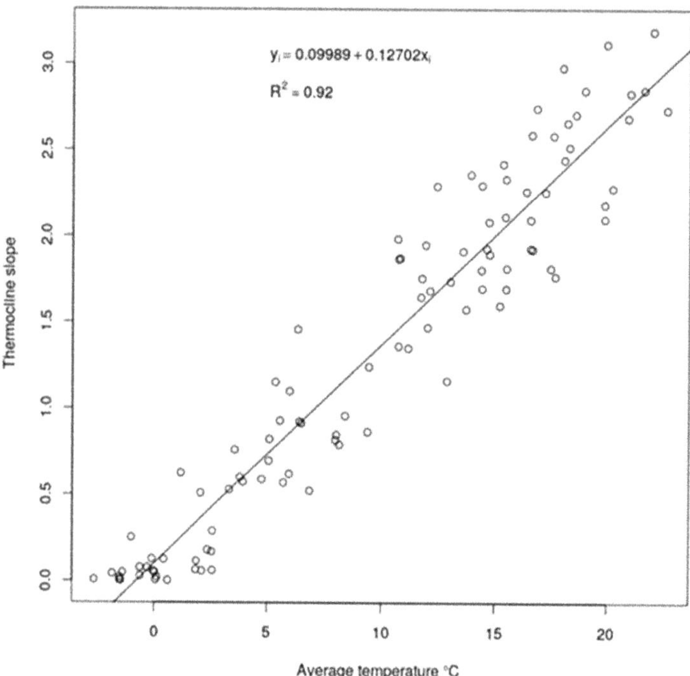

Figure 4. Relationship between the thermocline slopes and the average air temperature of current plus the preceding 14 days. Every dot is from a probe sampling campaign (from 2003 to 2015 included). The linear relationship is strong ($R^2 = 0.92$) and without any relevant bias in the plot, thus supporting the use of the average air temperature of the current and the last 14 days as a proxy of water stratification.

Table 3. Multiple linear regression (MLR) models of cyanobacteria biovolume (LOG CyanoBV) using the 14-days mean air temperature (T14) and one of the physical–chemical variables listed in Table S1 of SI (AN: Ammonium nitrogen; CD: Conductivity; TP: Total phosphorous; SD: Secchi disk depth; and pH). The statistical significance of each coefficient is indicated when below 0.1 (p-values: ***: 0.001; **: 0.01; *: 0.05; °, 0.1). For each model the coefficient of determination (R^2), the adjusted R^2 (R^2adj), and the F statistic are reported. LOG means ten-based logarithm.

No.	Linear Model	R^2	R^2 adj	F
1	LOG CyanoBV = 14.894 * + 0.052 T14 ** − 5.431 CD *	0.31	0.29	16.58
2	LOG CyanoBV = 0.658 + 0.096 T14 *** + 0.527 LOG TP °	0.29	0.27	15.45
3	LOG CyanoBV = 0.244 + 0.065 T14 *** + 0.178 pH	0.27	0.26	13.56
4	LOG CyanoBV = 1.895 *** + 0.068 T14 *** − 0.055 SD	0.27	0.25	14.05
5	LOG CyanoBV = 1.724 *** + 0.077 T14 *** + 0.109 LOG AN	0.27	0.25	13.53

Then, the performance of the selected two-variable models for cyanobacteria biovolume (based on T14 and TP) is detailed as a time series, from 2004 to 2014, in Figure 5 comparing the measured and predicted LOG CyanoBV. As shown in the figure, the model generally follows the temporal variability of cyanobacteria biovolume and is able to forecast the ordinary (seasonal) variability but undervalues or overestimates the extremes (the blooming or very low concentrations). The overall model capability is evaluated by the root mean square error and the obtained RMSE = 0.83 seems to be satisfactory given the model simplicity.

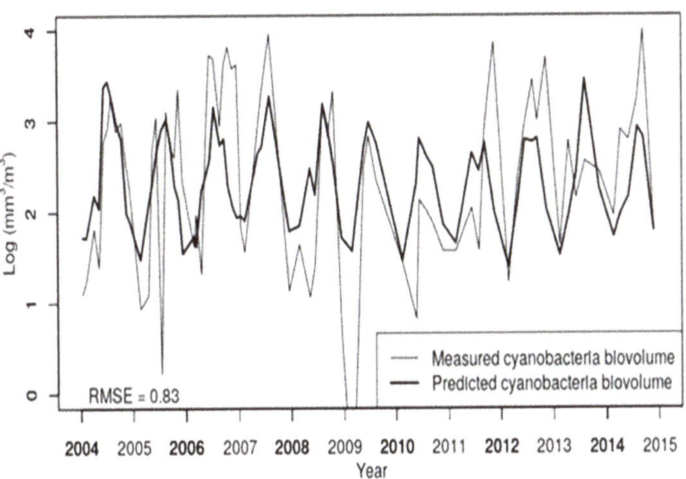

Figure 5. Measured and predicted cyanobacteria biovolumes at Lake Varese for all samples collected during the period 2004–2014. Predicted values are calculated using model No.2 of Table 3. RMSE indicates the root mean square error of predicted vs. measured values.

In order to check whether the explanatory power of the forecasting could be increased, models with three variables were also developed by keeping T14 as the pivotal variable, and combining with all variables from Table 1. The resulting models are shown in Table A4 of Appendix A. Almost no improvement in the explanatory power was detected in comparison to the two-variable models (Table 3), hence the models with three variables were not further considered. In addition, a possible improvement of the selected model (No.2, Table 3) was also tested, by adding meteorological parameters, such as wind speed, total precipitations, and photosynthetically active radiance (PAR). Since no improvement was observed (see explanations in Appendix A) the forecasting model with only two predictors (No.2 in Table 3) was selected as the final choice.

Finally, to verify the model robustness, model No.2 in Table 3 was further tested using an independent dataset (i.e., not used in model development) coming from sampling campaigns conducted by the JRC from the end of 2007 to the end of 2009 [49]. These samples were collected at the deepest locations of the lake and dates with no presence of cyanobacteria were excluded resulting in 67 nonzero records. Except conductivity (CD), that was provided by multiparametric probe, the remaining chemical–physical variables shown in Table 1 were available as surface samples. Concerning CyanoBV, the monitoring records were available at nine different depths. To be consistent with the previous dataset, the total CyanoBV was calculated using measurements in the euphotic zone, between the surface and 2.5 times the Secchi disk depth (Section 2.1.). Time series graphs of the predicted and the measured LOG CyanoBV are shown in Figure 6. In this case the RMSE value is 0.85, similar to the value obtained with the 2004–2014 dataset (RMSE = 0.83). Again, the model tends to overestimate the cyanobacteria biomass, but a similar effect was observed using the original dataset (see Figure 6), so the model performances are comparable using both the original and the validation dataset. This conclusion was confirmed also with a recent monitoring data for cyanobacteria bio-volume accomplished by the JRC in the summer of 2017 (unpublished data) (Figure A4 in Appendix A).

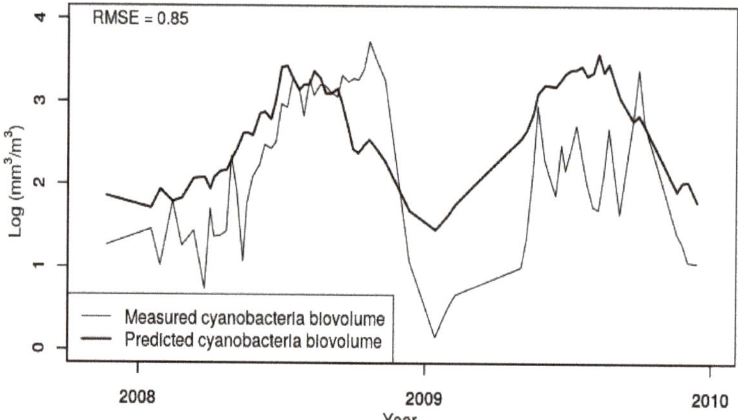

Figure 6. Validation of the model: Measured and predicted cyanobacteria biovolumes at Lake Varese using an independent dataset (not included in the MLR analysis) of water samples collected at Lake Varese during the period 2008–2010 by JRC. RMSE indicates the root mean square error of predicted vs. measured values.

4. Discussion

The aim of this study was to analyze the temporal patterns in cyanobacteria dominance in Lake Varese over a period of ten years (2004–2014) and to investigate the possibility to predict their abundance as a function of a set of meteorological data and water physical–chemical parameters. Overall, the picture of the phytoplankton community showed a change from 2006 with an increase of the cyanobacterial cells that accounted for more than the 50% of the total community, except in 2010. The community shift was also observed by the biovolume distribution, however, not so strongly dominated by the cyanobacteria particularly in 2008 and 2009 where the main contribution was due to the Bacillariophyta and Dynophyta, respectively as confirmed also by the other studies [50]. The different distribution in 2010 could be explained by a pilot study performed in 2009 to reduce the Phosphorus (P) loading by applying a lanthanum-modified bentonite clay to bind the P [32]. The authors showed a sharp reduction (more than 80%) of the P concentrations along the water column during 2009–2010 which may be the consequence of the dropped concentration of the cyanobacteria. However, the trend seems to be reverted with an increase of the cyanobacteria during the 2011–2014 years (Figure 2).

The cyanobacteria composition showed two distinct dominances; Chroococcales/Nostocales were mainly detected in the period from 2004 to 2009, while Oscillatoriales/Synechoccoccales increased after the 2010. In temperate regions, Oscillatoriales and most Chroococcales are associated with the increasing of temperature and water column stability; Nostocales, without the potential to fix nitrogen (Aphanizomenon), are associated with increasing TP concentration, while Synechoccoccales dominance is influenced by lower temperatures and water stability [4]. We could not have a clear picture of the dynamic distribution due to scarce periodicity of the sampling which would lead to a misinterpretation. After 2010, we observed an increase of the precipitation which may influence the shifting, however, Oscillatoriales have been reported to often dominate shallow polymictic eutrophic lakes showing cyclic successions between Microcystis (Chroococcales) and Planktothrix (Oscillatoriales). To get insight to the cyanobacterial biodiversity and spatio-temporal dynamic changes, it would need more frequent samplings such as weekly combined with the microscopy analysis and the metagenomics technique which allows a more deep analysis by sequencing the whole community (microbiome).

In the modeling studies, we did an attempt to develop an ANN using our data, but it resulted in an overfitted model while the selected MLR with just two predictors suggests that a general relationship has been captured as also confirmed by the validation step (see Figure 6), further supporting the robustness of the model as an accepted predictor. In our studies, we did not use the cell density as parameter (although available for the data set as shown in Figure 2) because the cell size can vary considerably within and between species, and toxin concentration relates more closely to the amount of dry matter in a sample than to the number of cells. In addition, cell identification by optical microscopy may underestimate the real value of cyanobacteria cells, indeed, smaller species may be (classified) reported as ultraplankton [51]. Therefore, the biovolume was selected as a more appropriate parameter. Furthermore, the biovolume measurements require less time for the microscopic analysis than cell identification and enumeration providing data with lower uncertainties. The parameters selected in this work such as the water temperature, DO, nutrient availability, and the water transparency measured as SD, play a key role in the occurrence of cyanobacterial blooms [19,20,22,23,26,27,52–54]. In addition to these factors known to be good predictors for cyanobacteria abundance, we tried to evaluate whether the air temperature could be considered as parameter. Indeed, air temperature is one of the main factors driving the evolution of water temperature in a lake [55] and influencing as well the vertical temperature profile and lake stratification [56], which are both important parameters for the occurrence of algal bloom outbreaks. In Lake Varese and generally in monomictic lakes, during spring, when air temperature rises, water stratification normally starts to build-up, stabilizes during summer until fall, when decreasing air temperatures break the stratification. Another factor which may lead to changes in water turbulence is the wind stress.

Taranu et al. has shown conflicting results on the response of cyanobacteria to climate and nutrients for dimictic (lakes that mix from top to bottom in two water mixing periods within a year) and polymictic lakes (mix from top to bottom for more than two water mixing periods) [23]. In dimictic lakes the stronger predictor was water-column stability, while for polymictic lakes, was nutrients loading. Journey et al., focusing on two monomictic lakes (mix from top to bottom during one mixing period per year), found strong correlations between cyanobacterial biovolumes and water stratification, while an opposite relationship was found with the nutrient levels [29]. In addition, it has been shown that the meteorological variables as air temperature, wind speed, and relative humidity, could be drivers of hypolimnetic anoxia, which is an indirect consequence of thermal stratification [28]. Indeed, temperature and stratification are cross-linked factors, with stratification forming and strengthening at higher air and water temperatures. In this paper, based on this correlation, we have demonstrated that 14-days average air temperature can be used as a proxy of the stratification strength for Lake Varese (Figure A3 in Appendix A and Figure 4). Indeed, the strongest correlation was found at 14 days (T14) preceding the current water temperature and the parameter T14. Despite wind speed being generally a physical variable influencing lake stratification [57], the lack of relationship with the thermocline slopes can be explained because the Lake Varese area usually does not experience strong wind speed. Based on that and considering that phosphorous is one of the most important factors for lake management [17], the model using T14 and total phosphorous (TP, model No.2 in Table 3) was chosen in this work as the best candidate to predict the Cyanobacteria biovolume at Lake Varese. In our analysis, the model using ammonium nitrogen (AN) as predictor (model No.5 in Table 3) was not considered further because AN is only one form of available nitrogen in water while it is known that all bioavailable forms of nitrogen (ammonium nitrogen, nitrate nitrogen, urea, and alanine nitrogen) influence the cyanobacteria abundance [58].

Nevertheless, the predictive power of this model is rather low, representing only about 30% of the total variability. The difficulty of predicting cyanobacteria blooms using physico–chemical environmental variables is a common problem highlighted also by previous studies [17,23,59]. On the other hand, as reported by Janssen et al., it is urgent and challenging to provide an algal bloom prediction at global level for the lakes [21]. The 14-days average air temperature together with total phosphorus can be, therefore, used to predict an algal outbreak two weeks in advance and, eventually, to adopt management actions to reduce their occurrence in monomictic and eutrophic shallow lakes.

The threshold for health alert in recreational waters is normally defined for cyanobacterial density in many countries. A density of 100,000 cyanobacterial cells per ml (which is equivalent to approximately 50 µg/L of chlorophyll-a if cyanobacteria dominate) is a guideline for a moderate health alert in recreational waters [60], although the regulated threshold varies at national level. The conversion of cyanobacteria cell density into biovolume is not simple, as measurements of the cyanobacteria genera/species are needed [41]. Each country is free to define the alert values for the presence of cyanobacteria either in cell abundance or in biovolume [61], only few countries defined thresholds in terms of cyanobacterial biovolume (Figure 7). The Netherlands and New Zealand defined a surveillance level for cyanobacetria biovolume below 2.5 mm^3/L (LOG CyanoBV = 3.34 mm^3/m^3) and 0.5 mm^3/L (LOG CyanoBV = 2.7 mm^3/m^3), respectively. An alert is set above these thresholds requiring weekly monitoring and issue warning to the public. Above 15 mm^3/L (LOG CyanoBV = 4.18 mm^3/m^3) and 10 mm^3/L (LOG CyanoBV = 4 mm^3/m^3) The Netherlands and New Zealand authorities set an action level, continue the monitoring, notify the public of a potential risk to health, and if potentially toxic taxa are present, consider testing samples for cyanotoxins. Germany defined a single threshold for surveillance and alert level at 1 mm^3/L (LOG CyanoBV = 3 mm^3/m^3), above which local authorities must publish warnings, discourage bathing, and consider temporary closure.

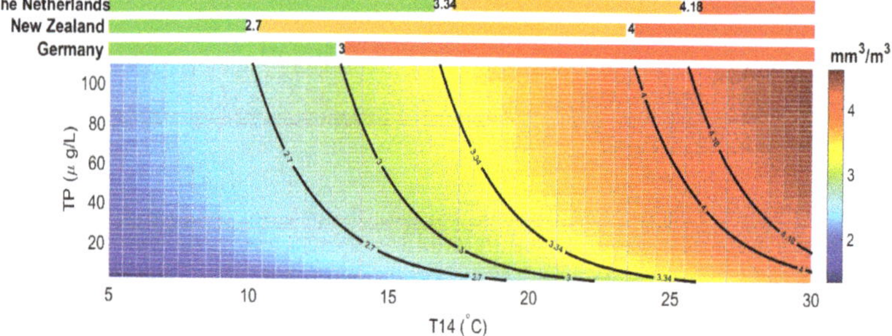

Figure 7. Estimated cyanobacteria biovolume calculated for Lake Varese using model No.2 of Table 3. The colored field represents LOG CyanoBV at different levels of the two predictors, T14 ranging between 5 and 30 °C, and TP ranging between 2.5 and 110 µg/L. The colored bars above the plot and the black lines refers to examples of threshold levels for cyanobacteria biovolume (surveillance in green, alert in orange, and action in red) of defined by legislation in The Netherlands, New Zealand, and Germany.

5. Conclusions

The main findings of this study can be summarized in the following points:

- In Lake Varese, a shift in the phytoplankton community distribution towards cyanobacteria dominance onwards 2010 was observed;
- This change may be related to changes in the nutrients, as well as precipitation patterns, as suggested by other studies, but more frequent samplings combined with the microscopy analysis and the metagenomics technique (microbiome) would allow a more conclusive analysis;
- Air temperature can be used as a good proxy of the lake surface water temperature and of the lake stratification;
- The 14 days mean air temperature showed the highest correlation with lake stratification strength derived by vertical water temperature profiles. At surface, this parameter is easily computable from weather forecast data, and together with total phosphorus continuous measured in situ,

can be used as an early warning tool to anticipate by two weeks the beginning of cyanobacteria blooms in Lake Varese.

This model can help predict and mitigate the impact of climate change on water and ecosystem resource management.

Author Contributions: Conceptualization, T.L.; data curation, A.B. and P.G.; investigation, N.C. and L.P.; methodology, N.C. and D.C.A.; project administration, T.L.; validation, L.P. and D.M.; writing—Original draft, N.C., D.C.A., L.P., D.M., and S.D.; writing—Review and editing, L.P., A.M., I.S., and T.L. All authors have read and agreed to the published version of the manuscript.

Funding: This research received no external funding.

Acknowledgments: We thank Dorota Napierska and Giovanni Strona for their critical review to the manuscript. We thank Chiara Facca for constructive discussion. We gratefully acknowledge the three anonymous reviewers for their helpful comments and suggestions on the manuscript. These studies have been performed within the Project Biocli4crisma funded by the European Commission Joint Research Centre (JRC) as an exploratory project.

Conflicts of Interest: The authors declare no conflict of interest.

Appendix A

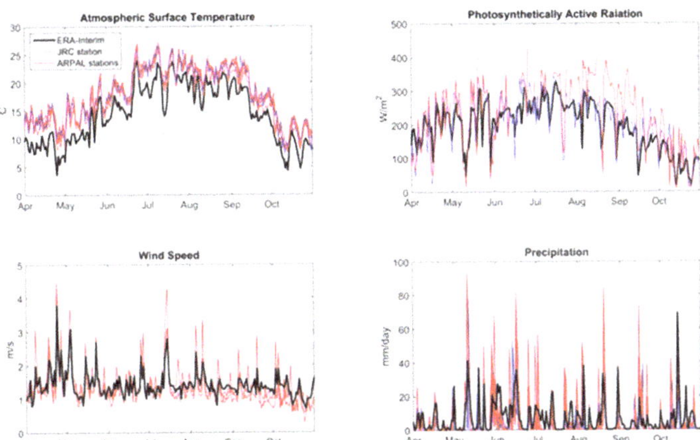

Figure A1. ERA-Interim reanalysis of daily mean data at Lake Varese (black) compared to ground observations in the area of Lake Varese for the period 2004–2014: Surface temperature; photosyntetically active radiation (PAR); wind speed; total precipitation.

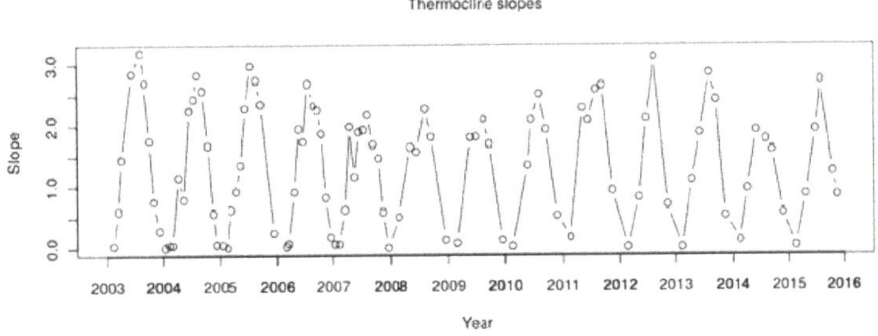

Figure A2. Thermocline maximum slopes (°C/m) calculated for Lake Varese, from 2003 to 2015.

Table A1. Total phytoplankton community expressed as cell density (cells/L) and biovolume (mm^3/m^3) for the period 2004–2014.

Year	Cell Density (Cells/L)	Biovolume (mm^3/m^3)
2004	2.11×10^7	1.43×10^4
2005	2.54×10^7	1.22×10^5
2006	2.65×10^7	6.01×10^4
2007	1.32×10^8	2.84×10^4
2008	1.60×10^8	1.80×10^4
2009	4.54×10^7	1.27×10^4
2010	7.85×10^7	2.73×10^4
2011	2.17×10^8	3.12×10^4
2012	2.04×10^8	2.16×10^4
2013	1.31×10^8	2.31×10^4
2014	8.48×10^7	2.27×10^4

Appendix B. Validation of T14

To verify whether the 14 days' time could be used from the weather forecast we compared the T14 values from the ERA-Interim reanalysis with T14 calculated using forecasted temperatures from the ECMWF Integrate Forecast System. Considering all sampling days of the LOG CyanoBV model dataset, we found a regression line with slope of 1.00 and a coefficient of determination (R^2) of 0.99, the T14 value with forecasted temperature is on average 0.7 °C larger than the T14 calculated using the reanalysis. Thus, we can assume that the 14 days average of forecasted atmospheric temperatures could be reliable enough for an early warning of cyanobacteria algal bloom outbreak, despite the forecast uncertainty increases with time.

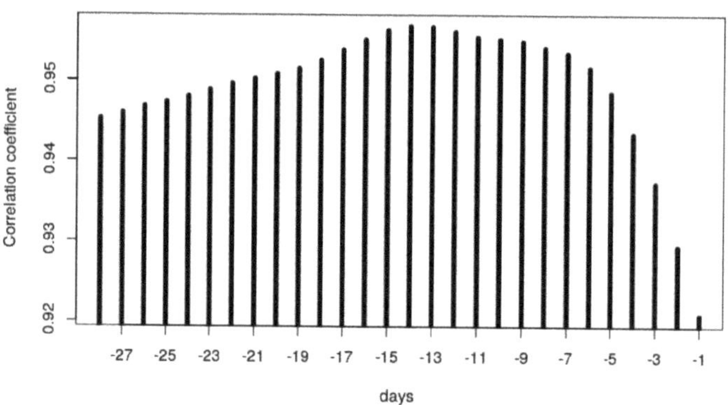

Figure A3. Relationship between the thermocline slope and the average atmospheric surface temperature of the current and the preceding 1–28 days. The maximum correlation is found for the preceding 14 days.

Appendix C. Two-Variable MLR Models

Since T14 and WT are strongly correlated (Pearson correlation coefficient is 0.98), WT was excluded from the performed trials. The ten-based logarithm of the CD, TP, DO, RS, and AN, was used in MLR because Log10 transformation best approximates the normal distributions for these variables. Table 1 shows the list of models for cyanobacteria biovolume (LOG CyanoBV) computed through the MLR method. The fraction of the total variation of measured LOG CyanoBV that can be explained by the regression equation (coefficient of determination, R^2) is ranged from 0.28 to 0.33, and a bit lower when taking into account the sample size (73) and the number of predictors (2), R^2adj ranged from 0.26 to 0.31.

The coefficients estimated for the atmospheric temperature (T14) are always statistically significant at the 1% or 0.1% level, while only the coefficient of CD in model No.1 was statistically significant at the 5% level. However, all the models listed in Table 1, are statistically significant according to the analysis of variance with F value higher than 3.12, which is the critical value for a *p*-value of 0.05, two predictor variables, and 73 samples.

Table A2. Two-variable multiple linear regression (MLR) models of cyanobacteria biovolume (LOG CyanoBV) using the 14-days mean atmospheric temperature (T14) and one of the physical–chemical variables (AN: Ammonium nitrogen; CD: Conductivity; TP: Total phosphorous; SD: Secchi disk depth; RS: Reactive silicates; DO: Dissolved oxygen; OS: Saturation oxygen percentage; and pH). The statistical significance of each coefficient is indicated when below 0.05 (*p*-values: ***: 0.001; **: 0.01; *: 0.05). For each model the coefficient of determination (R^2), the adjusted R^2 (R^2adj), and the F statistic are reported. LOG means 10-based logarithm.

No.	Linear Model	R^2	R^2 adj	F
1	LOG CyanoBV = 15.353 * + 0.055 T14 ** − 5.645 LOG CD *	0.33	0.31	17.16
2	LOG CyanoBV = 0.718 + 0.095 T14 *** + 0.463 LOG TP	0.31	0.29	15.41
3	LOG CyanoBV = 2.886 ** + 0.074 T14 *** − 1.324 LOG DO	0.3	0.28	14.95
4	LOG CyanoBV = 1.869 *** + 0.070 T14 *** − 0.065 SD	0.3	0.28	14.79
5	LOG CyanoBV = 1.561 *** + 0.079 T14 *** + 0.230 LOG RS	0.3	0.28	14.68
6	LOG CyanoBV = 1.759 *** + 0.079 T14 *** − 0.002 OS	0.29	0.27	14.01
7	LOG CyanoBV = −1.015 + 0.063 T14 ** + 0.329 pH	0.29	0.27	14.49
8	LOG CyanoBV = 1.616 *** + 0.077 T14 *** + 0.055 LOG AN	0.28	0.26	13.92

Table A3. List of water surface physical–chemical variables selected for the final MLR analysis. For each physical–chemical variable, the mean, median, range, and number of measurements above/below the level of quantification (LOQ) are reported.

		Mean	Median	Range	Above LOQ	Below LOQ
Total phosphorous (µg/L)	TP	40.4	25.5	2.5–110	69	9
Ammonium nitrogen (mg/L)	AN	0.11	0.048	0.0075–0.6	64	14
pH	pH	8.2	8.2	7.5–9.6	78	0
Conductivity (µS/cm 20 °C)	CD	256	257	175–310	78	0
Secchi disk depth (m)	SD	4	3.5	1.1–9.6	78	0

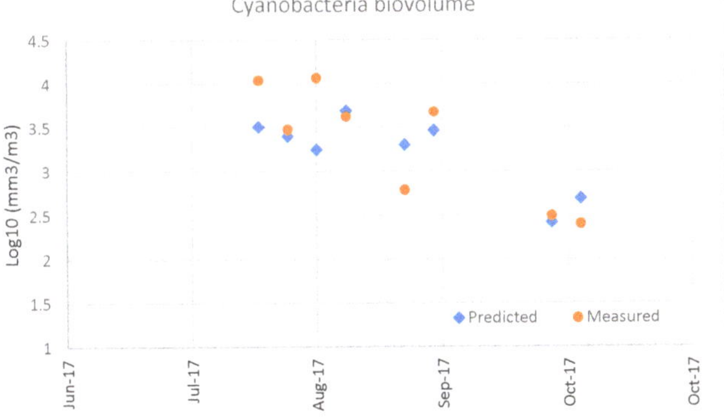

Figure A4. Validation of model No.2 in Table 1: Measured and predicted cyanobacteria biovolumes at Lake Varese using an independent dataset (not included in the MLR analysis) of water samples collected at Lake Varese during summer 2017. The T14 was calculated using forecasted temperatures at Lake Varese by the Global Forecast System (GFS) of the National Centers for Environmental Prediction (NCEP, http://www.emc.ncep.noaa.gov/).

Appendix D. Three-Variable MLR Models

Table A4. Three-variable multiple linear regression (MLR) models of cyanobacteria biovolume (LOG CyanoBV) using the 14-days mean atmospheric temperature (T14) and two of the physical–chemical variables (AN: Ammonium nitrogen; CD: Conductivity; TP: Total phosphorous; SD: Secchi disk depth; and pH). The statistical significance of each coefficient is indicated when below 0.1 (*p*-values: ***: 0.001; **: 0.01; *: 0.05; °, 0.1). For each model the coefficient of determination (R^2), the adjusted R^2 (R^2adj), and the F statistic are reported. LOG means 10-based logarithm.

No.	Linear Model	R^2	R^2 adj	F
1	LOG CyanoBV = 3.391 * + 0.074 T14 *** − 0.0097 CD * + 0.504 LOG TP °	0.33	0.31	12.41
2	LOG CyanoBV = 4.859 *** + 0.046 T14 ** − 0.0105 CD * − 0.067 SD	0.32	0.3	11.75
3	LOG CyanoBV = 4.401 *** + 0.052 T14 * + 0.0218 LOG AN − 0.010 CD *	0.31	0.28	74
4	LOG CyanoBV = 15.201 * + 0.053 T14 ** - 0.0236 pH − 5.481 LOG CD *	0.31	0.28	10.91
5	LOG CyanoBV = 0.918 + 0.091 T14 *** + 0.5077 LOG TP ° − 0.048 SD	0.3	0.27	10.53
6	LOG CyanoBV = −1.369 + 0.088 T14 *** + 0.2519 pH + 0.554 LOG TP °	0.3	0.27	10.42
7	LOG CyanoBV = 0.742 + 0.102 T14 *** + 0.1191 LOG AN + 0.532 LOG TP °	0.29	0.27	10.29
8	LOG CyanoBV = 2.070 *** + 0.076 T14 *** + 0.1702 LOG AN − 0.065 SD	0.28	0.25	9.49
9	LOG CyanoBV = 0.931 + 0.064 T14 *** + 0.1207 pH − 0.051 SD	0.27	0.24	9.3
10	LOG CyanoBV = 0.415 + 0.071 T14 ** + 0.1659 pH + 0.099 LOG AN	0.27	0.24	9

Appendix E. Testing an Improvement with Additional Meteorological Parameters

For a range of days from 1 to 28, the maximum squared wind speed (m^2/s^2), the average total precipitation (mm/day), and the PAR (W/m^2) were calculated (as for the average surface temperature). Adding wind speed (iteratively from 1 to 28 days) to the predictors of model No.2 (Table 3), the corresponding R^2 ranged from 0.29 to 0.31, thus no relevant improvement was detected. This is probably due to the generally weak winds at Lake Varese, not strong enough to break the lake stratification. Indeed, ERA-Interim data showed that annual mean wind speed is ranged between 1.82 and 2.02 m/s, i.e., light breeze for the period 2003–2015. Stronger winds, such as gentle breeze, between 3.3 and 5.2 m/s, were observed for 17–37 days/year, and only up to three days/year of moderate breeze, between 5.2 and 7.4 m/s. Similarly, for the average total precipitation, R^2 ranged from 0.29 to 0.30. On the opposite, an improvement was detected for PAR, where R^2 ranged from 0.37 to 0.40. Since PAR, calculated for 14 days, was correlated with T14 to a high degree (Pearson correlation coefficient is 0.82), including it in the model would be deemed as an overfitting.

References

1. Whitton, B.A.; Potts, M.E. *The Ecology of Cyanobacteria*; Springer: Dordrecht, The Netherlands, 2000.
2. Ho, J.C.; Michalak, A.M.; Pahlevan, N. Widespread global increase in intense lake phytoplankton blooms since the 1980s. *Nature* **2019**, *574*, 667–670. [CrossRef]
3. Sanseverino, I.; Antonio, D.C.; Pozzoli, L.; Dobricic, S.; Lettieri, T. Algal bloom and its economic impact. In *JRC Technical Report*; JRC: Ispra, Italy, 2016.
4. O'Neil, J.M.; Davis, T.W.; Burford, M.A.; Gobler, C.J. The rise of harmful cyanobacteria blooms: The potential roles of eutrophication and climate change. *Harmful Algae* **2012**, *14*, 313–334. [CrossRef]
5. Hudnell, H.K.; Dortch, Q. A Synopsis of Research Needs Identified at the Interagency, International Symposium on Cyanobacterial Harmful Algal Blooms (ISOC-HAB). In *Cyanobacterial Harmful Algal Blooms: State of the Science and Research Needs*; Hudnell, H.K., Ed.; Springer: New York, NY, USA, 2008; pp. 17–43.
6. Sivonen, K. Cyanobacterial toxins and toxin production. *Phycologia* **1996**, *35*, 12–24. [CrossRef]
7. Rastogi, R.P.; Madamwar, D.; Incharoensakdi, A. Bloom Dynamics of Cyanobacteria and Their Toxins: Environmental Health Impacts and Mitigation Strategies. *Front. Microbiol.* **2015**, *6*, 1254. [CrossRef] [PubMed]

8. Stauffer, B.A.; Bowers, H.A.; Buckley, E.; Davis, T.W.; Johengen, T.H.; Kudela, R.; McManus, M.A.; Purcell, H.; Smith, G.J.; Vander Woude, A.; et al. Considerations in Harmful Algal Bloom Research and Monitoring: Perspectives From a Consensus-Building Workshop and Technology Testing. *Front. Mar. Sci.* **2019**, *6*, 1–18. [CrossRef]

9. Beaulieu, M.; Pick, F.; Palmer, M.; Watson, S.; Winter, J.; Zurawell, R.; Gregory-Eaves, I. Comparing predictive cyanobacterial models from temperate regions. *Can. J. Fish. Aquat. Sci.* **2014**, *71*, 1830–1839. [CrossRef]

10. Ondenka, M. Correlations between several environmental factors affecting the bloom events of cyanobacteria in Liptovska Mara reservoir (Slovakia)—A simple regression model. *Ecol. Model. Elsevier* **2007**, *209*, 412–416. [CrossRef]

11. French, T.D.; Petticrew, E.L. Chlorophyll a seasonality in four shallow lakes (northern British Columbia, Canada) and the critical roles of internal phosphorous loading and temperature. *Hydrobiologia* **2007**, *575*, 285–299. [CrossRef]

12. Malek, S.; Syed Ahmad, S.M.; Singh, S.K.K.; Milow, P.; Salleh, A. Assessment of predictive models for chlorophyll-a concentration of a tropical lake. *BMC Bioinform.* **2011**, *12*, S12. [CrossRef]

13. Dimberg, P.H.; Olofsson, C.J. A comparison between regression models and genetic programming for predictions of chlorophyll-a concentrations in Northern Lakes. *Environ. Model Assess.* **2016**, *21*, 221–232. [CrossRef]

14. Malagó, A.; Bouraoui, F.; Grizzetti, B.; De Roo, A. Modelling nutrient fluxes into the Mediterranean Sea. *J. Hydrol. Reg. Stud.* **2019**, *22*, 100592.

15. Srivastava, A.; Ahn, C.Y.; Asthana, R.K.; Lee, H.G.; Oh, H.M. Status, Alert System, and Prediction of Cyanobacterial Bloom in South Korea. *Biomed Res. Int.* **2015**, *2015*, 8. [CrossRef] [PubMed]

16. Ghaffar, S.; Stevenson, R.J.; Khan, Z. Cyanobacteria Dominance in Lakes and Evaluation of Its Predictors: A Study of Southern Appalachians Ecoregion, USA. *Matec Web Conf.* **2016**, *60*, 02001. [CrossRef]

17. Downing, J.A.; Watson, S.B.; McCauley, E. Predicting Cyanobacteria dominance in lakes. *Can. J. Fish. Aquat. Sci.* **2001**, *58*, 1905–1908. [CrossRef]

18. Håkanson, L.; Bryhn, A.C.; Hytteborn, J.K. On the issue of limiting nutrient and predictions of cyanobacteria in aquatic systems. *Sci. Total Environ.* **2007**, *379*, 89–108. [CrossRef]

19. Beaulieu, M.; Pick, F.; Gregory-Eaves, I. Nutrients and water temperature are significant predictors of cyanobacterial biomass in a 1147 lakes data set. *Limnol. Oceanogr.* **2013**, *58*, 1736–1746. [CrossRef]

20. Huber, V.; Wagner, C.; Gerten, D.; Adrian, R. To bloom or not to bloom: Contrasting responses of cyanobacteria to recent heat waves explained by critical thresholds of abiotic drivers. *Oecologia* **2012**, *169*, 245–256. [CrossRef]

21. Janssen, A.B.G.; Janse, J.H.; Beusen, A.H.W.; Chang, M.; Harrison, J.A.; Huttunen, I.; Kong, X.; Rost, J.; Teurlincx, S.; Troost, T.A.; et al. How to model algal blooms in any lake on earth. *Curr. Opin. Environ. Sustain.* **2019**, *36*, 1–10. [CrossRef]

22. Kosten, S.; Huszar, V.L.M.; Bécares, E.; Costa, L.S.; Van Donk, E.; Hansson, L.A.; Jeppesen, E.; Kruk, C.; Lacerot, G.; Mazzeo, N.; et al. Warmer climates boost cyanobacterial dominance in shallow lakes. *Glob. Chang. Biol.* **2012**, *18*, 118–126. [CrossRef]

23. Taranu, Z.E.; Zurawell, R.W.; Pick, F.; Gregory-Eaves, I. Predicting cyanobacterial dynamics in the face of global change: The importance of scale and environmental context. *Glob. Chang. Biol.* **2012**, *18*, 3477–3490. [CrossRef]

24. Wagner, C.; Adrian, R. Cyanobacteria dominance: Quantifying the effects of climate change. *Limnol. Oceanogr.* **2009**, *54*, 2460–2468. [CrossRef]

25. Wells, M.L.; Trainer, V.L.; Smayda, T.J.; Karlson, B.S.O.; Trick, C.G.; Kudela, R.M.; Ishikawa, A.; Bernard, S.; Wulff, A.; Anderson, D.M.; et al. Harmful algal blooms and climate change: Learning from the past and present to forecast the future. *Harmful Algae* **2015**, *49*, 68–93. [CrossRef] [PubMed]

26. JÖHnk, K.D.; Huisman, J.E.F.; Sharples, J.; Sommeijer, B.E.N.; Visser, P.M.; Stroom, J.M. Summer heatwaves promote blooms of harmful cyanobacteria. *Glob. Chang. Biol.* **2008**, *14*, 495–512. [CrossRef]

27. Walsby, A.E.; Hayes, P.K.; Boje, R.; Stal, L.J. The selective advantage of buoyancy provided by gas vesicles for planktonic cyanobacteria in the Baltic Sea. *New Phytol.* **1997**, *136*, 407–417. [CrossRef]

28. Snortheim, C.A.; Hanson, P.C.; McMahon, K.D.; Read, J.S.; Carey, C.C.; Dugan, H.A. Meteorological drivers of hypolimnetic anoxia in a eutrophic, north temperate lake. *Ecol. Model.* **2017**, *343*, 39–53. [CrossRef]

29. Journey, C.A.; Beaulieu, K.M.; Bradley, P.M. Environmental Factors that Influence Cyanobacteria and Geosmin Occurrence in Reservoirs. In *Current Perspectives in Contaminant Hydrology and Water Resources Sustainability*; Bradley, P.M., Ed.; InTech: Rijeka, Croatia, 2013; pp. 27–55.

30. Zaccara, S.; Canziani, A.; Roella, V.; Crosa, G. A northern Italian shallow lake as a case study for eutrophication control. *Limnology* **2007**, *8*, 155–160. [CrossRef]

31. Guilizzoni, P.; Lami, A.; Ruggiu, D.; Bonomi, G. Stratigraphy of specific algal and bacterial carotenoids in the sediments of Lake Varese (N. Italy). *Hydrobiologia* **1986**, *143*, 321–325. [CrossRef]

32. Crosa, G.; Yasseri, S.; Nowak, K.-E.; Canziani, A.; Roella, V.; Zaccara, S. Recovery of Lake Varese: Reducing trophic status through internal P load capping. *Fundam. Appl. Limnol. Arch. Hydrobiol.* **2013**, *183*, 49–61. [CrossRef]

33. Morabito, G.; Hamza, W.; Ruggiu, D. Carbon assimilation and phytoplankton growth rates across the trophic spectrum: An application of the chlorophyll labelling technique. *J. Limnol.* **2004**, *63*, 33–43. [CrossRef]

34. Ruggiu, D. Osservazioni conclusive. Considerazioni generali sull'evoluzione a lungo termine dei popolamenti planctonici. Fitoplancton. In *C.N.R. Istituto Italiano di Idrobiologia, Ricerche sull'evoluzione del Lago Maggiore*; Aspetti limnologici. Campagna 1992 e Rapporto quinquennale 1988–1992; Commissione Internazionale per la protezione delle acque italo-svizzere: Torino, Italy, 1993; pp. 115–116.

35. Mosello, R.; Panzani, P.; Pugnetti, A.; Ruggiu, D. An assessment of the hydrochemistry of the eutrophic Lake Varese (N. Italy), coincident with the implementation of the first restoration measures. *Mem. Dell'istituto Ital. Idrobiol.* **1991**, *49*, 99–116.

36. Premazzi, G.; Cardoso, A.C.; Rodari, E.; Austoni, M.; Chiaudani, G. Hypolimnetic withdrawal coupled with oxygenation as lake restoration measures: The successful case of Lake Varese (Italy). *Limnetica* **2005**, *24*, 123–132.

37. Ambrosetti, W.; Barbanti, L.; Sala, N. Residence time and physical processes in lakes. *J. Limnol.* **2003**, *62*, 1–15. [CrossRef]

38. Premazzi, G.; Dalmiglio, A.; Cardoso, A.C.; Chiaudani, G. Lake management in Italy: The implications of the Water Framework Directive. *Lakes Reserv.* **2003**, *8*, 41–59. [CrossRef]

39. Balzamo, S.; Martone, C. *Metodi Biologici per le Acque Superficiali Interne*; Manuali e Linee Guida 111/2014; ISPRA: Rome, Italy, 2014.

40. Utermöhl, H. Zur Vervollkommnung der quantitativen Phytoplankton-Methodik. *Mitt. Int. Ver. Angew. Limnol.* **1958**, *9*, 1–38. [CrossRef]

41. Sun, J.; Liu, D. Geometric models for calculating cell biovolume and surface area for phytoplankton. *J. Plankton Res.* **2003**, *25*, 1331–1346. [CrossRef]

42. Dee, D.P.; Uppala, S.M.; Simmons, A.J.; Berrisford, P.; Poli, P.; Kobayashi, S.; Andrae, U.; Balmaseda, M.A.; Balsamo, G.; Bauer, P.; et al. The ERA-Interim reanalysis: Configuration and performance of the data assimilation system. *Q. J. R. Meteorol. Soc.* **2011**, *137*, 553–597. [CrossRef]

43. Bresnan, E.; Davidson, K.; Edwards, M.; Fernand, L.; Gowen, R.; Hall, A.; Kennington, K.; McKinney, A.; Milligan, S.; Raine, R.; et al. Impacts of climate change on harmful algal blooms. *MCCIP Science Review.* **2013**, 236–243. [CrossRef]

44. Paerl, H.W.; Paul, V.J. Climate change: Links to global expansion of harmful cyanobacteria. *Water Res.* **2012**, *46*, 1349–1363. [CrossRef]

45. Paerl, H.W.; Hall, N.S.; Calandrino, E.S. Controlling harmful cyanobacterial blooms in a world experiencing anthropogenic and climatic-induced change. *Sci. Total Environ.* **2011**, *409*, 1739–1745. [CrossRef]

46. Fiedler, P.C. Comparison of objective descriptions of the thermocline. *Limnol. Oceanogr.* **2010**, *8*, 313–325. [CrossRef]

47. Meriluoto, J.; Spoof, L.; Codd, G.A. *Handbook of Cyanobacterial Monitoring and Cyanotoxin Analysis*; Wiley: Chichester, UK, 2017.

48. Huisman, J.; Matthijs, H.; Visser, P. *Harmful Cyanobacteria*; Springer: Dordrecht, The Netherlands, 2005.

49. Nõges, P.; Nõges, T.; Ghiani, M.; Sena, F.; Fresner, R.; Friedl, M.; Mildner, J. Increased nutrient loading and rapid changes in phytoplankton expected with climate change in stratified South European lakes: Sensitivity of lakes with different trophic state and catchment properties. *Hydrobiologia* **2011**, *667*, 255–270. [CrossRef]

50. Nõges, P.; Nõges, T.; Ghiani, M.; Paracchini, B.; Grande, J.P.; Sena, F. Morphometry and trophic state modify the thermal response of lakes to meteorological forcing. *Hydrobiologia* **2011**, *667*, 241–254. [CrossRef]

51. Caron, D.A. Technique for enumeration of heterotrophic and phototrophic nanoplankton, using epifluorescence microscopy, and comparison with other procedures. *Appl. Environ. Microbiol.* **1983**, *46*, 491–498. [CrossRef] [PubMed]

52. Kimambo, O.N.; Gumbo, J.R.; Chikoore, H. The occurrence of cyanobacteria blooms in freshwater ecosystems and their link with hydro-meteorological and environmental variations in Tanzania. *Heliyon* **2019**, *5*, e01312. [CrossRef] [PubMed]

53. Takarina, N.D.; Wardhana, W. Relationship between cyanobacteria community and water quality parameters on intertidal zone of fish ponds, Blanakan, West Java. *Aip Conf. Proc.* **2017**, *1862*, 030114. [CrossRef]

54. Megard, R.O.; Settles, J.C.; Boyer, H.A.; Combs, W.S., Jr. Light, Secchi disks, and trophic states1. *Limnol. Oceanogr.* **1980**, *25*, 373–377. [CrossRef]

55. Piccolroaz, S.; Toffolon, M.; Majone, B. A simple lumped model to convert air temperature into surface water temperature in lakes. *Hydrol. Earth Syst. Sci.* **2013**, *17*, 3323–3338. [CrossRef]

56. Hadley, K.R.; Paterson, A.M.; Stainsby, E.A.; Michelutti, N.; Yao, H.; Rusak, J.A.; Ingram, R.; McConnell, C.; Smol, J.P. Climate warming alters thermal stability but not stratification phenology in a small north-temperate lake. *Hydrol. Process.* **2014**, *28*, 6309–6319. [CrossRef]

57. Kerimoglu, O.; Rinke, K. Stratification dynamics in a shallow reservoir under different hydro-meteorological scenarios and operational strategies. *Water Resour. Res.* **2013**, *49*, 7518–7527. [CrossRef]

58. Chaffin, J.D.; Bridgeman, T.B. Organic and inorganic nitrogen utilization by nitrogen-stressed cyanobacteria during bloom conditions. *J. Appl. Phycol.* **2013**, *26*, 299–309. [CrossRef]

59. Oh, H.M.; Ahn, C.Y.; Lee, J.W.; Chon, T.S.; Choi, K.H.; Park, Y.S. Community patterning and identification of predominant factors in algal bloom in Daechung Reservoir (Korea) using artificial neural networks. *Ecol. Model.* **2007**, *203*, 109–118. [CrossRef]

60. Chorus, I.; Bartram, J. *Toxic Cyanobacteria in Water: A Guide to Their Public Health Consequences, Monitoring and Management/edited by Ingrid Chorus and Jamie Bertram*; World Health Organization: Geneva, Switzerland, 1999.

61. Chorus, I. *Current Approaches to Cyanotoxin Risk Assessment, Risk Management and Regulations in Different Countries*; Umweltbundesamt: Dessau-Roßlau, Germany, 2012.

Article

Impact of Nutrient and Stoichiometry Gradients on Microbial Assemblages in Erhai Lake and Its Input Streams

Yang Liu [1,2,3], Xiaodong Qu [1,3], James J. Elser [4], Wenqi Peng [1,3], Min Zhang [1,3,*], Ze Ren [4,*], Haiping Zhang [1,3], Yuhang Zhang [1] and Hua Yang [5]

[1] State Key Laboratory of Simulation and Regulation of Water Cycle in River Basin, China Institute of Water Resources and Hydropower Research, Beijing 100038, China
[2] College of Hydrology and Water Resources, Hohai University, Nanjing 210098, China
[3] Department of Water Environment, China Institute of Water Resources and Hydropower Research, Beijing 100038, China
[4] Flathead Lake Biological Station, University of Montana, Polson, MT 59860, USA
[5] Yunnan Hydrology and Water Resources Bureau Dali Branch, Dali 671000, China
*   Correspondence: zhangmin@iwhr.com (M.Z.); renzedyk@gmail.com (Z.R.); Tel.: +86-010-6878-1946 (M.Z.)

Received: 23 July 2019; Accepted: 13 August 2019; Published: 17 August 2019

Abstract: Networks of lakes and streams are linked by downslope flows of material and energy within catchments. Understanding how bacterial assemblages are associated with nutrients and stoichiometric gradients in lakes and streams is essential for understanding biogeochemical cycling in freshwater ecosystems. In this study, we conducted field sampling of bacterial communities from lake water and stream biofilms in Erhai Lake watershed. We determined bacterial communities using high-throughput 16S rRNA gene sequencing and explored the relationship between bacterial composition and environmental factors using networking analysis, canonical correspondence analysis (CCA), and variation partitioning analysis (VPA). Physicochemical parameters, nutrients, and nutrient ratios gradients between the lake and the streams were strongly associated with the differences in community composition and the dominant taxa. Cyanobacteria dominated in Erhai Lake, while Proteobacteria dominated in streams. The stream bacterial network was more stable with multiple stressors, including physicochemical-factors and nutrient-factors, while the lake bacterial network was more fragile and susceptible to human activities with dominant nutrients (phosphorus). Negative correlations between bacterial communities and soluble reactive phosphorus (SRP) as well as positive correlations between bacterial communities and dissolved organic carbon (DOC) in the network indicated these factors had strong effect on bacterial succession. Erhai Lake is in a eutrophic state, and high relative abundances of Synechococcus (40.62%) and Microcystis (16.2%) were noted during the course of our study. CCA indicated that nutrients (phosphorus) were key parameters driving Cyanobacteria-dominated community structure. By classifying the environmental factors into five categories, VPA analyses identified that P-factor (total phosphorus (TP) and SRP) as well as the synergistic effect of C-factor (DOC), N-factor (NO_3^-), and P-factor (TP and SRP) played a central role in structuring the bacterial communities in Erhai Lake. Heterogeneous physicochemical conditions explained the variations in bacterial assemblages in streams. This study provides a picture of stream–lake linkages from the perspective of bacterial community structure as well as key factors driving bacterial assemblages within lakes and streams at the whole watershed scale. We further argue that better management of phosphorus on the watershed scale is needed for ameliorating eutrophication of Erhai Lake.

Keywords: stream-lake linkage; taxonomic; bacterial community; environmental change

1. Introduction

Microbial assemblages are fundamental components of aquatic ecosystems and play a key role in biogeochemical cycles in both lotic and lentic ecosystems [1,2]. Their high diversity, small size, and rapid generation have caused microbes to become the most sensitive aquatic organisms to environmental perturbations, especially to nutrient alteration [3–5]. Understanding these responses is increasingly important given that biogeochemical cycles have been dramatically altered by human activities [6], with a two-fold increase in nitrogen (N) availability [7,8] and approximately a four-fold increase in phosphorus (P) mobilization [9,10] relative to preindustrial times.

Freshwater ecosystems include networks of lakes and streams linked within catchments by downslope water flow and nutrient flux [11]. The spatial distribution of lakes and streams can influence various chemical and biological characteristics, while the interactions among aquatic habitats control, at least in part, their individual functioning [11,12]. Ecosystem metabolism is directly influenced by the relative balance of external loading of nutrients and dissolved organic carbon in lakes [13]. Stream ecosystems are primary receivers of nutrients and organic matter that export from terrestrial ecosystems through transport and storage of water, nutrients, and energy [14–16], and these nutrients and organic matter are transported by streams into lakes and marine environments [13]. In stream ecosystems, microbial assemblages primarily occur in the form of benthic biofilms [17], which play an important role in biogeochemical cycling, being responsible for organic matter mineralization, nutrient uptake, and transformation of contaminants [18]. Likewise, bacterial communities in lake ecosystems often have a close relationship with nutrient alteration. Data currently available suggest that bacterial community composition in freshwater ecosystems is controlled both by biogeography and environmental nutrient supplies, and that the relative importance of these controls shifts spatially from local to regional areas [11,19,20]. However, most studies of bacterial communities have focused only on single environment types, such as lakes or streams. Comparative information on the comparison of bacterial assemblages and their driving forces in lakes and input streams is still limited [16].

Erhai Lake is the second largest plateau lake and the seventh largest freshwater lake in China. This lake has valuable functions in supporting human life and regional development, including drinkable water supply, agricultural irrigation, climate regulation, tourism, and hydroelectric power generation [21–23]. Anthropogenic activities in the basin generate a large amount of nutrients and waste, which directly perturb biogeochemical cycles and accelerate whole-basin ecosystem deterioration [24]. Previous research on the bacterial communities in Erhai Lake basin focused only on streams [25] or lakes [26–29]. At the watershed scale, the ecosystems of Erhai Lake and its tributaries are highly connected by material and energy fluxes. To gain a better understanding of how nutrients are associated with variations in bacterial communities between the two kinds of ecosystems, we conducted an extensive bacterial survey using high-throughput ILLUMINA sequencing of 16S rRNA in Erhai Lake and its surrounding major tributaries. This study aimed to address four major questions regarding Erhai Lake and streams: (1) How do they differ in bacterial community structure? (2) How are co-occurrence patterns of bacterial taxa structured? (3) What are the major environmental factors associated with bacterial assemblages at the watershed scale? (4) How do these associations differ for the lake and the inflow streams? By understanding bacterial communities and its associations with human activities in the lake and its input streams, this study seeks to provide insight into the management of Erhai Lake watershed.

2. Materials and Methods

2.1. Study Area

Erhai Lake (25°25′–26°16′ N, 99°32′–100°27′ E) is the second largest freshwater lake in China. Erhai Lake watershed experiences a typical subtropical plateau monsoon climate with 85% of precipitation occurring from May to October and long-term average annual precipitation of 909 mm. The annual average air temperature is 15.7 °C (22 °C on average in summer). Erhai Lake watershed is located in the upper Mekong River tributaries with a total of 117 tributaries. Miju stream is the largest tributary of Erhai Lake originating in the northern mountains of Erhai Lake watershed, and there are also many small tributaries located in the western part of Erhai Lake. All of these tributaries pass through urbanized areas around the lake, which dramatically altered physical, chemical, and biological conditions of stream ecosystems. Moreover, the rapid urbanization around the lake has caused severe lake eutrophication and water contamination [27].

2.2. Sampling and Physicochemical Analyses

During August 2016, a total of nineteen samples spread evenly across Erhai Lake were collected, and seventeen samples were collected from six tributaries [Miju stream (MJ), Wanhuaxi stream (WH), Yangxi stream (YX), Jinxi stream (JX), Baishi stream (BS), and Taoxi (TX) stream]. According to the different level of human activities between upstream and downstream, two sampling sites were chosen in each stream.. Surface water samples for bacterial analyses were collected at a depth of 0.5 m in Erhai Lake. A total of 600 mL water was filtered with Whatman Nylon membrane filters (pore size: 0.2 µm), and these filters were immediately stored and transported in liquid N_2. Benthic biofilms were removed by vigorously scrubbing (with a sterilized nylon brush) a 4.5 cm-diameter area from the surface of a stone substrate with five replicates at each sampling site. After the slurry was rinsed with sterile water, 10 mL of the mixed slurry was filtered with 0.2 µm membrane filters, and these filters were immediately frozen in liquid N_2 in the field. All bacterial samples were stored at −80 °C until DNA extraction and subsequent analyses.

At each sampling site, water temperature (Temp), dissolved oxygen (DO), pH values, and conductivity (Cond) were measured in situ using a YSI Model 80 m (Yellow Springs Instruments, Yellow Springs, OH, USA). Altitude was measured using a Global Positioning System (GPS) unit (Triton 500, Magellan, Santa Clara, CA, USA). Water samples were acid-fixed and transported to the laboratory at 4 °C for chemical analyses. Water was filtered through Whatman Glass Fibre Filters to analyze dissolved organic carbon (DOC), soluble reactive phosphorus (SRP), nitrate (NO_3^-), and ammonium (NH_4^+). Total nitrogen (TN) was analyzed using ion chromatography after persulfate oxidation. NO_3^- was determined using ion chromatography, and NH_4^+ was analyzed using the indophenol colorimetric method. Total phosphorus (TP) and SRP were quantified using the ammonium molybdate method. DOC was analyzed using a Shimadzu TOC Analyzer (TOC-VCPH, Shimadzu Scientific Instruments, Columbia, MD, USA).

2.3. DNA Extraction, PCR, and Sequencing

The bacterial community was characterized by 16S rRNA gene sequencing. Bacterial genomic DNA was extracted using a PowerSoil DNA Isolation Kit (MoBio, Carlsbad, CA, USA) following the manufacturer's protocols. The 16S rRNA genes covering the V3 to the V4 regions were amplified using primers 806R-GGACTACHVGGGTWTCTAAT and 338F-ACTCCTACGGGAGGCAGCA (Invitrogen, Vienna, Austria) [25]. The PCR was performed according to the standard procedures of Applied Biosystems 2720 Thermal Cycler (ABI, Foster City, CA, USA). By using 1× Tris-Acetate-Edta buffer, amplified DNA samples were verified by 1.0% agarose gel electrophoresis and purified using the Gel Extraction Kit (Qiagen, Hilden, Germany). The final sequencing process was based on a MiSeq sequencing platform (Illumina, San Diego, CA, USA).

Raw sequence data (available from the National Center for Biotechnology Information, SRP167734) were processed using QIIME software [30]. The range of sequence reads per sample was 12,009–30,315. Quality filtering on merged sequences was performed. Sequences were discarded when these sequences did not meet the following criteria: sequence length < 200 bp, no ambiguous bases, and mean quality score > 20. Then, these sequences were compared with reference database (RDP Gold database) using the UCHIME algorithm to detect chimeric sequences [31], and the chimeric sequences were removed. After quality filtering, these sequences were grouped into operational taxonomic units (OTUs) using the clustering program VSERACH 1.9.6 [32] against the Silva 123 database based on the complete linkage algorithm at an identity level of 97% sequence similarity. The Ribosomal Database Program (RDP) classifier was used to assign a taxonomic category to all OTUs at a confidence threshold of 0.8. The RDP classifier uses the Silva 123 database, which has taxonomic categories predicted to the species level.

2.4. Data Analysis

We compared taxonomic profiles of the bacterial communities to assess the differences and the linkages between bacterial communities in lake and streams. To determine whether physicochemical factors, nutrients, organic carbon, and stoichiometric factors were significantly different between lake and streams, we conducted bootstrap *t*-tests using SPSS software (Version 12.0) (IBM, Armonk, NY, USA). The DOC:NO_3^-:TP (C:N:P) ratios were calculated based on stoichiometric homeostasis [33] with the following equation:

$$(x : y)_{\text{molar}} = \frac{x}{y} \frac{y_n}{x_n}$$

where x and y are nutrient concentrations, and x_n and y_n are the molar mass.

The spatial variability of nutrient concentrations, nutrient ratios, and bacterial community composition was visualized in ArcGIS 10.2 (ESRI, Redlands, CA, USA). The Multiple Response Permutation Procedure (MRPP) is a nonparametric procedure of testing the null hypothesis between two or more groups of entities in the software R (version 3.3.2) (Lucent Technologies, Murray Hill, NJ, USA). Heatmaps were applied to reveal the differences of OTUs (relative abundance higher than 0.1%, Heatplus and Gplots packages in R version 3.3.2). All pairwise Spearman's rank correlations between those OTUs and environmental variables were calculated in R (Hmisc package, version 4.0-1). Only robust ($r > 0.6$ or $r < -0.6$) and statistically significant (p-values < 0.01, p-values were adjusted using Benjaminin–Hochber method) were considered. Cytoscape (version 3.5.1) was used for network visualization. Nodes represented the OTUs, and edges represented Spearman's correlation relationship. All of the nodes were classified at the phylum level, and an edge-weighted spring-embedded network was applied to display the arrangement of the variables. Topological and node/edge metrics, including centralization, heterogeneity, characteristic path length, and clustering coefficient, were calculated. Modular structure analyses were conducted using the ClusterMaker app in Cytoscape. A total of 30 random networks with the same size of a real network were generated using Network Randomizer app (version 1.1.3) and topological parameters were calculated individually. The *t*-tests were employed to calculate the differences between topological parameters of the real network and the random networks using the standard deviations derived from the corresponding random network. Canonical correspondence analysis (CCA) was applied to analyze the bacterial communities with respect to various environmental factors using Vegan package 2.4 in R (version 3.3.2). Monte Carlo permutations and Mantel tests ($p < 0.05$) were used to select a set of environmental variables that had significant and independent effects on bacterial communities. Environmental factors with high partial correlation coefficients ($p < 0.05$, $r > 0.5$) and variance inflation factors > 20 were eliminated from the final CCA. To identify effects of different categories of environmental variables on bacterial communities, environmental factors were classified into five categories (Table 1). Variation partitioning analysis (VPA) was performed using the Vegan package (version 2.5-3) in R to determine the relative contributions of physicochemical-factor, N-factor (NO_3^-), P-factor (TP and SRP), C-factor (DOC), and the interactions of various factors.

Table 1. Environmental variables in Erhai Lake and streams.

Categories of Environmental Variables	Parameters	Lake			Stream			p (*t*-tests)
		Average	SD	CV	Average	SD	CV	
Physicochemical-factor	Altitude (m)	1964.00	0.00	0	2200.28	265.28	0.071	0.001 **
	Temp (°C)	22.93	0.90	0.039	16.68	1.99	0.121	0.000 **
	DO (mg/L)	5.50	0.44	0.079	5.10	0.59	0.120	0.077
	Cond (µs/cm)	287.28	6.13	0.021	115.29	52.46	0.738	0.000 **
	pH	9.53	0.07	0.007	8.36	0.82	1.537	0.000 **
P-factor	TP (mg/L)	0.04	0.01	0.146	0.06	0.02	0.327	0.011 *
	SRP (mg/L)	0.01	0.002	0.267	0.028	0.008	0.431	0.000 **
N-factor	TN (mg/L)	0.54	0.08	0.138	0.53	0.32	0.529	0.863
	NO_3^- (mg/L)	0.50	0.06	0.125	0.47	0.29	0.565	0.690
	NH_4^+ (mg/L)	0.017	0.004	0.259	0.016	0.003	0.471	0.621
C-factor	DOC (mg/L)	4.95	0.44	0.088	1.16	0.25	0.437	0.000 **
Nutrient ratios	DOC:NO_3^- (C:N)	11.626	1.282	0.110	4.366	2.597	0.595	0.001 **
	DOC:TP (C:P)	293.011	46.494	0.158	79.068	26.261	0.332	0.000 **
	NO_3^-:TP (N:P)	25.267	3.400	0.135	24.119	14.978	0.621	0.761

** indicates $p < 0.01$ and * indicates $p < 0.05$. SD: standard division, CV: coefficient of variation, DO: dissolved oxygen, Cond: conductivity, TP: total phosphorus, SRP: soluble reactive phosphorus, TN: total nitrogen, DOC: dissolved organic carbon.

3. Results

3.1. Environmental Conditions and Nutrients

Significant differences in various environmental variables and nutrient concentrations were observed. Environmental variables were divided into five categories, including physicochemical-factor, P-factor, C-factor, N-factor, and nutrient ratios (Table 1). Temp, Cond, pH, and DOC were significantly higher in Erhai Lake than in streams (Table 1, *t*-tests, $p < 0.01$). TP and SRP concentration were higher in streams than in the lake ($p < 0.01$). Streams were the major sources of nitrogen and phosphorus in Erhai Lake (Figure 1). The concentration of TP in MJ was higher than that in other tributaries (*t*-tests, $p < 0.01$), which caused P concentrations to be higher in northern areas of the lake than in central and southern areas. Although SRP was the major component of TP in streams (29.17~74.44%), it only was a small component in the lake (15.38~30.95%). Moreover, NO_3^- was the major component of TN in streams (57.09~93.78%) and the lake (86.18%~96.54%). In order to ensure the validity of data, we chose DOC, NO_3^-, and TP to calculate DOC:NO_3^-:TP (C:N:P). Both C:N and C:P ratios were higher in the lake than in streams (Table 1).

3.2. Bacterial Assemblages

After quality filtering, a total of 12,566 OTUs were identified. The relative abundance of different phyla dramatically differed in the lake relative to the streams. The proportion of shared OTUs was 27.9% between lake and streams, while the proportions of detected OTUs unique to Erhai Lake and streams were 31.1% and 41.0%, respectively. Except for the Bacteroidetes, all the other dominant phyla were significantly different between Erhai Lake and the streams. In the lake, the dominant phylum was Cyanobacteria (relative abundance of 43.8%), followed by Proteobacteria (15.1%) and Actinobacteria (14.8%) (Figure 2). The dominant phyla in streams were Proteobacteria (52.4%) and Cyanobacteria (30.9%) (Figure 2). Relative abundances of Proteobacteria and Thermi in streams were much higher than those in the lake (Figure 2, $p < 0.05$). Relative abundances of Cyanobacteria, Actinobacteria, Bacteroidetes, Verrucomicrobia, Chlorobi, and Planctomycetes were much higher in the lake than those in the streams (Figure 2, $p < 0.05$).

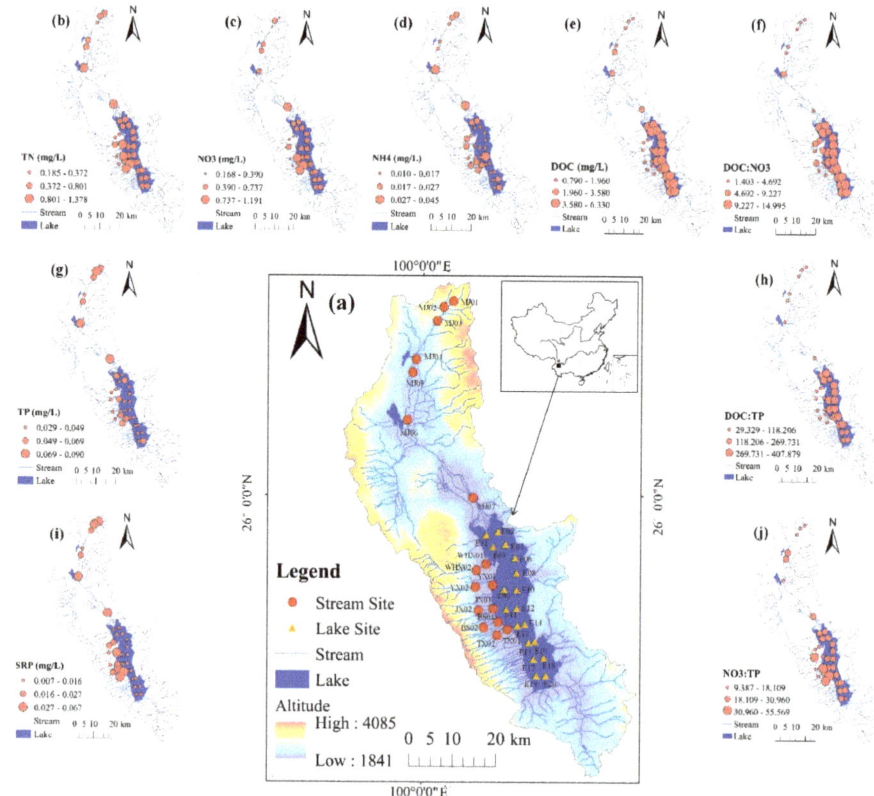

Figure 1. Study area and nutrient variables. (**a**) The distribution of sampling sites and (**b–j**) spatial patterns of nutrient variables in Erhai Lake and its tributaries. The map was created in ArcGIS 10.2 (http://desktop.arcgis.com/en/arcmap/) using ASTER GDEM data downloaded from the United States Geological Survey [ASTER GDEM is a product of the Ministry of Economy, Trade and Industry (METI) and the National Aeronautics and Space Administration (NASA)].

Unweighted Pair Group Method with Arithmatic Mean (UPGMA) clustering analyses illustrated the remarkable variations in bacterial community composition among the sampling sites (Figure 3a). In the lake, bacterial assemblages were clearly separated into two subgroups, and MRPP analyses demonstrated the significant difference between these two subgroups ($p < 0.05$). Subgroup 1 (EH09– EH11, EH14–EH18) had no dominant bacterial community under low nutrient concentrations (Figure 3a,b). In subgroup 2 (EH01–EH08, EH12, EH13, EH19, and EH20), Cyanobacteria (47.99–67.10%) was the dominant phyla under high phosphorus concentrations (Figure 3a,b). Bacterial community structure varied with spatial heterogeneity of environmental variables in streams. Heatmap clustering analyses also illustrated the variability of OTUs among sampling sites (Figure 4). Synechococcus (40.62%) and Microcystis (12.06%) made up the Cyanobacteria in Erhai Lake. Synechococcus (26.3%) dominated Cyanobacteria in streams, while Microcystis (0.23%) only was a small component of Cyanobacteria in streams.

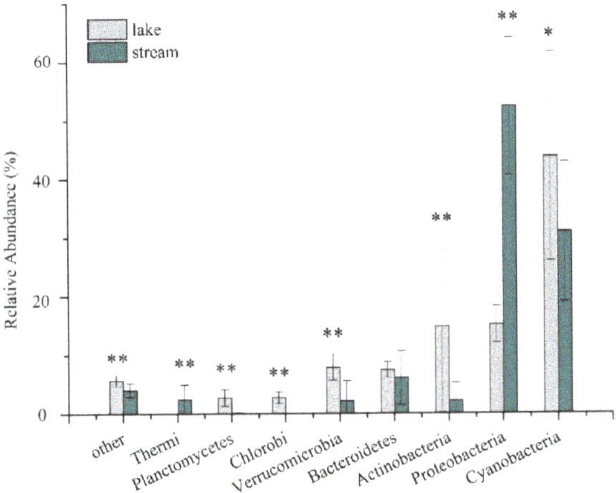

Figure 2. Phyla-level bacterial relative abundances in Erhai Lake and streams. Only phyla with a relative abundance > 1% in either streams or the lake are shown; "other" represents the unsigned operational taxonomic units (OTUs) and the phyla with a relative abundance < 1%. The comparison of lake and streams bacterial abundances was assessed by *t*-tests. ** indicates statistical significance at $p < 0.01$ and * indicates $p < 0.05$.

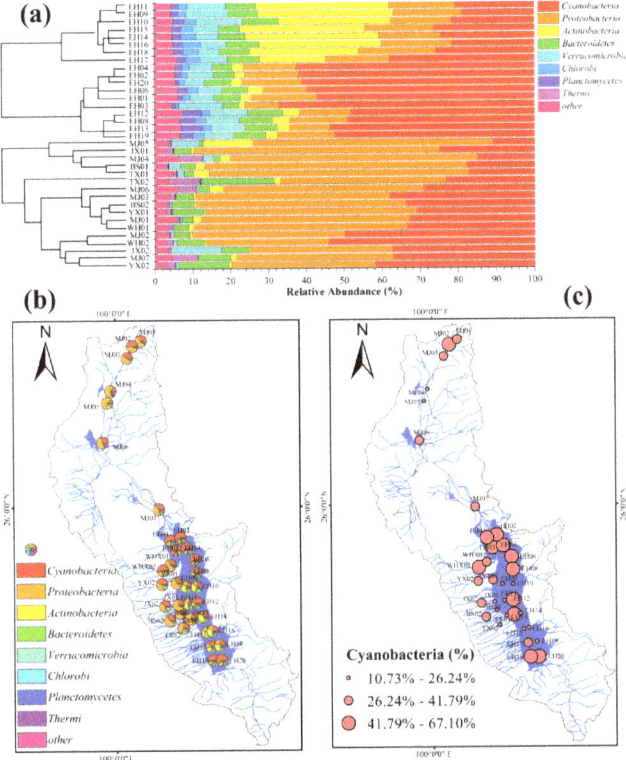

Figure 3. Composition of bacterial communities in Erhai Lake and streams (**a**) as well as the spatial variations in bacterial communities (**b**) and Cyanobacteria (**c**).

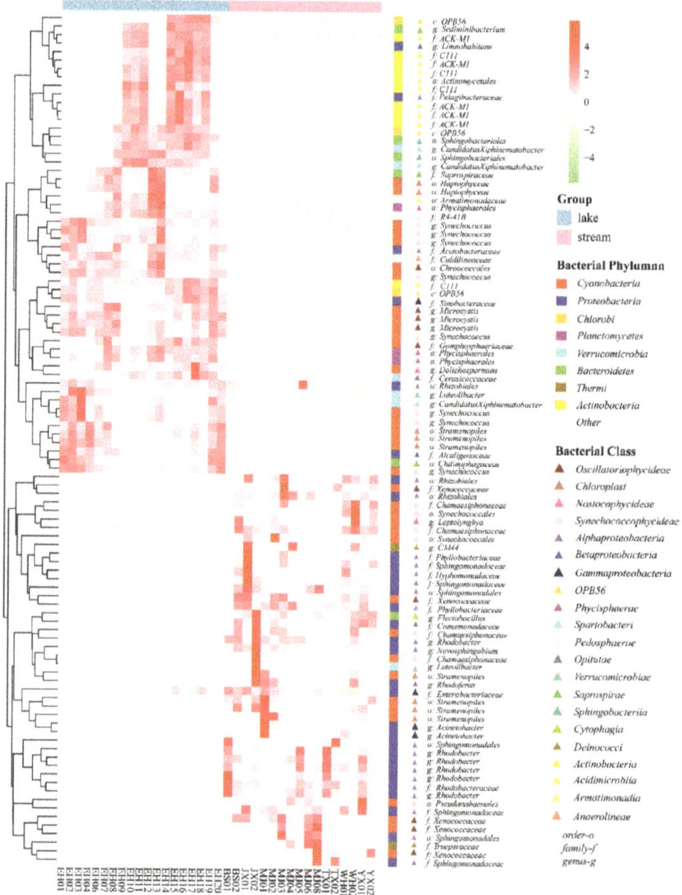

Figure 4. Heatmap showing the taxonomic differences of OTUs between Erhai Lake and streams based on Bray–Curtis distance. Bray–Curtis distances were calculated using relative abundances of OTUs (relative abundance > 0.1%).

3.3. Bacteria Co-Occurrence

OTUs (relative abundances > 0.01%) were used to build co-occurrence networks (Figure 5, $r > 0.6$ or $r < -0.6$, $p < 0.01$). The integrated bacterial network contained 511 nodes (i.e., OTUs) and 12 environmental parameters with 17,129 edges (significant interactions) (Figure 5a). The lake bacterial network contained 416 nodes (i.e., OTUs) and 5898 edges (Figure 5b), and the stream bacterial network contained 435 nodes with 2808 edges (Figure 5c). In total, 97.78% inter-genus correlations were positive within the integrated bacterial network. Significant correlations were mostly concentrated within Cyanobacteria, Proteobacteria, Verrucomicrobia, and Actinobacteria phyla. The positive relationship of OTUs in the lake bacterial network was 69.2% (Figure 5b), and in the streams network, it was 100% (Figure 5c). These results suggested that lake microorganisms have more competitive relationships than stream microorganisms due to the homogeneous habitat and the limited resources in the lake. In the integrated bacterial network, Proteobacteria had the highest accumulated betweenness centrality score of 0.67. The betweenness centrality is a measure of centrality of nodes in a network based on shortest paths. The higher the betweenness centrality scores of nodes are, the more associations over the network there are. C:P had the highest accumulated betweenness centrality score (0.065), followed by SRP (0.035), Temp (0.033), and DOC (0.032). Moreover, both C:P (31.76%) and DOC (46.58%) were positively associated with nodes belonging to Cyanobacteria, while SRP (36.62%) was negatively associated with nodes belonging to Cyanobacteria.

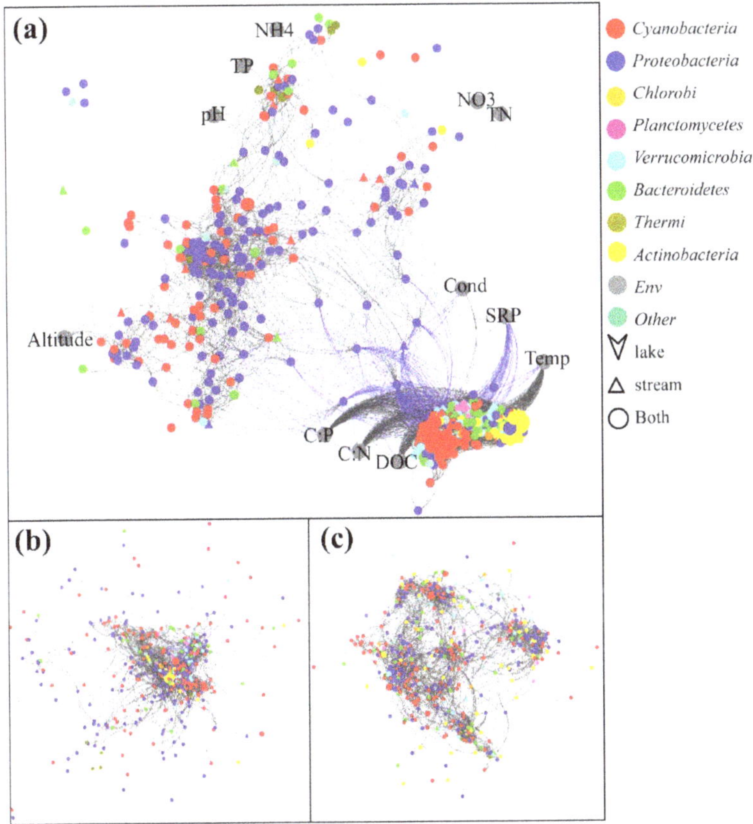

Figure 5. Co-occurrence networks of bacterial communities in integrated lake/stream (**a**), Erhai Lake (**b**), and streams (**c**). Nodes represent the OTUs. Edges represent Spearman's correlation relationships, and only strong and significant correlations (Spearman's $r > 0.6$ or $r < -0.6$, $p < 0.01$) are shown. The grey and the blue lines indicate positive and negative correlations, respectively.

Topological parameters were calculated to describe the interactions, the compactness, and the modularity among OTUs in the networks (Table 2). Among the three networks, the integrated lake/stream network showed the highest clustering coefficient and network centralization, while characteristic path length and modularity values were highest in the streams sub-network (Table 2). Bacterial communities in the three bacterial networks exhibited modular structures with modularity values ranging from 0.572 to 0.784 (Table 2). These results confirmed that all of these bacterial networks had "small world" properties and significant modular structures, because modularity, network centralization, clustering coefficient, and average path length were higher in real networks than in random networks (*t*-tests, $p < 0.05$).

Table 2. Topological parameters of real co-occurrence networks and corresponding random networks (some parameters of the random network are shown in mean ± SD. SD represent standard deviation). *t*-tests were used to compare the topological parameters between real and random networks.

Topological Parameter	Lake and Streams		Lake		Stream	
	Random	Real	Random	Real	Random	Real
Number of nodes	511	511	416	416	435	435
Average number of neighbors	65.840	67.037	28.351	28.351	12.906	12.906
Network Centralization	0.045 ± 0.003 *	0.232	0.038 ± 0.004 *	0.193	0.028 ± 0.008 *	0.044
Network Heterogeneity	0.116 ± 0.004 *	0.745	0.182 ± 0.006 *	1.128	0.272 ± 0.001 *	0.700
Characteristic Path Length	1.880 ± 0.001 *	2.654	2.060 ± 0.001 *	3.799	2.651 ± 0.002 *	5.134
Clustering Coefficient	0.125 ± 0.006 *	0.705	0.069 ± 0.001 *	0.538	0.029 ± 0.002 *	0.596
Modularity	0.439 ± 0.021 *	0.639	0.438 ± 0.011 *	0.572	0.446 ± 0.018 *	0.784

* indicates $p < 0.05$.

3.4. Environmental Factors Associated with the Bacterial Community

Spearman's rank correlation, CCA, and VPA were used to assess the potential effect of environmental variables on bacterial communities (Figures 6 and 7). C:P was significantly associated with DOC and TP, N:P was significantly associated with NO_3^- and TP, and C:N was only significantly associated with NO_3^- (Figure 6). We noted that both C:P and N:P showed the co-limitation of the numerator and the denominator, while C:N was limited by the denominator in Erhai Lake. In streams, both C:P and N:P were limited by the numerator, and C:N was limited by the denominator, because DOC significantly associated with C:P, and NO_3^- significantly associated with N:P and C:N (Figure 6). Moreover, TP, SRP, and C:P ratios were the predominant environmental factors closely associated with Cyanobacteria in the lake, and C:P ratios were also positively correlated with Verrucomicrobia ($p < 0.05$, Figure 6a). In streams, TN and NO_3^- were positively associated with Cyanobacteria, and N:P ratios were negatively associated with Cyanobacteria ($p < 0.05$, Figure 6b). Moreover, Actinobacteria positively associated with TN, NO_3^-, DOC, and N:P ($p < 0.05$).

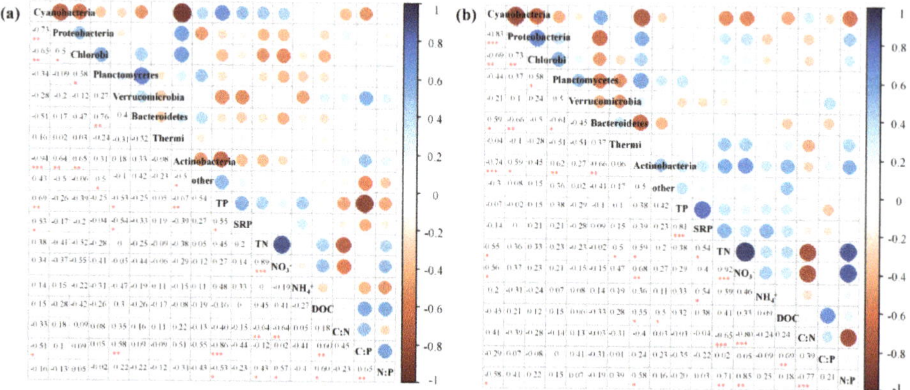

Figure 6. Spearman's rank correlation between nutrients and microbes at the phyla-level in (**a**) Erhai Lake and (**b**) streams. Blue nodes indicate positive associations and red nodes negative associations. Nodes are sized and present the degree of correlations. *** indicates statistical significance at $p < 0.001$, ** indicates $p < 0.01$ and * indicates $p < 0.05$.

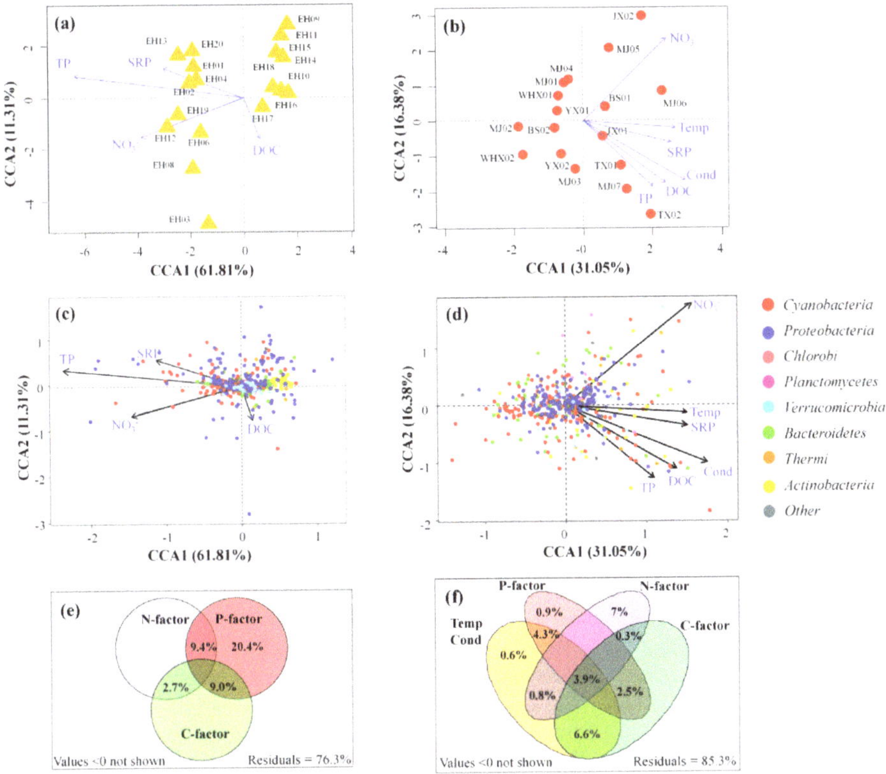

Figure 7. Canonical correspondence analysis (CCA) of bacterial communities and nutrient variables in Erhai Lake (**a, c, and e**) and streams (**b, d, and f**). All environmental variables had goodness of fit at the significant level $p < 0.05$ by envfit function. Variance partition analysis (VPA) determined the relative contributions of physicochemical factor, C-factors (DOC), N-factor (NO_3^-), P-factor (TP and SRP), and the interactions of these factors in Erhai Lake (**e**) and streams (**f**).

CCA and VPA were used to evaluate the relationship of environmental variables and bacterial communities (Figure 7). Some collinear variables were removed from CCA and VPA. NO_3^- was the predominant N-factor used in CCA and VPA, because NO_3^- was the major component of TN in Erhai watershed, and it positively associated with TN ($p < 0.01$, Figure 6). SRP did not dominate P-factor because SRP was a small component of TP in Erhai Lake; thus, TP and SRP constituted P-factor.

According to the Monte Carlo test and the Mantel test, TP, SRP, NO_3^-, and DOC were significantly associated with bacterial communities in Erhai Lake (Figure 7, two axes accounted for 73.12%, $p < 0.05$). The first axis was mainly defined by TP, SRP, and NO_3^- in the negative direction, and the second axis was mainly defined by DOC in the negative direction. Following an increase of dominant nutrient parameters, OTUs gradually changed from Actinobacteria-dominated to Cyanobacteria-dominated (Figure 7c). The sampling sites were divided into two subgroups at the first axis. The subgroup 1 (EH09–EH11 and EH14–EH18), located at the right of the diagram, primarily involved samples from middle to southern parts of the lake, while the sites of subgroup 2 (EH01–EH08, EH12, EH13, EH19, and EH20), located at the left of the diagram, included samples from the northern areas along with two sites near the lake mouth to the down streams (Figure 7a). Variance partitioning analyses (VPA) revealed that nutrient factors explained 23.7% of bacterial communities in Erhai Lake. Moreover, P-factor and the two interactions of C, N, and P (C × P and N × P) had high contributions to the overall explanation (Figure 7e). In addition, nutrients ratios (C:N, C:P, and N:P) explained 12.3% of bacterial communities in the lake (Figure S1).

Temp, Cond, NO_3^-, TP, SRP, and DOC were significantly associated with bacterial communities in streams (Figure 7b, two axes accounted for 47.43%, $p < 0.05$), and these variables defined the first axis. Bacterial communities were more evenly distributed, and no obvious pattern was identified (Figure 7d). Samples were divided into two subgroups based on axis 1 (Figure 7b): subgroup 1 (JX01, JX02, TX01, TX02, BS01, and MJ05–MJ07) and subgroup 2 (BS02, YX01, YX02, TX01, WHX01, and MJ01–MJ04). The samples of subgroup 1 (located at the right of the biplot) mainly represented sites experiencing higher levels of human disturbance or reference conditions in the streams, while the left direction represented the low levels of human activity (Figure 7b). N-factor, the interaction of physiochemical factor, C-factor, N-factor, and P-factor, had high contributions to the overall explanation of bacterial communities in the streams (Figure 7f). Nutrient ratios only explained 7.4% of bacterial communities in streams (Figure S1).

4. Discussion

For lacustrine bacterial assemblages, cells inflowing from tributaries can influence community structure [11,34]. These inputs partly explain the high level of overlap in identified OTUs between Erhai Lake and its input streams. However, according to similarity analyses of the bacterial communities, there was obvious spatial heterogeneity in the bacterial community structure between the lake and the streams, reflecting significant changes in relative abundances of various taxa. These results are consistent with previous studies of stream bacterial communities in connected lake and stream ecosystems [16,35]. The dramatic changes in bacterial community composition between the lake and the streams indicates significant variations in ecological function and nutrient metabolism [36]. In contrast to a previous study [26], our results showed extremely high relative abundances of Cyanobacteria in Erhai Lake. These high relative abundances of Cyanobacteria differed across sampling seasons, as Cyanobacteria always dominated Erhai Lake during summer [37]. We also found high abundances of Cyanobacteria in streams, while the dominant species of Cyanobacteria in streams were different from that in the lake. Because most of the Cyanobacteria are photoautotrophic, the increase in Cyanobacteria would likely correlate with changes in bacterial metabolism and biogeochemical processes [16] as well as with sampling season [26] and human impacts at downstream sites [25]. In the lake, the high relative abundance of Cyanobacteria dominated by Synechococcus (40.62%) and Microcystis (16.2%) attracted considerable attention due to the high risk of potential toxin producers [38] and the subsequent influence on drinking water supply, aquatic biodiversity, fisheries, water irrigation, recreation, and aesthetics [39].

Co-occurrence patterns of organisms were evaluated to assess community assembly rules and interactions in highly complex systems [40]. Co-occurrence network results revealed that bacterial networks in Erhai watershed had non-random, scale-free, and "small world" properties. The high degree of connectedness and the relatively high proportion of negative associations in the lake bacterial network indicated stronger competition relationships of OTUs in Erhai Lake. Moreover, Proteobacteria play the most important role as a transitional species from streams to lake (Figure 5). Several OTUs of Proteobacteria showed widespread connections with OTUs of lakes and streams bacterial networks. Those broad connections may serve as important connectors that strongly contribute to community stability and coexistence. Removal of those strongly connected components in a network often causes the collapse of structure and function in the ecosystem [40–42]. The topological parameters provided important information that helps us to understand community structure. Higher values of network centralization and heterogeneity indicated stronger connections of OTUs in Erhai Lake (Table 1), implying that a minor disturbance would more easily cause a large impact on the lake bacterial network [41–43]. Moreover, bacterial communities in these networks exhibited strong modular structures, and the stream bacterial networks exhibited a stronger modular structure than other networks (Table 1, modularity values > 0.4). In general, species interactions are more frequent and intense in a modular structure [43]. Notably higher modularity values of the stream networks related to the heterogeneous habitats were found in streams, which provide abundant niches for microorganisms [40]. Niche differentiation in stream biofilms is also highly associated with heterogeneous microhabitats, wider ranges of physiochemical conditions, and complex hydrological and hydraulic conditions, which are known to have significant effects on composition and function of bacterial communities [44–46]. Thus, heterogeneous microhabitats caused multiple factors affecting bacterial communities in streams.

UPGMA clustering of bacterial communities indicated an apparent spatial variation in bacterial communities in the lake and the streams. In general, physicochemical parameters, nutrients, and nutrient ratios could drive bacterial communities in freshwater ecosystems [34,47–51]. Our results demonstrated that N-factor (NO_3^-) (7%) and the interaction of multiple factors, including physicochemical-factor, N-factor, P-factor, and C-factor, drove the bacterial community structure on stream biofilms (Figure 7f). Streams are the primary recipient of xenobiotics and pollutants input from the watershed; thus, terrestrial soil, sewage, plant litter, and autochthonous material were the main sources of organic matter and nutrients in streams [52]. It has been widely reported that agriculture and urban development significantly increase nutrient concentration in streams [53]. These nutrients could accumulate on benthic biofilms and be used by bacterial communities [36,54–56]. In Erhai watershed, streams pass through mountain areas and urbanized areas, which causes significant differences of water quality in these heterogeneity ecosystems under different degrees of human activities perturbations [23]. Under these heterogeneity stream ecosystems, physicochemical-factor provides a diverse environment for bacterial communities [52,57], and DOC, NO_3^-, SRP, and TP provide available nutrients for bacterial growth [40,58].

Nutrients predominately explained the variation in lake bacterial communities, possibly due to the relatively uniform physical conditions [4,5]. Elevated inputs of phosphorus and nitrogen from its inflow streams have caused severe eutrophication and associated algae blooms in Erhai Lake. The seasonal succession of Cyanobacteria-dominated communities was highly correlated with nutrient input in the summer at Erhai Lake [51]. Our results demonstrated that P-factor, the interaction of N-factor and P-factor, as well as the interaction of N-factor, P-factor, and C-factor were highlighted as nutrient resources associated with spatial distribution and bacterial species in Erhai Lake during summer (Figures 5 and 7). P-factor (TP and SRP) was positively associated with Cyanobacterial communities (Figure 5), which is understandable because SRP is the most readily available form of phosphorus for autotrophs assimilation [59]. Lake bacterial community structure was positively associated with the SRP concentration in its inflow streams [49]. Thus, phosphorus loading from tributaries from sub-watersheds experiencing seasoning flooding may play an important role in supplying P resources to Erhai Lake. Moreover, low N:P ratios are often connected with an increasing load of P, and the synergy of N and P has historically been the key factor for harmful algal blooms in Erhai Lake [21,60–63]. The most likely mechanism for synergistic effects of N and P is that a single nutrient quickly induces limitation by the alternative nutrient [64]. Additionally, previous studies have shown that organic carbon plays an important role in regulating microbial assemblages and the food web of aquatic ecosystems [65–67]. In the present study, synergistic effects of C-factor and P-factor (C:P) and N-factor and P-factor (N:P) played crucial roles in regulating bacterial communities in Erhai Lake (Figure 7 and Figure S1). Thus, DOC was an important factor affecting bacterial community structures in Erhai Lake that could not be ignored, because high DOC content could provide enough carbon resource for fast-growing and competitive species [64]. TP and SRP were the key environmental factors that drove bacterial communities—especially Cyanobacterial-dominant communities—in Erhai Lake. Moreover, the synergistic effect of C-factor (DOC), N-factor (NO_3^-), and P-factor (TP and SRP) also played an important role in regulating the bacterial community structure in the Lake.

5. Conclusions

The bacterial communities of Erhai Lake and its tributaries are connected to and affected by terrestrial nutrient enrichment. The lake microorganism network showed stronger centralization and greater heterogeneity than the stream bacterial network, suggesting that lake bacterial communities were less stabile and resistant to human activities. In contrast, modularity values were higher in streams, implying the abundant niches of microorganisms under different degrees of human activities perturbation. The dramatic differences of bacterial community structure between the lake and the streams were mainly associated with SRP, DOC, C:P, and C:N. Through CCA and VPA analyses, we found that multiple factors, including physicochemical-factor, N-factor, P-factor, and C-factor, were the primary correlates of bacterial communities at the stream scale. However, the composition of lake bacterial communities was most closely associated with phosphorus concentrations and the interaction of N-factor, P-factor, and C-factor (C:P and N:P). These nutrient factors play an important role in structuring bacterial assemblages, stimulating rapid bacterial growth, and producing the Cyanobacteria blooms of Erhai Lake. Our results provide a further understanding of stream–lake linkages from the perspective of bacterial community structures and of the important effects of terrestrial nutrients on downstream ecosystems. These data indicate that it is important to enhance watershed management and decrease the pollution from agriculture and urban regions.

Supplementary Materials: The following are available online at http://www.mdpi.com/2073-4441/11/8/1711/s1, Figure S1: Canonical correspondence analysis (CCA) and Variance partition analysis (VPA) of bacterial communities and nutrient ratios in Erhai Lake (a, c, and e) and Streams (b, d, and f). Variance partition analysis (VPA) determined the relative contributions of C:N, C:P, and N:P.

Author Contributions: Y.L., X.Q., J.J.E., and M.Z., did the analyses, and prepared the manuscript; J.J.E. reviewed and revised the manuscript. W.P., X.Q., Z.R., and M.Z. designed the study. X.Q., Y.Z., H.Z. and H.Y. performed the field work and laboratory analysis; W.P. and J.J.E. gave suggestions during the whole work.

Funding: This study was funded by the State Key Laboratory of Simulation and Regulation of Water Cycle in River Basin (SKL2018CG02), and the National Natural Science Foundation of China (No. 51439007 and No. 41671048), and the IWHR Research and Development Support Program (WE0145B052018 and WE0145B532017).

Acknowledgments: We appreciate the anonymous reviewers for their valuable comments and efforts to improve this manuscript.

Conflicts of Interest: The authors declare no conflict of interest.

References

1. Elser, J.J.; Goldman, C.R. Zooplankton effects on phytoplankton in lakes of contrasting trophic status. *Limnol. Oceanogr.* **1991**, *36*, 64–90. [CrossRef]

2. Loreau, M. Microbial diversity, producer-decomposer interactions and ecosystem processes: A theoretical model. *Proc. R. Soc. B Biol. Sci.* **2001**, *268*, 303–309. [CrossRef] [PubMed]

3. Peter, H.; Sommaruga, R. Shifts in diversity and function of lake bacterial communities upon glacier retreat. *ISME J.* **2016**, *10*, 1545–1554. [CrossRef] [PubMed]

4. Mieczan, T.; Adamczuk, M.; Pawlik-Skowronska, B.; Toporowska, M.; Pawlik, S. Eutrophication of peatbogs: Consequences of P and N enrichment for microbial and metazoan communities in mesocosm experiments. *Aquat. Microb. Ecol.* **2015**, *74*, 121–141. [CrossRef]

5. Mieczan, T.; Adamczuk, M.; Tarkowska-Kukuryk, M.; Nawrot, D. Effect of water chemistry on zooplanktonic and microbial communities across freshwater ecotones in different macrophyte-dominated shallow lakes. *J. Limnol.* **2015**, *75*, 262–274. [CrossRef]

6. Martínez-Alonso, M.; Méndez-Álvarez, S.; Ramírez-Moreno, S.; González-Toril, E.; Amils, R.; Gaju, N. Spatial Heterogeneity of Bacterial Populationsin Monomictic Lake Estanya (Huesca, Spain). *Microb. Ecol.* **2008**, *55*, 737–750. [CrossRef] [PubMed]

7. Galloway, J.N.; Townsend, A.R.; Erisman, J.W.; Bekunda, M.; Cai, Z.; Freney, J.R.; Martinelli, L.A.; Seitzinger, S.P.; Sutton, M.A. Transformation of the Nitrogen Cycle: Recent Trends, Questions, and Potential Solutions. *Science* **2008**, *320*, 889–892. [CrossRef]

8. Peñuelas, J.; Sardans, J.; Rivas-ubach, A.; Janssens, I.A. The human-induced imbalance between C, N and P in Earth's life system. *Glob. Chang. Biol.* **2011**, *18*, 3–6. [CrossRef]

9. Bennett, E.M.; Carpenter, S.R.; Caraco, N.F. Human Impact on Erodable Phosphorus and Eutrophication. *Bioscience* **2001**, *51*, 227–234. [CrossRef]

10. Falkowski, P.G.; Fenchel, T.; Delong, E.F. The microbial engines that drive Earth's biogeochemical cycles. *Science* **2008**, *320*, 1034–1039. [CrossRef] [PubMed]

11. Crump, B.C.; Adams, H.E.; Hobbie, J.E.; Kling, G.W. Biogeography of bacterioplankton in lakes and streams of an arctic tundra catchment. *Ecology* **2007**, *88*, 1365–1378. [CrossRef]

12. Kipphut, G.W.; O'Brien, W.J.; Kling, G.W.; Miller, M.M. Integration of lakes and streams in a landscape perspective: The importance of material processing on spatial patterns and temporal coherence. *Freshw. Biol.* **2000**, *43*, 477–497.

13. Williamson, C.E.; Dodds, W.; Kratz, T.K.; Palmer, M.A. Lakes and streams as sentinels of environmental change in terrestrial and atmospheric processes. *Front. Ecol. Environ.* **2008**, *6*, 247–254. [CrossRef]

14. Levi, P.S.; Starnawski, P.; Poulsen, B.; Baattrup-Pedersen, A.; Schramm, A.; Riis, T. Microbial community diversity and composition varies with habitat characteristics and biofilm function in macrophyte-rich streams. *Oikos* **2017**, *126*, 398–409. [CrossRef]

15. McInerney, P.J.; Rees, G.N. Co-invasion hypothesis explains microbial community structure changes in upland streams affected by riparian invader. *Freshw. Sci.* **2017**, *36*, 297–306. [CrossRef]

16. Ren, Z.; Wang, F.; Qu, X.; Elser, J.J.; Liu, Y.; Chu, L. Taxonomic and Functional Differences between Microbial Communities in Qinghai Lake and Its Input Streams. *Front. Microbiol.* **2017**, *8*, 2319. [CrossRef] [PubMed]

17. Findlay, S.E.G. Stream microbial ecology. *J. N. Am. Benthol. Soc.* **2010**, *29*, 170–181. [CrossRef]

18. Dopheide, A.; Lear, G.; He, Z.; Jizhong, Z.; Lewis, G.D. Functional Gene Composition, Diversity and Redundancy in Microbial Stream Biofilm Communities. *PLoS ONE* **2016**, *10*, e0123179. [CrossRef]

19. Ji, B.; Qin, H.; Guo, S.; Chen, W.; Zhang, X.; Liang, J. Bacterial communities of four adjacent fresh lakes at different trophic status. *Ecotoxicol. Environ. Saf.* **2018**, *157*, 388–394. [CrossRef]

20. Horton, D.J.; Theis, K.R.; Uzarski, D.G.; Learman, D.R. Microbial community structure and microbial networks correspond to nutrient gradients within coastal wetlands of the Laurentian Great Lakes. *FEMS Microbiol. Ecol.* **2019**, *95*, fiz033. [CrossRef]

21. Li, J.; Lars-Anders, H.; Kenneth, P. Nutrient Control to Prevent the Occurrence of Cyanobacterial Blooms in a Eutrophic Lake in Southern Sweden, Used for Drinking Water Supply. *Water* **2018**, *10*, 919. [CrossRef]

22. Ni, Z.; Wang, S.; Zhang, M. Sediment amino acids as indicators of anthropogenic activities and potential environmental risk in Erhai Lake, Southwest China. *Sci. Total. Environ.* **2016**, *551*, 217–227. [CrossRef]

23. Feng, S.; Zhang, L.; Wang, S.; Nadykto, A.B.; Xu, Y.; Shi, Q.; Jiang, B.; Qian, W. Characterization of dissolved organic nitrogen in wet deposition from Lake Erhai basin by using ultrahigh resolution FT-ICR mass spectrometry. *Chemosphere* **2016**, *156*, 438–445. [CrossRef]

24. Chen, A.; Lei, B.; Hu, W.; Wang, H.; Zhai, L.; Mao, Y.; Fu, B.; Zhang, D. Temporal-spatial variations and influencing factors of nitrogen in the shallow groundwater of the nearshore vegetable field of Erhai Lake, China. *Environ. Sci. Pollut. Res.* **2018**, *25*, 4858–4870. [CrossRef]

25. Qu, X.; Ren, Z.; Zhang, H.; Zhang, M.; Zhang, Y.; Liu, X.; Peng, W. Influences of anthropogenic land use on microbial community structure and functional potentials of stream benthic biofilms. *Sci. Rep.* **2017**, *7*, 15117. [CrossRef]

26. Dai, Y.; Yang, Y.; Wu, Z.; Feng, Q.; Xie, S.; Liu, Y. Spatiotemporal variation of planktonic and sediment bacterial assemblages in two plateau freshwater lakes at different trophic status. *Appl. Microbiol. Biotechnol.* **2016**, *100*, 4161–4175. [CrossRef]

27. Xiong, W.; Xie, P.; Wang, S.; Niu, Y.; Yang, X.; Chen, W. Sources of organic matter affect depth-related microbial community composition in sediments of Lake Erhai, Southwest China. *J. Limnol.* **2014**, *73*, 310–323. [CrossRef]

28. Li, B.; Chen, H.; Li, N.; Wu, Z.; Wen, Z.; Xie, S.; Liu, Y. Spatio-temporal shifts in the archaeal community of a constructed wetland treating river water. *Sci. Total. Environ.* **2017**, *605*, 269–275. [CrossRef]

29. Liu, Y.; Zhang, J.; Zhao, L.; Li, Y.; Yang, Y.; Xie, S. Aerobic and nitrite-dependent methane-oxidizing microorganisms in sediments of freshwater lakes on the Yunnan Plateau. *Appl. Microbiol. Biotechnol.* **2015**, *99*, 2371–2381. [CrossRef]

30. Caporaso, J.G.; Kuczynski, J.; Stombaugh, J.; Bittinger, K.; Bushman, F.D.; Costello, E.K.; Fierer, N.; Peña, A.G.; Goodrich, J.K.; Gordon, J.I.; et al. QIIME allows analysis of high-throughput community sequencing data. *Nat. Methods* **2010**, *7*, 335–336. [CrossRef]

31. Edgar, R.C.; Haas, B.J.; Clemente, J.C.; Quince, C.; Knight, R. UCHIME improves sensitivity and speed of chimera detection. *Bioinformatics* **2011**, *27*, 2194–2200. [CrossRef]

32. Edgar, R.C. Search and clustering orders of magnitude faster than BLAST. *Bioinformatics* **2010**, *26*, 2460–2461. [CrossRef]

33. Downing, J.A.; McCauley, E. The nitrogen: Phosphorus relationship in lakes. *Limnol. Oceanogr.* **1992**, *37*, 936–945. [CrossRef]

34. Zhulidov, A.V.; Kämäri, J.; Robarts, R.D.; Pavlov, D.F.; Rekolainen, S.; Gurtovaya, T.Y.; Meriläinen, J.J.; Lugovoy, V.V.; Pavlov, D. Long-term dynamics of water-borne nitrogen, phosphorus and suspended solids in the lower Don River basin (Russian Federation). *J. Water Clim. Chang.* **2011**, *2*, 201–211. [CrossRef]

35. Crump, B.C.; Hobbie, J.E. Synchrony and seasonality in bacterioplankton communities of two temperate rivers. *Limnol. Oceanogr.* **2005**, *50*, 1718–1729. [CrossRef]

36. Peter, H.; Romaní, A.M.; Tranvik, L.J.; Ylla, I. Different diversity-functioning relationship in lake and stream bacterial communities. *FEMS Microbiol. Ecol.* **2013**, *85*, 95–103.

37. Cao, J.; Hou, Z.; Li, Z.; Chu, Z.; Yang, P.; Zheng, B. Succession of phytoplankton functional groups and their driving factors in a subtropical plateau lake. *Sci. Total. Environ.* **2018**, 1127–1137. [CrossRef]

38. Liao, J.; Zhao, L.; Cao, X.; Sun, J.; Gao, Z.; Wang, J.; Jiang, D.; Fan, H.; Huang, Y. Cyanobacteria in lakes on Yungui Plateau, China are assembled via niche processes driven by water physicochemical property, lake morphology and watershed land-use. *Sci. Rep.* **2016**, *6*, 36357. [CrossRef]

39. Liu, W.; Li, S.; Bu, H.; Zhang, Q.; Liu, G. Eutrophication in the Yunnan Plateau lakes: The influence of lake morphology, watershed land use, and socioeconomic factors. *Environ. Sci. Pollut. Res. Int.* **2012**, *19*, 858–870. [CrossRef]

40. Freedman, Z.B.; Zak, D.R. Atmospheric N Deposition Alters Connectance, but not Functional Potential among Saprotrophic Bacterial Communities. *Mol. Ecol.* **2015**, *24*, 3170–3180. [CrossRef]

41. Montoya, J.M.; Pimm, S.L.; Solé, R.V. Ecological networks and their fragility. *Nature* **2006**, *442*, 259–264. [CrossRef] [PubMed]

42. Saavedra, S.; Stouffer, D.B.; Uzzi, B.; Bascompte, J. Strong contributors to network persistence are the most vulnerable to extinction. *Nature* **2011**, *478*, 233–235. [CrossRef] [PubMed]

43. Newman, M.E.J. Modularity and community structure in networks. *Proc. Nati. Acad. Sci. USA* **2006**, *103*, 8577–8582. [CrossRef] [PubMed]

44. Brenda, L.H.; Jane, C.M.; Mary, E.W. Early bacterial and fungal colonization of leaf litter in Fossil Creek. *J. N. Am. Benthol. Soc.* **2009**, *28*, 383–396.

45. Fazi, S.; Amalfitano, S.; Pernthaler, J.; Puddu, A. Bacterial communities associated with benthic organic matter in headwater stream microhabitats. *Environ. Microbiol.* **2005**, *7*, 1633–1640. [CrossRef] [PubMed]

46. Findlay, R.H.; Yeates, C.; Hullar, M.A.J.; Stahl, D.A.; Kaplan, L.A. Biome-Level Biogeography of Streambed Microbiota. *Appl. Environ. Microbiol.* **2008**, *74*, 3014–3021. [CrossRef]

47. Zhang, X.; Yan, Q.; Yu, Y.; Dai, L. Spatiotemporal pattern of bacterioplankton in Donghu Lake. *Chin. J. Oceanol. Limnol.* **2014**, *32*, 554–564. [CrossRef]

48. Zwirglmaier, K.; Keiz, K.; Engel, M.; Geist, J.; Raeder, U. Seasonal and spatial patterns of microbial diversity along a trophic gradient in the interconnected lakes of the Osterseen Lake District, Bavaria. *Front. Microbiol.* **2015**, *6*, 1168. [CrossRef]

49. Zhao, H.; Zhang, L.; Wang, S.; Jiao, L. Features and influencing factors of nitrogen and phosphorus diffusive fluxes at the sediment-water interface of Erhai Lake. *Environ. Sci. Pollut. Res. Int.* **2018**, *25*, 1933–1942. [CrossRef]

50. Zhang, L.; Xu, K.; Wang, S.; Wang, S.; Li, Y.; Li, Q.; Meng, Z. Characteristics of dissolved organic nitrogen in overlying water of typical lakes of Yunnan Plateau, China. *Ecol. Indic.* **2018**, *84*, 727–737. [CrossRef]

51. Zhu, R.; Wang, H.; Chen, J.; Shen, H.; Deng, X. Use the predictive models to explore the key factors affecting phytoplankton succession in Lake Erhai, China. *Environ. Sci. Pollut. Res. Int.* **2018**, *25*, 1283–1293. [CrossRef]

52. Tello, A.; Corner, R.; Telfer, T. How do land-based salmonid farms affect stream ecology? *Environ. Pollut.* **2010**, *158*, 1147–1158. [CrossRef]

53. Matthaei, C.D.; Piggott, J.J.; Townsend, C.R. Multiple stressors in agricultural streams: Interactions among sediment addition, nutrient enrichment and water abstraction. *J. Appl. Ecol.* **2010**, *47*, 639–649. [CrossRef]

54. Peter, H.; Ylla, I.; Gudasz, C.; Romani, A.M.; Sabater, S.; Tranvik, L.J. Multifunctionality and diversity in bacterial biofilms. *PLoS ONE* **2011**, *6*, e23225. [CrossRef]

55. Mulholland, P.J.; Helton, A.M.; Poole, G.C.; Hall, R.O.; Hamilton, S.K.; Peterson, B.J.; Tank, J.L.; Ashkenas, L.R.; Cooper, L.W.; Dahm, C.N.; et al. Stream denitrification across biomes and its response to anthropogenic nitrate loading. *Nature* **2008**, *452*, 202–205. [CrossRef]

56. Valett, H.M.; Thomas, S.A.; Mulholland, P.J.; Webster, J.R.; Dahm, C.N.; Fellows, C.S.; Crenshaw, C.L.; Peterson, C.G. Endogenous and exogenous control of ecosystem function: N cycling in headwater streams. *Ecology* **2008**, *89*, 3515–3527. [CrossRef]

57. Wilhelm, L.; Besemer, K.; Fragner, L.; Peter, H.; Weckwerth, W.; Battin, T.J. Altitudinal patterns of diversity and functional traits of metabolically active microorganisms in stream biofilms. *ISME J.* **2015**, *9*, 2454–2464. [CrossRef]

58. Ernst, A.; Deicher, M.; Herman, P.M.J.; Wollenzien, U.I.A. Nitrate and Phosphate Affect Cultivability of Cyanobacteria from Environments with Low Nutrient Levels. *Appl. Environ. Microbiol.* **2005**, *71*, 3379–3383. [CrossRef]

59. Correll, D. Phosphorus: A rate limiting nutrient in surface waters. *Poult. Sci.* **1999**, *78*, 674–682. [CrossRef]

60. Vrede, T.; Ballantyne, A.; Algesten, G.; Gudasz, C.; Lindahl, S.; Brunberg, A.K.; Mille-Lindblom, C.; Mille-Lindblom, C. Effects of N:P loading ratios on phytoplankton community composition, primary production and N fixation in a eutrophic lake. *Freshw. Biol.* **2009**, *54*, 331–344. [CrossRef]

61. Yu, G.; Jiang, Y.; Song, G.; Tan, W.; Zhu, M.; Li, R. Variation of Microcystis and microcystins coupling nitrogen and phosphorus nutrients in Lake Erhai, a drinking-water source in Southwest Plateau, China. *Environ. Sci. Pollut. Res.* **2014**, *21*, 9887–9898. [CrossRef]

62. Nikolai, S.J.; Dzialowski, A.R. Effects of internal phosphorus loading on nutrient limitation in a eutrophic reservoir. *Limnologica* **2014**, *49*, 33–41. [CrossRef]

63. Ding, S.; Chen, M.; Gong, M.; Fan, X.; Qin, B.; Xu, H.; Gao, S.; Jin, Z.; Tsang, D.C.; Zhang, C. Internal phosphorus loading from sediments causes seasonal nitrogen limitation for harmful algal blooms. *Sci. Total. Environ.* **2018**, *625*, 872–884. [CrossRef]

64. Frost, P.C.; Elser, J.J. Effects of light and nutrients on the net accumulationand elemental composition of epilithon in boreal lakes. *Freshw. Biol.* **2002**, *47*, 173–183. [CrossRef]

65. Jansson, M.; Persson, L.; De Roos, A.M.; Jones, R.I.; Tranvik, L.J. Terrestrial carbon and intraspecific size-variation shape lake ecosystems. *Trends Ecol. Evol.* **2007**, *22*, 316–322. [CrossRef]

66. Dong, X.-P.; Donoghue, P.C.J.; Cheng, H.; Liu, J.-B. Fossil embryos from the Middle and Late Cambrian period of Hunan, south China. *Nature* **2004**, *427*, 237–240. [CrossRef]

67. Ren, Z.; Qu, X.; Peng, W.; Yu, Y.; Zhang, M. Nutrients Drive the Structures of Bacterial Communities in Sediments and Surface Waters in the River-Lake System of Poyang Lake. *Water* **2019**, *11*, 930. [CrossRef]

MDPI

St. Alban-Anlage 66

4052 Basel

Switzerland

Tel. +41 61 683 77 34

Fax +41 61 302 89 18

www.mdpi.com

Water Editorial Office

E-mail: water@mdpi.com

www.mdpi.com/journal/water

www.ingramcontent.com/pod-product-compliance
Lightning Source LLC
LaVergne TN
LVHW071001170726
843515LV00006B/1044